A Practical Approach to Motor Vehicle Engineering and Maintenance

A Practical Approach to Motor Vehicle Engineering and Maintenance

Second Edition

Derek Newbold
Formerly of Hinckley College

Allan Bonnick, M.Phil, B.Sc, C.Eng, M.I.Mech.Eng
Formerly Principal Lecturer at Eastbourne College of Arts and Technology

ELSEVIER
BUTTERWORTH
HEINEMANN

AMSTERDAM • BOSTON • HEIDELBERG • LONDON • NEW YORK • OXFORD
PARIS • SAN DIEGO • SAN FRANCISCO • SINGAPORE • SYDNEY • TOKYO

Elsevier Butterworth-Heinemann
Linacre House, Jordan Hill, Oxford OX2 8DP
30 Corporate Drive, Burlington, MA 01803

First published by Arnold 2000
Reprinted 2002, 2003, 2004
Second edition 2005

British Library Cataloguing in Publication Data
A catalogue record for this book is available from the British Library

Library of Congress Cataloguing in Publication Data
A catalogue record for this book is available from the Library of Congress

ISBN 0 7506 6314 6

For information on all Elsevier Butterworth-Heinemann
publications visit our website at http://books.elsevier.com

Typeset by Integra Software Services Pvt. Ltd, Pondicherry, India
www.integra-india.com
Printed and bound in Great Britian

Contents

Foreword

Although this book has been written mainly for those who wish to gain an NVQ in motor vehicle work, it is equally suitable for a wide variety of people who are undertaking the City & Guilds, BTEC as well those who just want to know about cars to enable them to undertake their own servicing with a greater degree of confidence. It can be used in schools, colleges and garage workshops as each task is being undertaken.

It covers the fundamental principles for each system found in the motor vehicle. It is by no means exhaustive, but it does allow the student to take simple steps in understanding how each system works.

The NVQ qualification is not as daunting as many people think. The main problem seems to be in the gaining of evidence and the assembling of a portfolio, so, with this in mind, there are a number of exercises at the end of each section for the student to complete. The evidence required by the NVQ assessor from the student will be gained gradually and built into a student portfolio. Each completed job sheet (suitably signed after the job has been checked) should show evidence of what you have done, how you did it and that it was completed to a satisfactory standard. When the student feels confident enough to complete the task alone then he or she may request an assessment. Examples of assessments are given in the book to show what might be required by the assessor.

The text has a number of words and sentences that are highlighted when they first appear in the text. These are mostly key words that will help the student to remember what is essential from the text. No matter how much the student 'waffles' when answering questions, unless he or she understands the subject and uses the key words, it will be very difficult for the assessor to give marks. Remember the key words and you will be half-way there. Some of the words used are specific to the motor vehicle and to the NVQ. We have tried to explain those relating to the motor vehicle in the text and those relating to the NVQ in a glossary at the end of the book. There are a number of practical assignments and learning tasks in each chapter. If these are undertaken in a realistic way, they will enable the student to complete the task repeatedly, up to an acceptable standard, without supervision, which is the requirement of the NVQ.

The workshop job sheet should contain all the elements of the performance criteria as requested in the NVQ. They should be signed by both the person undertaking the task (you) and the assessor/supervisor to say the task has been completed satisfactorily.

The illustrations have been selected to give the maximum amount of support when learning about new topics. We advise students to attempt the learning tasks when they have completed the related section of their training and education.

This book is based on many years of teaching and helping students and apprentices who have gone on to become successful and valued mechanics. Some of them have become owners of their own motor vehicle workshops, each in turn has appreciated the training that was given to them in their early years. Our hope is that this book will enable a wider variety of people to achieve their hopes and ambitions.

Derek Newbold
Allan Bonnick

Foreword to second edition

In order to take account of feedback from practising teachers and to allow for developments in technology, education and training this 2nd edition includes coverage of developments such as; variable value timing, on board diagnostics, computer controlled systems etc. maintenance aspects are covered and a range of learning tasks and self assessment question is included.

Allan Bonnick
2005

Acknowledgements

Thanks are due to the following companies who supplied information and in many cases permission to reproduce photographs and diagrams:

- Audi UK Ltd
- Bowers Metrology Group (Moore & Wright)
- Champion Spark Plugs Ltd
- Crypton Technology Group
- Cummins Engine Company Inc.
- Delmar Publishers Inc.
- Dunlop Holdings Ltd
- Ford Motor Company UK Ltd
- Haynes Publications
- Honda UK Ltd
- KIA Cars Ltd
- Lucas Automotive Ltd
- Rover Group
- Vauxhall Motors Ltd
- Volvo Group (UK)
- Toyota GB Ltd
- Zahnradfabrik Friedrichshafen AG

Special thanks are due to the following for supplying much useful information, and with permission to reproduce pictures and diagrams from:

The Automotive Chassis: Engineering principles 2nd edition, J Reimpell & H Stoll (Vogel-Buchverlag, Würzburg, 1995)

If we have used information or mentioned a company name in the text, not listed here, our apologies and acknowledgements.

Introduction to the retail motor industry and Health & Safety at Work

Motor vehicle maintenance and repair

This introductory section contains:

- Details of the motor vehicle maintenance and repair industry;
- An introduction to Health & Safety at Work.

The motor vehicle maintenance and repair industry – the garage industry

By taking a few examples of aspects of modern life, it is possible to gain an insight into areas of activity where motor vehicle maintenance and repair plays an important part. One has only to consider the effect of a vehicle breakdown in any of these areas to gain an appreciation of the part that motor vehicle technicians play in the day-to-day operation of society when they maintain and repair vehicles.

In order for these activities to take place, the vehicles must be serviced and maintained at regular intervals and, in the event of a breakdown, action must be taken to clear the road and repair the vehicle so as to restore it to good working condition, as quickly as possible. In the UK it is the motor vehicle repair and maintenance industry that performs the bulk of vehicle maintenance and repair work. It is for trainees and students, preparing for work in this industry, that this book is designed.

Some details about the type of work involved in repair and maintenance of vehicles

There are approximately 30 000 000 vehicles of various types in use in the UK in 2004 and a large vehicle repair and maintenance industry exists to provide the necessary services. The

Area of activity	Type of activity	Type of vehicle
Personal transport	Getting to work. Taking children to school. Going on holiday. Visiting friends. Going shopping	Cars, people carriers, motorcycles, scooters and mopeds
Public transport	Activities as for personal transport, but much more important in towns and cities where traffic congestion causes delays	Buses, coaches
Good vehicles	Movement of food, fuel, materials for industry. Bringing food and materials into the country. Exporting manufactured goods, etc.	Trucks, vans, tankers, articulated vehicles
Emergency services	Fire and rescue services. Ambulances. Doctors. Movement of blood supplies and human organs. AA, RAC and other forms of roadside assistance	Cars, vans, fire engines, rescue vehicles, mobile cranes
Armed forces	Defence of the realm. Action overseas. Maintenance of services in times of need	Trucks, armoured vehicle, tanks, fuel tankers, tank transporters
Postal and other delivery services	Communications for business Private correspondence. Delivery of mail order goods	Vans, trucks
Fuel deliveries	Fuel deliveries to service stations and fuel depots. Domestic fuel supplies	Tankers of various sizes

majority of motor cars are repaired and maintained in retail garages and businesses vary in size, from large-sized vehicle dealerships employing several people in a range of occupations to small one man type businesses. Buses tend to be cared for in specialised workshops operated by Local Authorities and specialised bus companies. The repair and maintenance of heavy goods vehicles is often carried out in garage workshops that are owned by transport companies.

The garage industry employs several hundred thousand people in a range of occupations. The work is interesting, often demanding – both physically and mentally. There are many opportunities for job satisfaction. For example, it is most rewarding to restore a vehicle to full working condition after it has suffered some form of failure. There are opportunities for promotion, technicians often progressing to service managers and general managers, or to run their own companies. The Office of Fair Trading (OFT) report for the year 2000 shows that more than 50% of cars were more than 5 years old. These cars require annual MOT inspections and there is a tendency for them to be serviced and repaired in the independent sector of the industry. The OFT report shows that there are approximately 16 000 independent garages and 6 500 garages that are franchised to one or more motor manufacturers. Evidently there are plenty of opportunities for employment.

Health and safety

It is the responsibility of every person involved in work to protect their own safety and that of any other persons who may be affected by their activities. At the basic level this means that everyone must work in a safe manner and know how to react in an emergency. Personal cleanliness issues such as regular washing, removal of substances from skin, use of barrier creams, use of protective clothing and goggles and gloves, etc. are factors that contribute to one's well being. Behaving in an orderly way, not indulging in horseplay, learning how to employ safe working practices and helping to keep the workplace clean and tidy, are all ways in which an individual can contribute to their own and others health and safety at work.

There are various laws and regulations that govern working practice in the motor vehicle repair industry, in the UK – the main ones are:

- Health & Safety at Work Act 1974;
- The Factories Act 1961;
- The Offices, Shops and Railway Premises Act 1963.

Some of the other regulations that relate to safety in motor vehicle repair and maintenance are:

- The Control of substances Hazardous to Health (COSHH) Regulations 1988;
- Regulations about the storage and handling of flammable liquids;
- The Grinding Wheel Regulations.

Health and safety laws are enforced by a factory inspector from the Health and Safety Executive (HSE) or an environmental health officer from the local council.

Safety policy

As stated at the beginning of this section, each individual has a responsibility to work safely and to avoid causing danger to anyone else. This means that each individual must know how to perform their work in a safe way and how to react with other people in the event of an accident or emergency. In any establishment, however small, safety planning must be performed by a competent person, who must then familiarise all others engaged in the enterprise with the plans that have been devised, to ensure that all health and safety issues are properly covered. These plans are a set of rules and guidelines for achieving health and safety standards, in effect a policy.

Every motor vehicle repair business that employs five or more people must write down their policy for health and safety and have it to hand for inspection by the HSE.

Motor vehicle repair and maintenance health and safety topics

Readers are advised to purchase a copy of the publication "Health & Safety in Motor Vehicle Repair – HSG67" available from HMSO for approximately £5.50. The HSE website: www.hse.gov.uk is also a valuable source of information. Several pages are devoted to motor vehicle repair and the home page offers the user the opportunity to select 'your industry', a click on 'Motor vehicle repair' brings up many opportunities to study a variety of health and safety topics.

This publication (HSG67) states that:

"*Most accidents involve trips and falls or poor methods of lifting and handling; serious injuries often resulting from these apparently simple causes. Accidents involving vehicles are frequent and cause serious injuries and deaths every year. Work on petrol tanks in particular causes serious burns, hundreds of fires and some deaths each year ... There is also widespread potential for work related ill-health in garages. Many of the substances used require careful storage, handling and control.*"

(HSE (2004) Health & safety in Motor Vehicle Repair. HMSO)

From this you might think that motor vehicle work is hazardous but, if work is carried out properly and safety factors are always considered, it is possible to work without injury to anyone. Safety training and related skills training must figure prominently in any course of education and training that aims to prepare people for work in motor vehicle maintenance and repair. The following descriptions are intended to highlight some of the everyday safety issues.

The information contained in this book does not attempt to provide full coverage of all safety and health issues. The information provided merely draws attention to some safety issues, and is not intended to cover all areas of safety. Readers should ensure that any education course, training scheme, apprenticeship or other training arrangement, does contain all necessary safety training.

Lifting equipment

The types of lifting equipment that are commonly used in vehicle repair workshops are: jacks of various types, axle stands, vehicle hoists (lifts), floor cranes and vehicle recovery equipment. Hydraulic jacks are used to raise the vehicle. For work to be done underneath the vehicle, the vehicle must be on level ground, and the wheels that are remaining on the ground must be chocked or have the handbrake applied. Axle stands must be placed in the correct positions before any attempt is made to get under the vehicle. Figure A.1.1 shows the hydraulic jack placed at a suitable jacking point. This is important: i) to ensure that the jacking point is secure from the safety point of view; and ii) to prevent damage to the vehicle.

Hydraulic jacks must be maintained in good condition. The safe working load must be clearly marked on the jack. Axle stands must be of good quality and of a load carrying capacity that is correct for the vehicle being supported on them. The proper pins that allow for height adjustment should be attached to the stands and the stands must be kept in good condition.

Vehicle hoists

The two-post vehicle hoist shown in Fig. A.1.2 is an example of a type that is widely used in

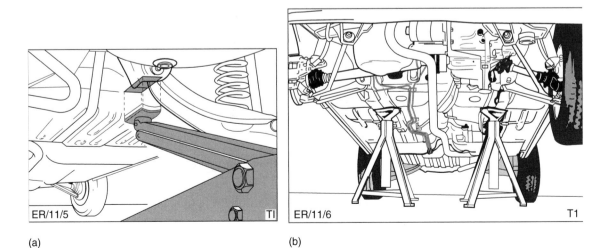

(a)   (b)

Fig. A.1.1 (a) Hydraulic jack positioned at jacking point, (b) Axle stands in position prior to working under the vehicle. (Reproduced with the kind permission of Ford Motor Company Limited.)

the garage industry. The HSE guide (HSG67) states:

"Careful attention should be paid to manufacturers' recommendations when using two-post hoists – vehicle chassis, sub frame and jacking points should be in good condition; support arms should be set to the correct height before the vehicle is raised. The weight distribution of the vehicle being lifted and the effect of the removal of major components should be constantly evaluated."

Comment

Hoists, in common with other equipment, should not be used until the method of operation is properly understood. For trainees this means that proper training must be provided and trainees should not attempt to use hoists until they have been trained and given permission to do so. Part of that training will be a demonstration of the procedure for locating jacking points and positioning the support arm pads. This ensures that the vehicle will be stable when lifted and the support arm pads prevent damage to the vehicle structure. The weight distribution of the vehicle on the hoist will be affected according to the work being performed. For example, if the hoist is being used during a transmission unit change, the removal of the transmission unit will greatly affect weight distribution and this must be allowed for as the work progresses.

Four post hoists. The HSE guide says that:

"Four-post hoists should have effective 'dead man's' controls, toe protection and automatic chocking. Toe traps should also be avoided when body straightening jigs are fitted. Raised platforms should never be used as working areas unless proper working balconies or platforms with barrier rails are provided."

Comment

The dead man's control is designed to avoid mishaps during the raising and lowering of the hoist. Toe protection is to prevent damage to the feet when the hoist is lowered to the ground, and the automatic chocking is to prevent the vehicle from being accidentally rolled off the raised hoist. Body straightening jigs are used in vehicle body repair shops and their use requires special training. The point about working above floor level and the use of balconies should be noted.

In general, hoists make work on the underside of a vehicle less demanding than lying on one's back. A point to note is that hard hats should be worn when working under a vehicle on a hoist. The area around the hoist must be kept clear and a check should always be made when lowering or raising at hoist to ensure that no person, vehicle or other object is likely to be

Fig. A.1.2 Wheel free. Post hoist. (Reproduced with the kind permission of Ford Motor Company Limited.)

harmed by the operation. Hoists must be frequently examined to ensure that they are maintained in safe working order.

Electrical safety

Safety in the use of electrical equipment must figure prominently at all stages of training.

> ### Learning task
>
> As an exercise to develop knowledge in this area you should now read pages 4 and 5 of HSG67 and compare that with the training and tuition that you are receiving. Also visit the HSE motor vehicle repair website and study what it has to say about electrical safety. Check with your supervisor to make sure that you know what to do in the event of an electrical emergency in your present circumstances.

Compressed air equipment

Compressed air is used in a number of vehicle repair operations such as tyre inflation, pneumatic tools, greasing machines and oil dispensers, etc. The compressed air is supplied from a compressor and air cylinder and this is equipment that must be inspected regularly by a competent person. The usual procedure is for the insurance company to perform this work. Air lines, hoses, tyre inflaters and pressure gauges, couplings, etc. should be inspected at regular intervals to ensure that they are kept in working order. Inflation of the split rim type of commercial vehicle tyres requires special attention and the tyres should only be inflated behind a specially designed guard. An HSE booklet HSG62 covers many aspects of health and safety for those engaged in tyre and exhaust fitting work.

> ### Learning task
>
> Read page 6 of HSG67 and note particularly items 43 and 45. Visit the HSE website to see what it has to say that is relevant to motor vehicle repair.

Petrol fires

Referring back to the statement at the beginning of this section we have:

"Work on petrol tanks in particular causes serious burns, hundreds of fires and some deaths each year."

Comment

Petrol gives off a flammable vapour that is heavier than air; this means that it will settle at a low level and spread over a wide area. Petrol vapour is also invisible. The vapour can be ignited by a flame or spark at some distance from any visible sign of the liquid. Great care must be exercised when dealing with petrol and the appropriate fire extinguisher should be close to hand when performing any task involving petrol or any other flammable substance. Petrol may only be stored in containers specified in the Petroleum Spirit Regulations. Advice on this is readily available from the local Fire Authority. In the event of spillage on clothes, the clothes should be removed for cleaning because petrol vapour can accumulate inside the clothing which can result in serious burns in the event of ignition.

> ### Learning task
>
> Read page 8 of HSG67. What does HSG67 say about the fuel gauge sender unit? Discuss this with your tutor/trainer, or supervisor.

Brake and clutch linings

"There is also widespread potential for work related ill-health in garages. Many of the substances used require careful storage, handling and control."

(HSG67)

Some brake and clutch linings contain asbestos and to guard against harmful effects special measures must be employed. These measures include wearing a mask and using an appropriate vacuum cleaner to remove dust and preventing asbestos dust from getting into the surrounding air.

> ### Learning task
>
> Read page 9 of HSG67 and then describe to your tutor/trainer the type of vacuum cleaner that should be used to remove brake lining or clutch lining dust. Also explain the purpose of wetting any dust that may have been left on the floor.

Oils and lubricants

Oils and lubricants used in vehicles contain chemicals to change their properties and make them suitable for use in many vehicle applications. Many of the chemicals used are harmful to health if not handled properly.

> ### Learning task
>
> Read page 12 of HSG67 and discuss it with your tutor/trainer or supervisor. Protective clothing, suitable gloves and barrier creams help to reduce risk. Most training organisations have illustrated booklets and posters that highlight the effects of over exposure to substances like used engine oil. Read item 72 on page 12 of HSG67 and discuss with your trainer/supervisor what the self-inspection should consist of.

Used oils must only be disposed of in approved ways. Special containers are available for receiving oil as it drained from a vehicle. Vehicle repair workshops must have facilities to store used oil prior to disposing of it through approved channels. Used oil has a market value and some companies collect it for reprocessing.

> ### Learning task
>
> Visit the website: http://ehsni.gov.uk study what this says about disposal of used oil from garages and private homes. Make a note of the main points.

"Accidents involving vehicles are very frequent and cause serious injuries and deaths every year."

(HSG67)

Comment

Work that involves movement of vehicles both inside and outside of the workshop can be dangerous if not handled properly. Vehicles should only be moved by authorised persons and great care must be taken when manoeuvring them.

> ### Learning task
>
> Study the sections of HSG67 that deal with rolling road dynamometers and brake testers, moving and road testing vehicles. Visit the HSE

website and note the entries that deal with this topic. Discuss your findings with your colleagues, fellow students and workmates.

Other topics

The booklet HSG67 draws attention to many topics relating to health and safety in motor vehicle repair and maintenance. Two of these topics are vehicle valeting and use of steam cleaners and water pressure cleaners.

Comment

Vehicle valeting tasks, such as cleaning the exterior and interior of the vehicle, removing stains from upholstery, etc. often involve the use of chemical substances. Cleaning of components during servicing and repair procedures also entails the use of chemicals, and steam and high-pressure water cleaners.

> ### Learning task
>
> Read pages 14 and 15 of HSG67. Note the points about COSHH. Note down the procedures that are used in your workplace for dealing with COSHH. Make a note of the types of steam and water pressure cleaning apparatus and any other cleaning tanks that are used in any place where you are involved in practical work. Note any special cleaning substances that are used. Visit the HSE website and study the motor vehicle repair section that deals with COSHH.

The intention in this section has been to draw attention to some aspects of Health and Safety. Health and safety are aspects of working life that one must be constantly aware of throughout working life. Many aspects of work, such as keeping a workplace clean and tidy, and working methodically, contribute to health and safety and also form part of an orderly approach to tasks that assist one in the performance of complex tasks such as fault diagnosis.

Organization of the firm

A company, or firm as it is often called, will often display its structure in the form of an organization chart. For any group of people who are engaged in some joint activity it is necessary to have someone in charge and for all members of the 'team' to know what their role is and which person they should speak to when they need advice. Just as a football team is organized and each person is given a position on the field so a business is organized so that people can work as a team.

The 'line chart' (Fig. A.1.3) shows how the work is divided up into manageable units, or areas of work, and how the personnel in those areas relate to each other. In this example the managing director is the 'boss'. Under the managing director come the accountant, the service manager, the parts department manager and the sales manager; these are known as 'line' managers. Below these line managers, in the hierarchical structure that is used in many businesses, are supervisors (foremen/women) and then come the technicians, clerks, etc.

Policy

A policy is a set of rules and guidelines that should be written down so that everyone in an organization knows what they are trying to achieve. Policy is decided by the people at the top of an organization, in consultation with whoever else they see fit. It is important for each employee to know how policy affects them and it should be suitably covered in a contract of employment. For example, every firm must have a 'safety policy'. It is the duty of every employee to know about safety and firms are subject to inspection to ensure that they are complying with the laws that relate to health and safety at work, and similar legislation.

Discipline

Standards of workmanship, hours of work, relations with other employees, relations with customers and many other factors need to be overseen by members of the firm who have some authority because, from time to time, it may be necessary to take steps to change some aspect. For example, it may be the case that a particular technician is starting to arrive late for work; it will be necessary for someone to deal with this situation before it gets out of hand and this is where the question of 'authority' arises.

Authority

Authority is power deriving from position. The organization chart illustrates who has authority (power) over whom. The vertical position on the chart of a member of staff indicates that they are in charge of, and have power over, the employees lower down the chart. The service manager in a garage has authority in relation to the operation of the workshop; in addition they are responsible for the satisfactory performance of the workshop side of the business.

Accountability

We all have to account for actions that we take. This means having to explain why we took a particular line of action. We are all responsible for the actions that we take. This means that we did certain things that led to some outcome. By having to explain why something 'went wrong', i.e. by having to account for our behaviour, we

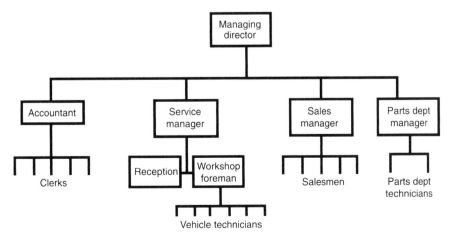

Fig. A.1.3 A garage business line organization chart

are made to examine our responsibility. We are 'responsible' for seeing that we get to work on time, for making sure that we do a job properly and safely, and for helping to promote good working relationships.

Delegating

It is often said that you can delegate (pass down) authority but not responsibility. So, if you are a skilled technician who has a trainee working with you and you have been given a major service to do, it will be your responsibility to ensure that the job is done properly. You will have had some authority given to you to instruct the trainee, but the responsibility for the quality of the work remains with you.

Having introduced some ideas about business organization and the working relationships that are necessary to ensure success, we can turn our attention to the business activities that are the immediate concern of the service technician. In order to place some structure on this section it is probably a good idea to start with 'reception'.

Reception

Reception in a garage is the main point of contact between customers and the services that the workshop provides. It is vitally important that the communication between the customers and reception and between reception and the workshop is good. It is at reception that the customer will hand over the vehicle and it is at reception that it will be handed back to the customer. The reception engineer is a vital link in the transaction between the customer and the firm and it is at this point that friction may occur if something has gone wrong. Staff in reception need to be cool headed, able to think on their feet and equipped with a vocabulary that enables them to cope with any situation that may 'crop up'. From time to time they may need to call on the assistance of the service manager and it is vitally important that there are good relations and clear lines of communication in this area.

In many garages there will be an area set aside for reception and the staff placed there will have set duties to perform. In the case of small garages employing, perhaps, one or two people reception may take place in the general office and the person who is free at the time may

perform the function. In either case, certain principles apply and it is these that are now addressed.

Customer categories

Because a good deal of skill and tact is required when dealing with customers it is useful to categorize them as follows:

- informed
- non-informed
- new
- regular

Informed customers

Informed customers are those who are knowledgeable about their vehicle and who probably know the whole procedure that the vehicle will follow before it is returned to them. This means that the transaction that takes place between them and the receptionist will largely consist of taking down details of the customer's requirements and agreeing a time for the completion of the work and the collection of the vehicle. Depending on the status of the customer, i.e. new or regular, it will be necessary to make arrangements for payment. It may also be necessary to agree details for contacting the customer should some unforeseen problem arise while work on the vehicle is in progress, so that details can be discussed and additional work and a new completion time agreed.

Non-informed customers

The non-informed customer probably does not know much about cars or the working procedures of garages. This type of customer will need to be treated in quite a different way. The non-informed customer will – depending on the nature of the work they require for their vehicle – need to have more time spent on them. A proportion of this extra time will be devoted to explaining what the garage is going to do with their vehicle and also what the customer has to do while the vehicle is in the care of the garage. The latter part will largely consist of making arrangements for getting the customer to some chosen destination and for collecting the vehicle when the work is completed. On handing the car back to the 'non-informed' customer it will probably be necessary to describe, in a non-patronizing and not too technical way, the work that has been done and what it is that they are being asked to pay for.

New customers

A 'new' customer may be informed or non-informed and this is information that should be 'teased' out during the initial discussions with them. Courtesy and tact are key words in dealing with customers and they are factors that should be uppermost in one's mind at the new-customer-introduction stage. The extent of the introductory interview with the new customer will depend on what it is that they want done and the amount of time that they and the firm have to give to the exercise. If it is a garage with several departments it may be advisable to introduce the new customer to relevant personnel in the departments that are most likely to be concerned with them. Customers may wish to know that their vehicle will be handled by qualified staff and it is common practice for firms to display samples of staff qualification certificates in the reception area. At some stage it will be necessary to broach the subject of payment for services and this may be aided by the firm having a clearly stated policy. Again, it is not uncommon for a notice to be displayed which states the 'terms of business'; drawing a client's attention to this is relatively easy.

Regular customers

Regular customers are valuable to a business and they should always be treated with respect. It hardly needs saying that many of the steps that are needed for the new customer introduction will not be needed when dealing with a regular customer. However, customers will only remain loyal (regular) if they are properly looked after. It is vitally important to listen carefully to their requests and to ensure that the work is done properly, and on time and that the vehicle is returned to them in a clean condition.

Workshop activities

Reception will have recorded the details of the work that is to be done on the vehicle, and they will have agreed details of completion time, etc. with the workshop. At some point in the process a 'job card' will have been generated. The instructions about the work to be done must be clear and unambiguous. In some cases this may be quite brief, for example, 10 000 mile service. The details of the work to be performed will be contained in the service manual for the particular vehicle model. In other cases it may be rather general, for example, 'Attend to noisy wheel bearing. Near side front.' Describing exactly what work is required may entail further investigation by the technician. It may be that the noise is caused by the final drive. The whole thing thus becomes much more complicated and it may be necessary to conduct a preliminary examination and test of the vehicle before the final arrangements are made for carrying out the work.

Once the vehicle has been handed over to the workshop with a clear set of instructions about the work to be done, it becomes the responsibility of the technician entrusted with the job and their colleagues to get the work done efficiently and safely and to make sure that the vehicle is not damaged. This means that the workshop must have all necessary interior and exterior protection for the vehicle such as wing and seat covers, etc. Figure A.1.4 shows a suggested layout for a service bay.

Records

As the work proceeds a record must be made of materials used and time spent because this will be needed when making out the invoice. In large organizations the workshop records will be linked to stock control in the parts department and to other departments, such as accounts, through the company's information system which will probably be computer based.

Quality control

Vehicle technicians are expected to produce work of high quality and various systems of checking work are deployed. One aspect of checking quality that usually excites attention is the 'road test'. It is evident that this can only be performed by licensed drivers and it is usually restricted to experienced technicians. A road test is an important part of many jobs because it is probably the only way to ensure that the vehicle is functioning correctly. It is vitally important that it is conducted in a responsible way. Figure A.1.5 illustrates the point that there must be a good clear road in front of and behind the vehicle.

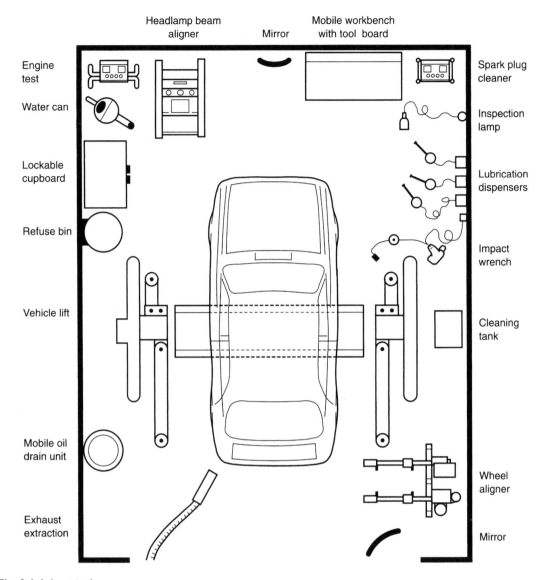

Headlamp beam aligner

Mirror

Mobile workbench with tool board

Engine test

Water can

Lockable cupboard

Refuse bin

Vehicle lift

Mobile oil drain unit

Exhaust extraction

Spark plug cleaner

Inspection lamp

Lubrication dispensers

Impact wrench

Cleaning tank

Wheel aligner

Mirror

Fig. A.1.4 A service bay

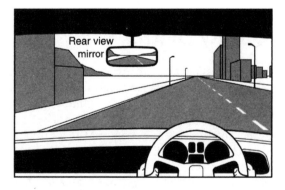

Fig. A.1.5 The front and rear view of the road when road testing

Rear view mirror

Returning the vehicle to the customer

On satisfactory completion of the work, the vehicle will be taken to the customer's collection point; the covers and finger marks and other small blemishes should be removed. In the meantime the accounts department will, if it is a cash customer, have prepared the account so that everything is ready for completion of the transaction when the customer collects the vehicle.

1
Engines and lubrication

Topics covered in this chapter

Engine construction
Four stroke and two stroke cycles
Valve timing – variable valve timing
Combustion
Lubrication and lubrication systems

1.1 Light vehicles

Vehicles classified in this category have a laden mass of less than three tonnes. A wide range of different body shapes and sizes are used in this category from simple two-seater cars to mini-buses and small trucks.

1.2 Layout of components

The power units used in the light vehicle can be fitted in a number of different places (Fig. 1.1). The source of power is provided by an **internal combustion engine**. The petrol or spark ignition (SI) engine is the most common, although the diesel or compression ignition (CI) engine is becoming more widely used. These are of the conventional design, where the pistons move up and down in the cylinders. Several other designs have been used; for example the Wankel engine used by Mazda in the RX7 and by Norton in their motor cycle. Another method of propulsion is electricity; the electric vehicle is gaining popularity as there is very little pollution of the atmosphere and it can

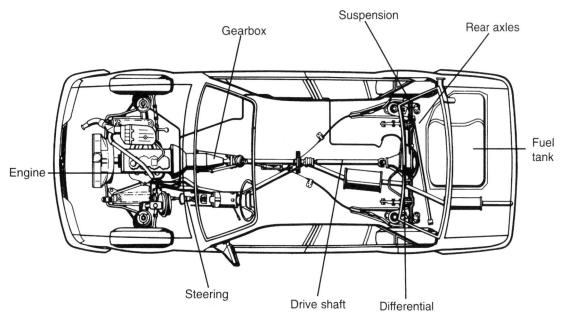

Fig. 1.1 Front engine RWD with integral body mountings

therefore meet the more stringent regulations coming into force each year.

The layout of the main components may conform to one of the following:

- **front engine front wheel drive** (FWD)
- **front engine rear wheel drive** (RWD)
- **mid-engine rear wheel drive**
- **rear engine rear wheel drive** (Fig. 1.2)
- one of the above but **four wheel drive** (4WD)

1.3 Location of major components

> *Learning task*
>
> Make a simple sketch in plan view of a front engine FWD vehicle. Label the main components: engine, clutch, gearbox, drive shafts, driving wheels, radiator, fuel tank.

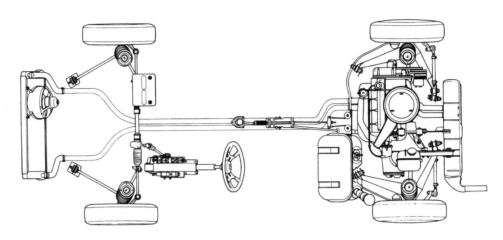

Fig. 1.2 Rear engine RWD ideal layout for a two seater coupe

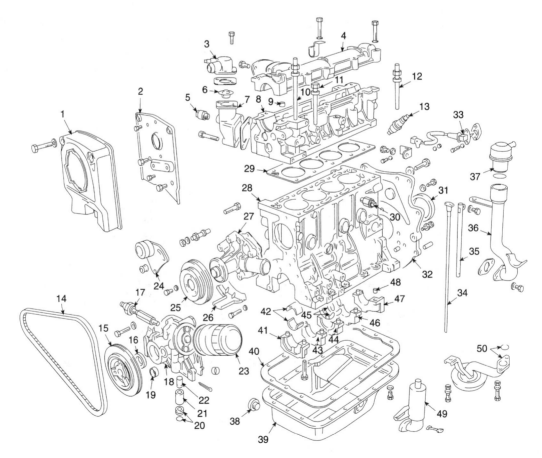

Fig. 1.3 Exploded view of engine block components

1 – Timing belt cover
2 – Timing belt cover back plate
3 – Thermostat outlet
4 – Camshaft cover
5 – Coolant thermistor
6 – Thermostat
7 – Thermostat housing
8 – Cylinder head
9 – Dowel
10 – Cylinder head stud – long
11 – Cylinder head bolt
12 – Cylinder head stud – short
13 – Spark plug
14 – Alternator/water pump belt
15 – Crankshaft pulley
16 – Oil seal
17 – Oil pressure switch
18 – Oil pump
19 – Oil pump plug
20 – Plug and 'O' ring
21 – Oil pressure relief valve spring
22 – Oil pressure relief valve plunger
23 – Oil filter cartridge
24 – Timing belt tensioner
25 – Water pump pulley
26 – Deflector
27 – Water pump
28 – Cylinder block
29 – Cylinder head gasket

30 – Knock sensor
31 – Crankshaft rear oil seal
32 – Gearbox adapter plate
33 – Crankshaft sensor
34 – Dipstick
35 – Dipstick tube
36 – Oil filler tube
37 – Oil filler cap
38 – Sump plug
39 – Oil sump
40 – Oil sump gasket
41 – Front main bearing cap
42 – Main bearing shells
43 – Intermediate main bearing cap
44 – Centre main bearing cap
45 – Thrust washers
46 – Intermediate main bearing cap
47 – Rear main bearing cap
48 – Dowel
49 – Oil separator
50 – Oil strainer and 'O' ring

Learning task

By using the information shown on Fig. 1.3 identify from Fig. 1.4 the following numbered components: 4, 5, 15, 27, 28, 29, 31, 32, 39, 41.

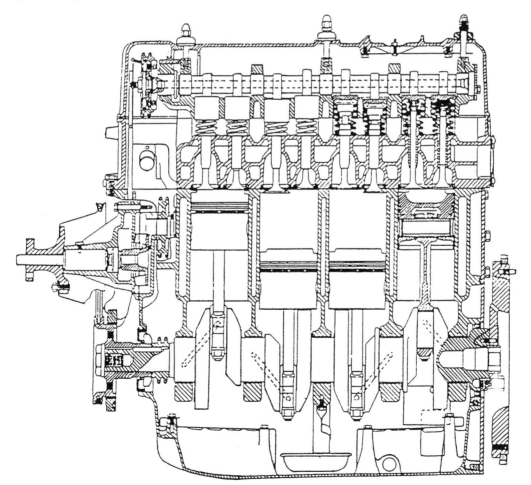

Fig. 1.4 Four cylinder in-line engine cross-section view

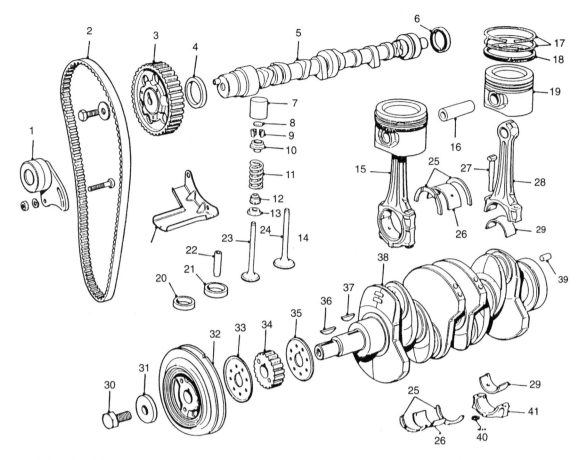

Fig. 1.5 Exploded view of internal engine components

1 – Timing belt tensioner
2 – Timing belt
3 – Camshaft gear
4 – Camshaft front oil seal
5 – Camshaft
6 – Camshaft rear oil seal
7 – Tappet
8 – Shim
9 – Cotters
10 – Cup
11 – Spring
12 – Valve stem seal
13 – Seat
14 – Deflector
15 – Connecting rod and piston – LH
16 – Gudgeon pin
17 – Compression rings
18 – Oil control ring
19 – Piston
20 – Exhaust valve seat insert
21 – Inlet valve seat insert
22 – Valve guide
23 – Exhaust valve
24 – Inlet valve
25 – Crankshaft washers

26 – Main bearing shell
27 – Connecting rod bolt
28 – Connecting rod – RH
29 – Big-end bearing shell
30 – Pulley bolt
31 – Washer
32 – Crankshaft pulley
33 – Gear flange
34 – Crankshaft gear
35 – Gear flange
36 – Pulley and gear key
37 – Oil pump key
38 – Crankshaft
39 – Flywheel dowel
40 – Connecting rod nut
41 – Connecting rod cap

Learning task

By using the information shown on Fig. 1.5, identify from Fig. 1.6 the following numbered components: 5, 7, 11, 15, 16, 19, 23, 32, 38, 39.

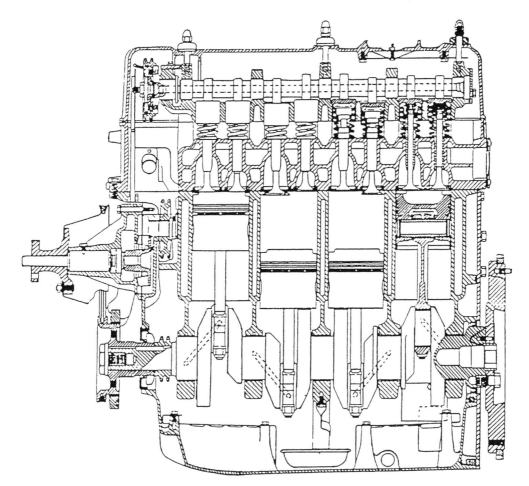

Fig. 1.6 Cross-sectioned view

1.4 Functional requirements of engine components

These are shown in Table 1.1.

> *Learning tasks*
>
> 1. Identify the properties of the main components (what they are made of) from the resources in the motor vehicle section in the library. The main components are: the cylinder head/block, pistons/rings, crankshaft/bearings, inlet/exhaust valves.
> 2. Give one advantage for the use of each material.

1.5 Definition of terms used in motor vehicle engines

These are given in Table 1.2.

1.6 Cycles of operation

The basic function of an engine is to convert chemical energy (the fuel) into mechanical energy and to produce usable power and torque (this is the ability to turn the driving wheels and move the vehicle). The spark ignition (SI) engine operates on the principle of the Otto four-stroke cycle or the Clerk two-stroke cycle of operations.

Table 1.1   Functional requirements of engine components

Component	Function or purpose
Camshaft cover	Encloses the camshaft and valve mechanism
Camshaft carrier	Locates the camshaft and provides upper bearing locations
Camshaft	Operates the valves and provides a means of driving the auxiliaries
Cylinder head	Provides for (in some cases) the combustion space, inlet and exhaust gas ports, supports the valve gear and part of the cooling system passages
Cam followers	Provide a bearing surface between the rotating camshaft and valve stem
Collets	Lock the valve spring to the valve
Valve spring	Keeps the valves closed, but allows them to be opened when required
Valve	Controls the flow of mixture into and out of the cylinder
Camshaft gear	Provides the drive to the camshaft at half crankshaft speed
Tensioner	Ensures the drive belt is maintained at a constant tension
Timing belt	Transmits the drive from the crankshaft to the camshaft without slipping
Pistons	Transmit the pressure of combustion down to the con-rod
Piston rings	These form a gas- and oil-tight seal
Gudgeon pin	Connects the piston to the con-rod
Connecting rod	Connects the piston to the crankshaft and converts the straight line motion of the piston to the rotary motion of the crankshaft
Crankshaft	Transmits the power from all the cylinders to the flywheel and transmission
Shell bearings	These form the bearing surface for the crankshaft journals

Table 1.2   Definitions of motor vehicle engine terms

Term	Definition
OHV	Over head valve
OHC	Over head camshaft
TDC	Top dead centre
BTDC	Before top dead centre
ATDC	After top dead centre
BDC	Bottom dead centre
BBDC	Before bottom dead centre
ABDC	After bottom dead centre
Bore	Diameter of the cylinder
Stroke	Distance moved by piston from TDC to BDC
Cylinder capacity	Stroke of the piston multiplied by the cross-sectional area of the bore
Swept volume (SV)	Volume created by the area of the bore times the stroke
Clearance volume (CV)	Volume left by piston when at TDC
Compression ratio (CR)	Ratio of swept volume (SV) and clearance volume (CV) against clearance volume (CV). This is expressed as $\dfrac{(SV + CV)}{CV}$
Engine capacity	Capacity of all the cylinders
IVO	Inlet valve opens
IVC	Inlet valve closes
EVO	Exhaust valve opens
EVC	Exhaust valve closes
Valve lead	The amount in crankshaft degrees the valves open before TDC or BDC
Valve lag	The amount in crankshaft degrees the valves close after TDC or BDC
Valve overlap	The amount in degrees the valves are open together measured at TDC
Valve timing	The point at which the valves should open in relation to piston/crankshaft movement
Valve clearance	The distance between the camshaft and valve stem, allows the valve to close
Ignition timing	The time at which the distributor opens the points in relation to the pistons
Combustion	This is the burning of the petrol and air mixture in the cylinder
Energy	Capacity for doing work
Power	Rate of doing work

Four-stroke cycle of operations

To complete the cycle of operations, four strokes of the piston are used. This involves two complete revolutions of the crankshaft, the inlet and exhaust valves being mechanically opened and closed at the correct times. Starting with the piston at TDC and the crankshaft rotating clockwise (looking from the front of the engine), the strokes operate as follows.

First stroke

With the inlet valve open and the exhaust valve closed the piston moves in a downwards direction drawing in a mixture of petrol vapour and air. This is called the **induction** stroke.

Second stroke

The piston moves up with both valves closed, thus compressing the mixture into the combustion chamber at the top of the cylinder. This is the **compression** stroke.

Third stroke

At the end of the compression stroke a spark occurs at the sparking plug. This ignites the mixture which burns very rapidly heating the gas to a very high temperature which also raises its pressure. This forces the piston down the cylinder and is called the **power** stroke.

Fourth stroke

As the piston begins to rise the exhaust valve opens and the spent gases are forced out of the cylinder. This is called the **exhaust** stroke. At the end of this stroke the exhaust valve closes and the inlet valve opens.

This cycle of induction, compression, power and exhaust operates on a continuous basis all the time the engine is running. As can be seen the complete cycle of operations of a four-stroke engine occupies two complete revolutions of the crankshaft. The SI engine draws into the cylinder a mixture of petrol and air which is compressed and burnt. The CI engine draws air only into the cylinder which is compressed to a very high pressure. This also raises its temperature and when fuel is sprayed into the combustion chamber it self-ignites. The four-stroke cycle is illustrated in Fig. 1.7.

> *Learning task*
>
> Read about the diesel four-stroke engine and describe in your own words the cycle of operations.

Two-stroke cycle of operations

By using both sides of the piston the four phases (induction, compression, power, and exhaust) are completed in two strokes of the piston and one revolution of the crankshaft. No valves are used as ports in the cylinder are covered and uncovered by the piston as it moves up and down the cylinder. When describing how this type of engine works it is best to look at what is happening above the piston and then below the piston.

First stroke (piston moving down the cylinder)

Events above the piston

The expanding gases which have been ignited by the spark plug force the piston down the cylinder. About two-thirds of the way down the **exhaust port** is uncovered and the burning gases leave the cylinder. As the piston continues to move downwards, the **transfer port** is uncovered; this allows a fresh mixture into the cylinder.

Events below the piston

The decending piston covers the **inlet port**. The air and fuel trapped in the crankcase is compressed.

Second stroke (piston moving up the cylinder)

Events above the piston

The transfer port is closed first, quickly followed by the closing of the exhaust port. Further movement of the piston compresses the mixture now trapped in the upper part of the cylinder.

Events below the piston

As the piston moves upwards, the depression created in the crankcase draws a fresh mixture in through the inlet port as it is uncovered by the piston.

The two-stroke cycle is shown in Fig. 1.8.

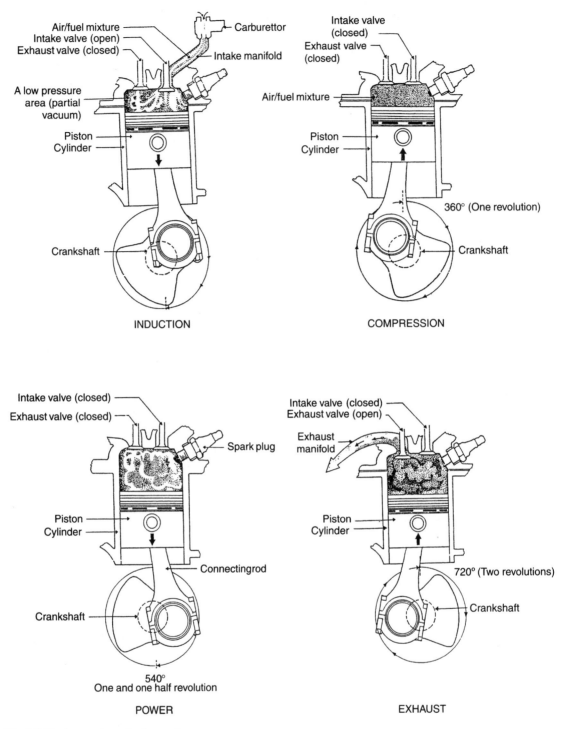

Fig. 1.7 The four-stroke cycle of operations

Two-stroke compression ignition engine

First stroke (piston moving upwards)

The cylinder is filled with air under pressure from the pressure charger. The piston rises covering the inlet ports; the **exhaust valves** are also closed. The air is compressed and fuel is sprayed into the cylinder. It mixes rapidly with the air until self-ignition occurs near TDC.

Second stroke (piston moving downwards)

The rapidly expanding gases force the piston downwards. The exhaust valves are arranged

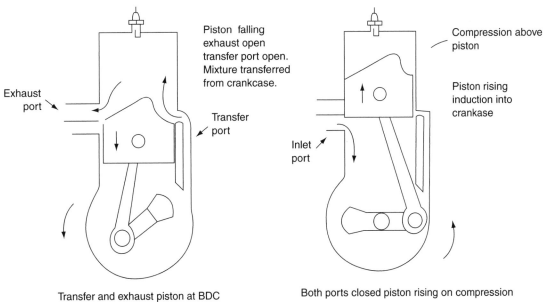

Fig. 1.8 The two-stroke cycle of operations

to open just before the piston uncovers the inlet ports. A new charge of air is forced into the cylinder through the open inlet ports forcing the spent gases out of the open exhaust valves and filling the cylinder with a fresh charge of air.

With the diesel two-stroke it is necessary to pressure charge the engine as there is no actual induction stroke. With the short port-opening period it is essential to fill the cylinder with a large mass of air to create the compression pressure and temperature rise to self-ignite the fuel when it is injected. This gives the following advantages:

- high power-to-weight ratio
- higher engine speeds
- simpler in construction
- good scavenging of exhaust gases

Its main disadvantages are:

- higher fuel consumption
- lower **volumetric efficiency**
- less complete combustion
- good **scavenging** of exhaust gases

Figure 1.9 shows the two-stroke CI engine.

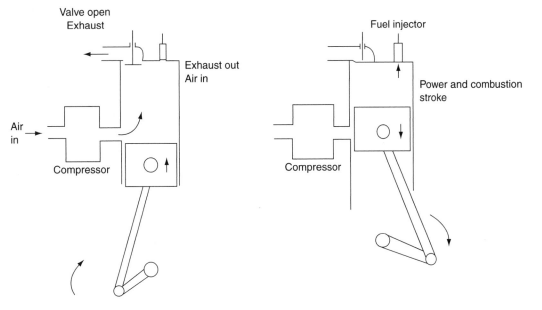

Fig. 1.9 Cross-section of a two-stroke diesel engine

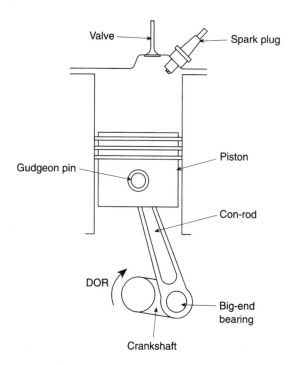

Fig. 1.10 Single-cylinder engine

1.7 The process of combustion

The process of combustion in a diesel engine differs from that in a petrol engine and therefore the two must be considered separately.

Combustion of the petrol/air mixture

Air and fuel are drawn into the cylinder and compressed into the combustion chamber by the rising piston. Just before TDC a spark at the spark plug ignites the mixture which burns rapidly across the combustion chamber in a controlled manner.

Combustion of the diesel/air mixture

Air only is drawn into the cylinder and compressed by the rising piston. The compression pressure, and therefore the temperature of the air, is very much higher in the CI engine than in the SI engine. Fuel in the form of very fine droplets is injected into the cylinder; towards the end of the compression stroke this fuel heats up and self-ignites. This causes a very rapid temperature and pressure rise forcing the piston down on its power stroke. The amount of fuel injected will determine the power developed by the engine.

Figure 1.10 shows the layout of a single-cylinder engine. The **reciprocating** motion of the piston is converted to **rotary** motion of the crankshaft by the connecting rod.

As the piston moves downwards the connecting rod is forced to move the crankshaft in a clockwise direction. In this way the linear (straight line) motion of the piston is converted into rotary motion of the crankshaft

Capacity, swept volume, compression ratio

The following questions show how engine **capacities** are worked out mathematically.

1. The swept volume of a cylinder in a four-cylinder engine is $298 \, cm^3$. Calculate the total volume of the engine.

   $$\text{Total volume} = \text{volume of 1 cylinder} \times 4$$
   $$= 298 \times 4$$
   $$= 1192 \, cm^3$$

2. The swept volume of a cylinder in a six-cylinder engine is $330 \, cm^3$. Calculate the total volume of the engine.

   $$\text{Total volume} = \text{volume of 1 cylinder} \times 6$$
   $$= 330 \times 6$$
   $$= 1980 \, cm^3$$

3. The total volume of a four-cylinder engine is 1498 cm^3. Calculate the swept volume of one cylinder.

$$\text{Volume of 1 cylinder} = \frac{\text{total volume}}{4}$$
$$= \frac{1498}{4}$$
$$= 374.5 \, \text{cm}^3$$

4. The cross-sectional area (CSA) of the piston crown is 48.5 cm^2 and the stroke is 12 cm. Calculate the swept volume of the cylinder and the capacity of the engine if it has six cylinders.

$$\text{Swept volume} = \text{CSA} \times \text{length of stroke}$$
$$= 48.5 \times 12$$
$$= 582 \, \text{cm}^3$$
$$\text{Capacity} = \text{SV} \times \text{no. of cylinders}$$
$$= 582 \times 6$$
$$= 3492 \, \text{cm}^3$$

Learning tasks

1. Calculate the SV of an engine in the workshop by removing a cylinder head and measuring the bore diameter and stroke. Check your results by looking in the manufacturer's manual.
2. What advantages are there to having an over-size cylinder (the diameter of the cylinder larger than the stroke)?
3. What are the constructional differences between the combustion chamber of an SI engine and a direct injection CI engine?
4. What safety aspects should be observed when working with petrol or diesel?

1.8 Cylinder arrangements and firing orders

There are three arrangements which may be used for an engine.

- **In-line engine** The cylinders are arranged in a single row, one behind the other. They may be vertical, as in most modern light vehicles, horizontal as used in coaches where the engine is positioned under the floor, or inclined at an angle to allow for a lower bonnet line.

- **Vee engine** The cylinders are arranged in two rows at an angle to one another. The angle for two-, four- and eight-cylinder engines is usually 90°. For six- and twelve-cylinder engines the angle is usually 60°. This is illustrated in Fig. 1.11.

- **Opposed piston or cylinder engine** This is where the cylinders are at an angle of 180° apart and usually positioned horizontally (see Fig. 1.12).

Learning task

Take a look at each type of engine and draw up a simple list of the main advantages and disadvantages, e.g. is it easier to work on for the mechanic? Does it allow for a lower bonnet line? Is the exhaust system easier to arrange? If so, what advantage/disadvantage is there in this?

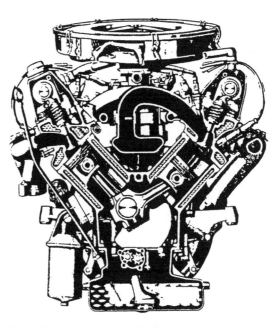

Fig. 1.11 A high-performance V eight-cylinder engine

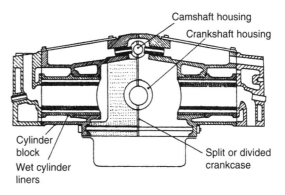

Fig. 1.12 Horizontally opposed cylinders with divided crankcase

Firing orders

When considering multi-cylinder engines and firing orders, the power strokes should be spaced at equal intervals to give the smoothest possible running of the engine. Each interval is equal to the number of degrees per cycle of operation. This will be 720° for a four-stroke engine. This is then divided by the number of cylinders, e.g. $720/4 = 180°$. Therefore the **firing interval** for a four-cylinder in-line engine will be **180°** and that for a six-cylinder in-line engine will be $720/6 = 120°$. The firing order is determined by two things.

- The position of the cylinders and the cranks on the crankshaft (this determines the *possible* firing orders).
- The arrangement of the cams on the camshaft (this must be in accordance with *one* of the possible firing orders).

The arrangements on the crankshaft are such that the **pistons** on a four-cylinder in-line engine are moved in pairs, e.g. numbers 1 and 4 form one pair and 2 and 3 form the other pair. This means that when number 1 is moving down, on its power stroke, number 4 will also be moving down, but on its induction stroke. Depending on the firing order, when number 2 piston moves upwards it will either be on its exhaust or compression stroke, number 3 will be on its compression or exhaust stroke.

From this then we can see that there are two possible firing orders for a four-cylinder in-line engine. These are 1342 or 1243, both of which are in common use today. Table 1.3 below shows the events in each of the cylinders for the two firing orders. The reasons for using more than one cylinder are very complex but in simple terms they are as follows.

- A multi-cylinder engine has a higher power-to-weight ratio than a single-cylinder engine.

- With multi-cylinder engines there are more power strokes for the same number of engine revolutions. This gives fewer fluctuations in torque and a smoother power output.
- A better acceleration is achieved due to smaller moving parts and more firing impulses.
- The crankshaft is balanced better; the crankshaft of a single-cylinder engine cannot be perfectly balanced. Very good balance is obtained with six or more cylinders.
- The piston crown cannot be adequately cooled on large single-cylinder engines; as the piston gets larger the centre of the crown becomes more difficult to keep cool.

Figure 1.13 shows firing orders for a range of different engines.

Speed relationship between crankshaft and camshaft

The movement ratio between the crankshaft and the camshaft is *always* 2:1 on all four-stroke engines. This can be simply determined either by counting the number of teeth on each gear or by measuring the diameter of the gears and dividing the driven gear by the driving gear. From this it will be seen that the camshaft gear is *always* twice as large as the crankshaft gear.

Learning task

Describe two methods you could use to determine the TDC position of number 1 cylinder on its firing stroke.

Cylinder arrangements

In-line cylinders

The four-cylinder in-line is the most popular design in Europe. It has the advantages of having easy access for its size and providing enough power for most applications. The larger six-cylinder engines provide for better acceleration and give better engine balance, and smoother running.

Horizontally opposed cylinders

In this layout the engine has little secondary imbalance giving very smooth running and long engine life. It also has a lower centre of gravity allowing for a lower bonnet line.

Table 1.3 Firing orders: 1342 and 1243

Cylinder number	1		2		3		4	
1st Stroke	I	I	C	E	E	C	P	P
2nd Stroke	C	C	P	I	I	P	E	E
3rd Stroke	P	P	E	C	C	E	I	I
4th Stroke	E	E	I	P	P	I	C	C

P – Power; I – Induction; E – Exhaust; C – Compression

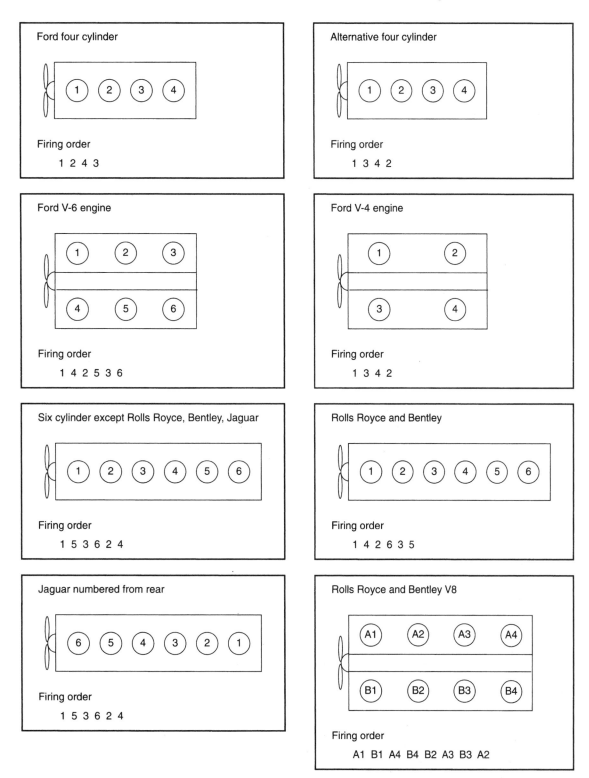

Fig. 1.13 Firing order diagrams for different engines

Vee cylinder arrangements

With this layout the engine is more compact than an in-line engine of the same number of cylinders and the vee 6, 8 and 12 are well balanced.

1.9 Valve-timing diagrams

The valve timing of an engine is set to give the best possible performance. This means that the

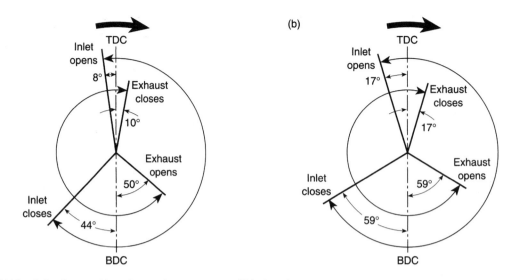

Fig. 1.14 Valve-timing diagrams (a) medium-performance engine, (b) high-performance engine

valves must be opened and closed at very precise times. The traditional way of showing exactly when the valve opens and closes is by the use of a valve-timing diagram (Fig. 1.14). As can be seen the valves are opened and closed in relation to the number of degrees of movement of the crankshaft. When comparing the diagrams for the petrol engine of medium and high performance cars, it will be noticed that the high performance car has larger valve opening periods, especially the closing of the inlet valve which is later. This is so that at high operating speeds the increased lag allows as much pressure energy as possible to be generated in the cylinder by the incoming air and fuel charge, prior to its further compression by the rising piston. There is also an increase in the value of valve overlap for the high performance engine. This means that at TDC both inlet and exhaust valves will be open together for a longer period of time giving a better breathing of the engine at these higher engine speeds (Fig. 1.15).

In the two-stroke petrol engine port timing is the equivalent to valve timing. It must take into account the time lapse before the ports are either fully opened or fully closed, and also the inertia effect of the incoming and outgoing flows of the crankcase and cylinder gases.

In the two-stroke diesel engine the main point to be noticed in comparison with that for a two-stroke petrol engine is that the exhaust event need no longer be symmetrical. This is made possible by the use of mechanically operated poppet valves. An early opening of the exhaust valves initiates (begins) thorough scavenging of

the exhaust gases just before the air inlet ports are uncovered by the piston. The exhaust valves are timed to close just before the air inlet ports to ensure the cylinder is fully charged with fresh air (Fig. 1.16).

Learning task

1. Find information for a sports engine and draw a valve-timing diagram.
2. What would be the result of fitting a cam belt one tooth out?

Valve timing: lead, lag and overlap

The opening and closing of the valves is pre-set by the position and shape of the cam lobes on the camshaft. Their position relative to the movement of the piston is, however, set by the correct positioning of the chain or belt connecting the camshaft to the crankshaft.

The effects on the engine if the valves were set to open **too early** would be loss of power and 'popping back' in the inlet manifold. If the valves were set to open **too late** the effects would be loss of power, overheating, poor starting and exhaust backfiring.

In both cases it could be possible for a valve to hit the piston as it passes TDC on the end of the exhaust stroke and start of the inlet stroke. This is known as valve overlap when shown on the valve-timing diagram.

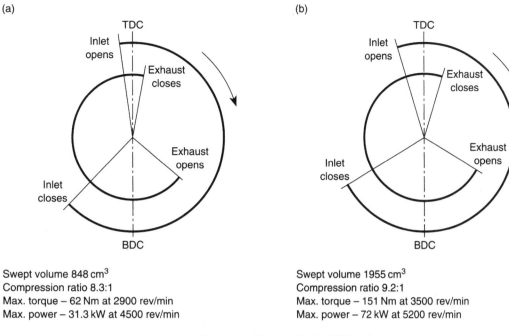

(a)

Swept volume 848 cm^3
Compression ratio 8.3:1
Max. torque – 62 Nm at 2900 rev/min
Max. power – 31.3 kW at 4500 rev/min

(b)

Swept volume 1955 cm^3
Compression ratio 9.2:1
Max. torque – 151 Nm at 3500 rev/min
Max. power – 72 kW at 5200 rev/min

Fig. 1.15 Valve-timing diagrams (a) four-cylinder OHV engine, (b) four-cylinder OHC engine

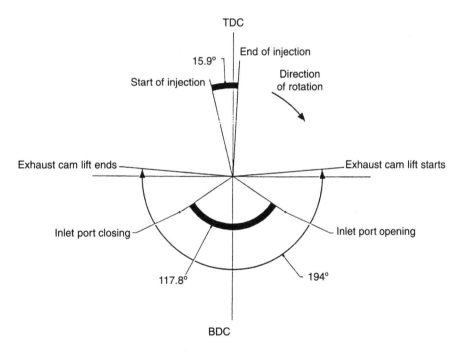

Fig. 1.16 A typical two-stroke CI engine port-timing diagram

1.10 Variable

Valve timing

Valve timing is normally a compromise in design that aims to achieve good performance across the operating speed range of the engine. Inlet valves open a few degrees of crank rotation before top dead centre and close several degrees after bottom dead centre, as shown in Figs 1.14 and 1.15. The period, at the piston top dead centre position, when both valves are open together, is known as valve overlap. It is normally fixed. However, early opening and late closing of the inlet valve has an unhelpful effect on low engine speed emissions and design developments, such as variable valve timing, have been introduced to improve emissions and other aspects of engine performance.

Variable valve timing

Variable valve timing is a development that has been enabled by the use of electronic control which permits valve timing to be changed while the engine is operating, to suit low speed, intermediate speed and high speed operation. The variations in inlet valve timing are approximately as follows:

- **Low speed** – inlet valves opened later to improve idling performance;
- **Intermediate speed** – inlet valves opened a few degrees earlier to take advantage of manifold design and thus improve cylinder filling and performance;
- **High speed** – a larger degree of early opening of the inlet valves.

The amount of variation in the timing may be limited to approximately 10°

The aim of the system is to reduce harmful exhaust emissions and improve engine performance.

Operation of a variable valve timing system

Helical spline inside hydraulic piston

Figure 1.17 shows the elements of a mechanical/hydraulic device that provides a means of varying valve timing. The drive end of the camshaft C is equipped with external splines that form a helical thread. Surrounding this helical splined portion of the camshaft is a hydraulically operated control piston A. The hollow interior of this control piston is splined to fit over the corresponding splines on the camshaft. The exterior of the control piston is also splined so that it can slide axially inside the camshaft drive gear B. Axial movement of the control piston A, relative to the camshaft, will cause the camshaft to rotate by a few degrees relative to the direction of rotation of the camshaft, thus advancing or retarding inlet valve opening. Control of the hydraulic pressure applied to the piston A is achieved by computer control of the solenoid operated valve D. Oil pressure is generated by the main engine oil pump and when pressure is released from the hydraulic piston the oil is

returned to the sump via the control valve. Because the cam profile is constant, this type of device serves to change valve opening and closing times but the valve period (angle between inlet valve opening and closing) remains constant.

Other variable valve-timing systems currently in use also rely on similar technology. One type provides camshaft rotation by means of a hydraulically operated paddle arrangement on the camshaft which means that the valve opening and closing is advanced or retarded and the valve period remains constant. Another type has separate cams for low and high speed operation and hydraulically operated mechanism effects the change from one set of cams to the other, by computer control, while the engine is operating.

Ignition timing

Correct setting of the ignition timing is vitally important, and, as described earlier, ignition takes place as the piston nears TDC towards the end of the compression stroke. Accuracy is necessary to gain the best power output and economy from an engine.

1.11 Combustion in a petrol engine

As the piston reaches TDC the fuel is ignited by the spark at the spark plug and the burning process of the mixture begins. As the gases rapidly expand the piston is forced down the cylinder on the power stroke. The speed of the flame front must not exceed the speed of the power stoke. Figure 1.18 shows the combustion process occurring in a wedge-shaped combustion chamber of a petrol engine.

1.12 Combustion in a diesel engine

Combustion in a petrol engine originates (begins) at the spark plug and then progresses across the combustion chamber in a controlled manner. In the case of the diesel engine, combustion of the fuel is initiated (started) by the heat of the air in the

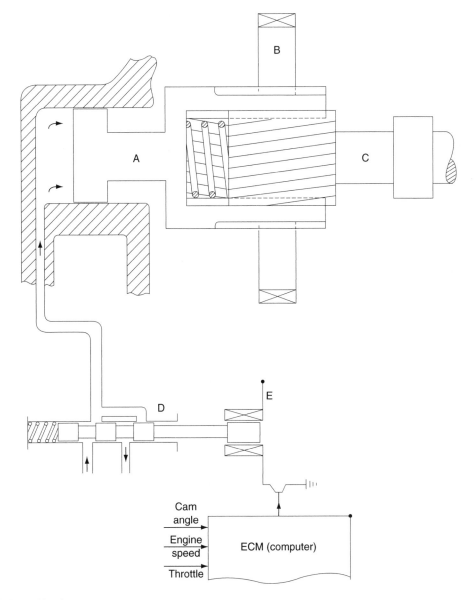

Fig. 1.17 A variable valve-timing system

chamber. As the droplets of fuel pass through the air they absorb the heat, and, if the temperature is high enough, the fuel will **vaporize** and **ignite**. Wide distribution of the fuel during the heating phase means that the burning process (combustion) starts at many different points in the chamber.

In direct injection systems, once ignition has started, most of the burning will tend to concentrate in zones fairly close to the injector. These zones must be fed with air in order to sweep away the burnt gases and supply the oxygen necessary for complete combustion. Any lack of oxygen in the combustion region will lead to black smoke in the exhaust.

A common and essential objective of all CI or diesel engine combustion systems is to achieve the maximum degree of mixing of the fuel, in the form of very fine droplets, with the air. This happens during injection of the fuel into the combustion chamber.

Injection will occur in the period approximately 15° BTDC on the compression stroke to approximately 10° ATDC. Mixing can be achieved by giving the air movement, either by shaping the inlet port, or by masking the inlet valve to give rotary movement to the incoming air charge about the axis of the cylinder. This movement of the air is called **swirl**.

There are two main methods of introducing fuel into the combustion chamber:

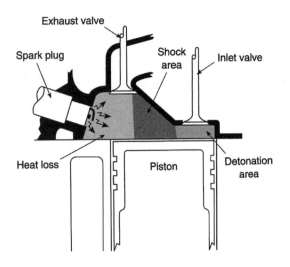

Fig. 1.18 Combustion zones in a wedge-shaped combustion chamber

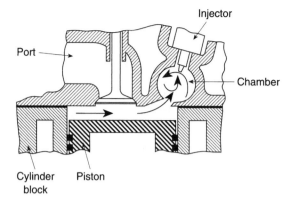

Fig. 1.20 Indirect injection

- direct, and
- indirect.

Direct injection

In the 'direct injection' system (Fig. 1.19) the fuel is injected directly into the combustion chamber which is formed in the piston crown. The air is made to rotate in this cavity at 90° to the incoming swirl by the squeezing out of the air from between the cylinder head face and the piston crown as the piston approaches the end of its compression stroke. This rapid movement of the air is called **turbulence**.

Maximum cylinder pressures are high, causing diesel knock, rough running and higher exhaust smoke. However, easier starting, no starting aids, high thermal efficiency and fairly constant torque output are the main

advantages. Also, because of the low surface area/volume ratio giving low heat losses, a characteristic of the system is a considerable saving in fuel giving good fuel consumption results. A disadvantage is a reduction in volumetric efficiency, due to the necessity of giving the incoming air the swirling movement as it passes into the cylinder. This effect can be largely overcome by the use of the 'indirect injection' system.

Indirect injection

In this arrangement the required movement of the air is made by transferring it, towards the end of the compression stroke, from the cylinder space into a small chamber (usually located in the cylinder head) via a restricting throat, arranged so as to give rapid rotation of the air in the chamber. The fuel is injected into the chamber at a point where the passage of air past the tip of the injector will give the maximum degree of mixing. This is shown in Fig. 1.20.

1.13 Phases of combustion

There are three distinct periods or phases.

First phase – ignition delay period

This is the time taken between the start of the injection of the fuel to the commencement of combustion. During this important period, the injected fuel particles are being heated by the hot air to the temperature required by the fuel to self-ignite.

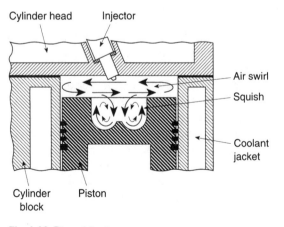

Fig. 1.19 Direct injection

Second phase – pressure rise or flame spread

The flame spread causes a sharp pressure rise due to the sudden combustion of the fuel that was injected during the first phase. The rate of pressure rise governs the extent of the combustion knock (diesel knock).

Third phase – direct or controlled burning

Direct burning of the fuel as it enters the chamber gives a more gradual pressure rise. The rate of combustion during this phase is directly controlled by the quantity of fuel injected into the cylinder. Combustion and expansion of the gases takes place as the piston descends on its power stroke producing a sustained torque on the crankshaft during the time the gases are burning.

Learning tasks

1. From what you have learned so far describe in your own words the main differences between combustion in a petrol engine and combustion in a diesel engine.
2. Examine the combustion chamber of a direct injection engine and make a sketch of the main components, e.g. piston positioned in the cylinder at TDC, the inlet and exhaust valves positioned in the cylinder head.

1.14 The main components used in the construction of the engine

Camshafts

As we have seen, the function of the camshaft is to open the valves at the correct time in the cycle of operations of the engine. It is also used as a drive for various auxiliary units such as the distributor, fuel pump and oil pump.

The position of the camshaft can be in the cylinder block (often termed as side-mounted). The main advantage of this arrangement is

that the timing is not disturbed when the cylinder head is removed. An alternative position is on the top of the cylinder head (termed the over head cam or OHC). This has the advantage of there being a considerable reduction in components that are required to transmit the movement of the camshaft to open the valves.

Figure 1.21 shows a typical camshaft for a four-cylinder engine. The camshaft driving gear or sprocket is located on the shaft by means of a woodruff key or dowel peg to ensure correct fitting, and therefore correct timing, and to give a positive location of the driving gear (see Fig. 1.22).

Camshaft drives

Several methods are employed to transmit the drive from the crankshaft to the camshaft, these are chain, gear and toothed belt (Fig. 1.23). The most common one in use on modern OHC engines is the toothed belt drive. This has the advantages of being silent in operation, requiring no lubrication and being fairly easy to remove and replace.

Learning tasks

1. From the manuals of over head camshaft engines, make a simple sketch of how the camshaft and crankshaft timing marks should be lined up.
2. When next in the workshop remove the timing belt from an engine. Rotate the crank shaft and re-align the timing marks, replace the belt and apply the correct tension.

Valves

Inlet valve

This is made from high tensile alloy steels, e.g. those containing nickel, chromium and molybdenum.

Exhaust valve

This is also made from high tensile alloy steels, for example, those containing alloys of cobalt chromium and silicon chromium, silicon chromium austenitic steel, all of which resist oxidation, corrosion and wear. Under full power it can reach temperatures of around 650 °C. For extreme operating conditions the valve stem is

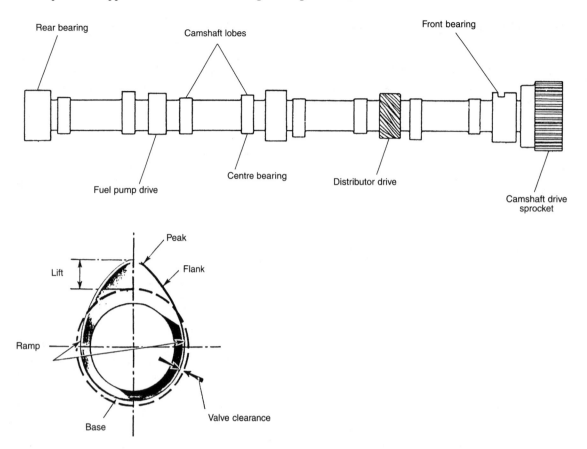

Fig. 1.21 The camshaft. The shape of a camshaft lobe is designed to give required performance with least strain on valve gear

made hollow and partly filled with sodium, which is a very soft metal having a melting point of approximately 98 °C. Under running conditions it is molten, and in splashing from end to end of the valve stem it assists the transfer of heat from the hot valve head to the valve stem.

Several types of valve have been used in the past and are to be found in current production engines. The most common are the poppet valve on the four-stroke engine as shown in Fig. 1.5 and the reed valve on the two-stroke motor cycle engine. The other types to have been tried are the rotary valve and the sleeve valve.

Valve stem seals

Because a clearance is necessary between the valve stem and the guide, valve stem seals are fitted to prevent excessive oil from passing down the stem and into the combustion chamber or exhaust manifold. They are most commonly fitted on the inlet valves as this is on the suction side of the combustion chamber and the oil is more readily drawn into the cylinder. As it is burnt it causes blue smoke to be passed out of the exhaust into the atmosphere. See Fig. 1.25(a).

Valve springs

The purpose of the valve spring is to close the valve. They also prevent the valve from

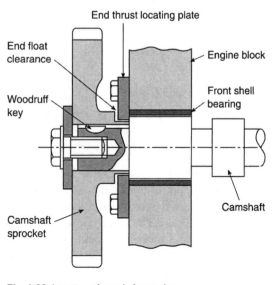

Fig. 1.22 Location of camshaft sprocket

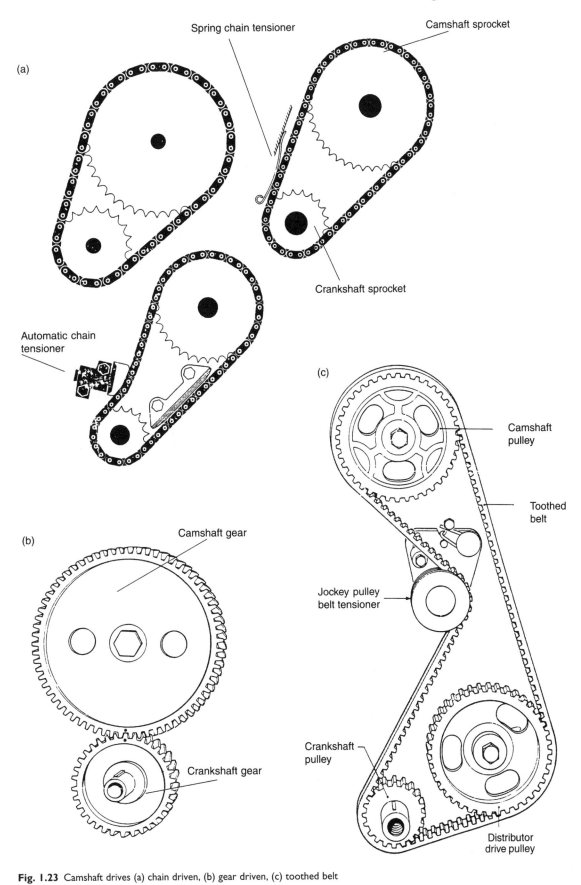

Fig. 1.23 Camshaft drives (a) chain driven, (b) gear driven, (c) toothed belt

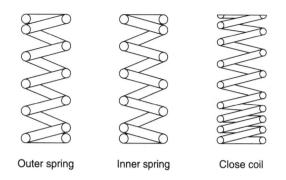

Outer spring   Inner spring   Close coil

Fig. 1.24 Valve springs

bouncing open at the wrong time in the engine cycle (Fig. 1.24). Always fit close coils towards the valve head. The springs are made from either plain high-carbon steel or a low alloy chromium-vanadium steel.

Valve guides

These are usually made of cast iron and are a press fit in the cylinder head, although bronze is sometimes used, particularly for exhaust valves, because of its better heat conducting properties. If the cylinder head is made of cast iron then the guide will be part of the same casting.

Operation of the valves

A number of different methods are used to open and close the valves, the most common methods are:

- **pushrod and rocker** used where the camshaft is positioned in the cylinder block;
- **rocker, lever or finger** used on OHC engines;
- **direct acting** used where the camshaft acts directly on the tappet;
- **hydraulic operation** used where the valve clearance is automatically taken up during normal running of the engine.

These are shown in Figs. 1.25(a)–(c) and Fig. 1.26.

Valve clearance and adjustment

A number of methods are used to position the camshaft so that the valve clearance can be correctly set. The follower must be resting on the lowest part of the cam profile, i.e. the base of the cam, before any adjustment can take place. To determine this position one of the following methods may be used.

- Turn the engine until the valve to be adjusted is fully open and rotate the crankshaft a further complete turn. The follower will now be on the lowest part of the cam and the clearance can be correctly set.
- The 'rocking method' is a quicker way of setting the valve clearances although it is not quite so accurate. Rotate the engine until the valves on number four cylinder are just changing over, i.e. the exhaust valve is just closing and the inlet valve is just opening. Numbers one and four pistons will now be at TDC, the valves on number one cylinder will be fully closed and can therefore be adjusted. To adjust the valve on number four cylinder rotate the engine one full turn (360°). The same method can be used to set numbers two and three cylinders.
- The 'rule of nine' can also be used. Number the valves from the front of the engine one to eight. Rotate the crankshaft until number one valve is fully open; number eight valve will be fully closed and can therefore be adjusted, $1 + 8 = 9$. In this way if the numbers are made to add up to nine the valves can be accurately set using the following method:

number 1 valve fully open, set number 8	$1 + 8 = 9$
number 2 valve fully open, set number 7	$2 + 7 = 9$
number 3 valve fully open, set number 6	$3 + 6 = 9$
number 4 valve fully open, set number 5	$4 + 5 = 9$
number 5 valve fully open, set number 4	$5 + 4 = 9$
number 6 valve fully open, set number 3	$6 + 3 = 9$
number 7 valve fully open, set number 2	$7 + 2 = 9$
number 8 valve fully open, set number 1	$8 + 1 = 9$

The above three methods can only be used on four-cylinder, four-stroke, in-line engines. For other layouts and cylinder arrangements, reference will need to be made to the manufacturer's manual to give the correct procedure to follow.

Hydraulic tappet operation

When the valve is closed, oil from the engine lubrication system passes through a port in the tappet body, through four grooves in the plunger and into the cylinder feed chamber. From the feed chamber the oil flows through a non-return valve (ball type) into the pressure chamber (A in Fig. 1.26).

The load of the cylinder spring enables the plunger to press the rocker arm against the valve, eliminating any free play.

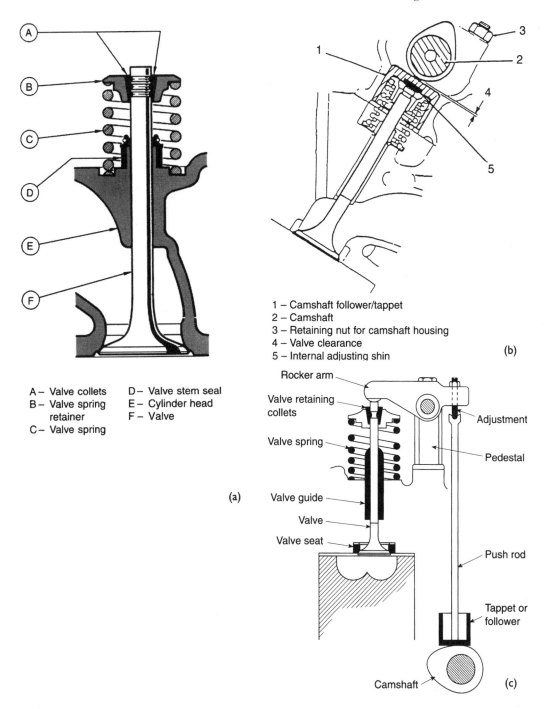

1 – Camshaft follower/tappet
2 – Camshaft
3 – Retaining nut for camshaft housing
4 – Valve clearance
5 – Internal adjusting shin

(b)

A – Valve collets
B – Valve spring retainer
C – Valve spring

D – Valve stem seal
E – Cylinder head
F – Valve

(a)

Rocker arm

Valve retaining collets

Valve spring

Valve guide

Valve

Valve seat

Adjustment

Pedestal

Push rod

Tappet or follower

Camshaft

(c)

Fig. 1.25 (a) A typical OHV assembly, (b) Direct acting type, (c) Push rod valve operation

As the cam lifts the follower the pressure in the pressure chamber rises, causing the non-return valve to close the port feed chamber. Since the oil cannot be compressed it forms a rigid connection between the tappet body, cylinder and plunger so that these parts rise together to open the valve.

The clearance between the tappet body and the cylinder is accurately controlled to allow a specific amount of oil to escape from the pressure chamber. Oil will only pass along the cylinder bore when the pressure is high during valve opening. After the valve has closed, this loss of oil will produce a small amount of free play; and also there will not be any pressure available in the pressure chamber. Oil from the feed chamber can then flow through the non-return valve into the pressure chamber so that the tappet

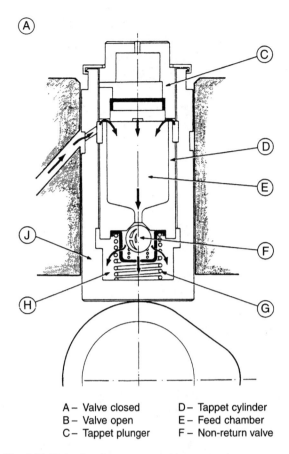
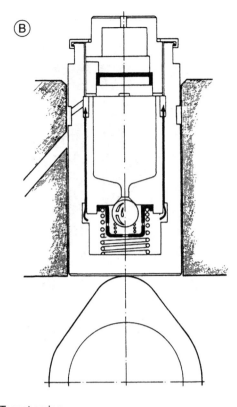

A – Valve closed   D – Tappet cylinder   G – Tappet spring
B – Valve open   E – Feed chamber   H – Pressure chamber
C – Tappet plunger   F – Non-return valve   J – Tappet body

Fig. 1.26 Hydraulic valve tappet assembly – sectional view

cylinder can be raised by the pressure of the spring to eliminate the play in the system before the valve is operated again.

The amount of oil flowing into the pressure chamber will be slightly more than the amount of oil lost when the tappet has to expand due to the increased play (wear). When the tappet has to be compressed due to the expansion of the valve slightly less than the amount of oil lost will flow into the pressure chamber.

> *Note:* If the engine is started up after standing unused for a lengthy period of time a chattering noise may be heard from the valve operating system. This is normal and will disappear after a few seconds when the tappets are pressurized with oil.

Cylinder head

Function

This forms the cover that is fitted on the top of the cylinder block. It may contain the combustion chamber which is formed in the space remaining when the piston reaches the top of its stroke. On some designs this space is formed in the top of the piston and the cylinder head is flat. The head will also contain the spark plugs, inlet and exhaust valves together with their operating mechanism, a number of ports which allow gases into and out of the cylinders, bolt holes to enable components to be bolted to the head and the head to be attached to the cylinder block. It may also contain part of a water jacket which forms part of the cooling system. The materials from which the cylinder head is made are often the same as the cylinder block, e.g. cast iron or aluminium. The cylinder head is illustrated in Fig. 1.27.

Cylinder head tightening sequence

To avoid distorting either the cylinder head or the head gasket the **tightening sequence** (order) must be followed. The torque settings must also be correct to prevent over-stretching of

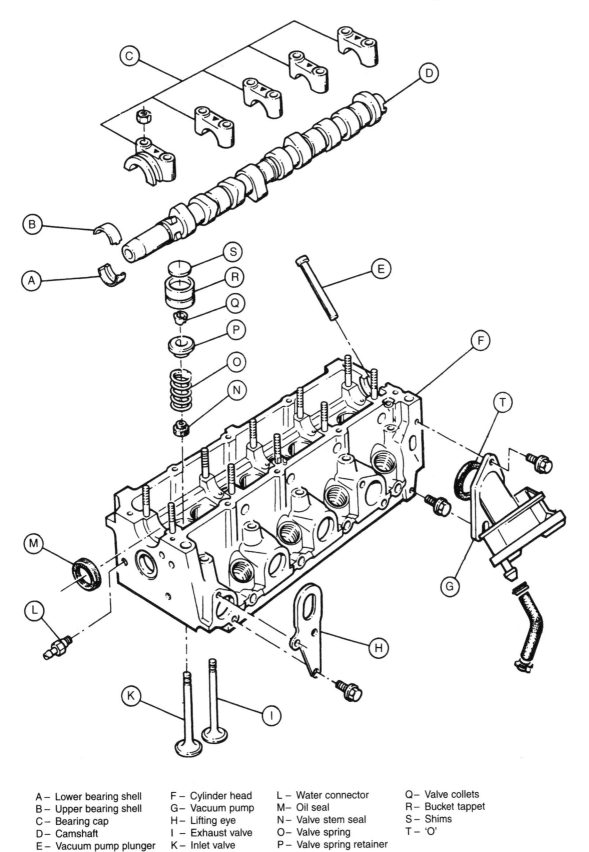

A – Lower bearing shell
B – Upper bearing shell
C – Bearing cap
D – Camshaft
E – Vacuum pump plunger

F – Cylinder head
G – Vacuum pump
H – Lifting eye
I – Exhaust valve
K – Inlet valve

L – Water connector
M – Oil seal
N – Valve stem seal
O – Valve spring
P – Valve spring retainer

Q – Valve collets
R – Bucket tappet
S – Shims
T – 'O'

Fig. 1.27 Cylinder head – exploded view

the bolts and studs or stripping of the threads. The manufacture's manual must always be referred to and the procedure followed; this will ensure that the above problems are not encountered when reassembling the cylinder head to the engine. In some cases it may be necessary to fit new bolts or nuts as these become stretched and cannot be used a second time. Figure 1.28 shows a typical tightening sequence.

Learning tasks

1. Why is the inlet valve larger in diameter than the exhaust valve?
2. What are the reasons for using hydraulic tappets?
3. Under what circumstances would it be necessary to **decoke** a cylinder head? Name any special tools or equipment that you would use.
4. Give the correct procedure for checking and adjusting the valve clearances for an OHC engine.
5. How would you check the serviceability of the valves and their seats?
6. When refitting the cylinder head, why is it necessary to tighten the bolts/nuts in the correct sequence? What would happen if this sequence was not followed?
7. Why is it necessary to have a gap between the valve stem and operating mechanism? What would happen if the engine was operated with no valve clearance?
8. Remove and refit a cylinder head from an engine, make out a job sheet and record any special tools used, faults found and safety procedures that were followed. Draw up an operations schedule for the task.

Piston

The main function of the piston is to provide the movable end of the cylinder, so as to convert the expansion of the burning gases on the power stroke into mechanical movement of the piston, connecting rod and crankshaft. On some types of engines the piston crown is designed to a specific shape instead of being flat. This allows for the shape of the combustion chamber to be included in the piston crown instead of in the cylinder head, and may also have an effect on the flow of gases into and out of the cylinder. Figure 1.29 shows an example of the shape of the combustion chamber being included in the piston crown.

Piston skirt

Several shapes are used in the manufacture of the lower part of the piston, called the piston skirt, e.g.

- **Solid skirt** used in both CI and SI high speed engines, where heavy loadings may be placed on the piston;
- **Split skirt** where small clearances are used to reduce piston slap when the engine is cold;
- **Slipper type** which is used to reduce the weight of the piston by cutting away the bottom of the non-thrust sides of the piston skirt; at the same time it reduces the area in contact with the cylinder wall, and also allows for a reduction in the overall height of the engine as BDC is now closer to the crankshaft.

A piston skirt is illustrated in Fig. 1.30. When cold, the piston head is smaller in diameter than the skirt. When the engine is operating at its normal temperature the piston head expands more than the skirt due to its being closer to

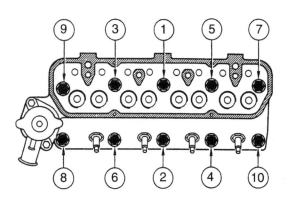

Fig. 1.28 Cylinder head bolt tightening sequence

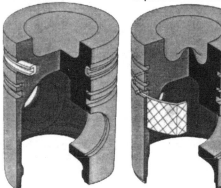

Fig. 1.29 Combustion chambers

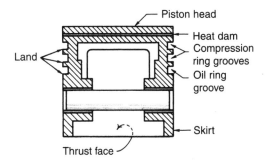

Fig. 1.30 Piston skirt

the very hot gases and also the fact that there is a greater volume of metal at this point.

Piston rings

The piston ring (Fig. 1.31) seals the gap left between the piston and the cylinder wall. Made from high-grade centrifugally cast iron, it is split to enable the ring to be assembled onto the piston. Some rings may be coated on their outer edge with chromium to give better wear characteristics and longer life. Normally three rings are fitted. The **top compression ring** takes most of the compression pressure and forms the first defence against the heat and escaping gases. It may be stepped so that it misses the ridge that tends to form in the cylinder bore at TDC. The second is also a compression ring that completes the sealing against compression loss. The third ring is the **oil control ring**. It is this ring that removes the excess oil from the cylinder wall, passing it back to the sump through holes drilled in the oil control ring groove of the piston. This ring may be made up from a number of steel rails that have radiused chromium plated edges. A crimped spring fitted in the ring groove next to the piston expands the rails against the cylinder wall. This type are commonly fitted where the piston comes very close to the oil in the sump, i.e. short stroke engines, or where some wear has taken place in the cylinder bore but not enough to warrant reboring the engine and fitting new pistons and rings.

Connecting rod

The con-rod connects the piston to the crankshaft. Its action converts the linear (straight line) movement of the piston into the rotary movement of the crankshaft. It is attached to the piston by the gudgeon pin via the **little-end** and to the crankshaft journal by the **big-end**. They are manufactured in the shape of an 'H' as this gives the greatest resistance to the stresses under which it operates whilst at the same time being as light as possible.

Cylinder block

The cylinder block (Fig. 1.32) contains the pistons, liners, crankshaft together with its bearings and sometimes the camshaft. It may also contain the oil pump and galleries to direct the oil to the bearings and the water jacket for the cooling system. It is normally

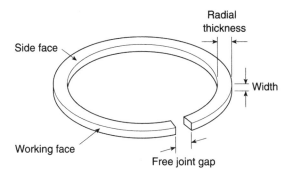

Fig. 1.31 Piston ring

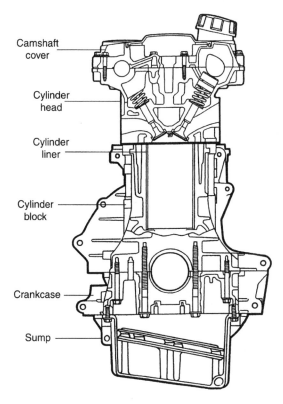

Fig. 1.32 Cylinder block and crankcase assembly

cast in a mould using **cast iron** or **aluminium** and machined to fine tolerances.

Cylinder liners

Cylinder liners are used to allow the engine block to be manufactured from a different kind of material, for example, an aluminium block and cast iron liners. The liners, being made of cast iron, have a much better wear resistance than many other materials. The block can be made from a lighter material than cast iron therefore saving weight. Two of the most common types of liners used are:

- a dry-type liner which forms a lining in the cylinder and is a press fit in the block;
- a wet liner, which forms the cylinder and is in direct contact with the coolant; with seals between the liner and the cylinder block at the top and bottom they are held in position by the cylinder head.

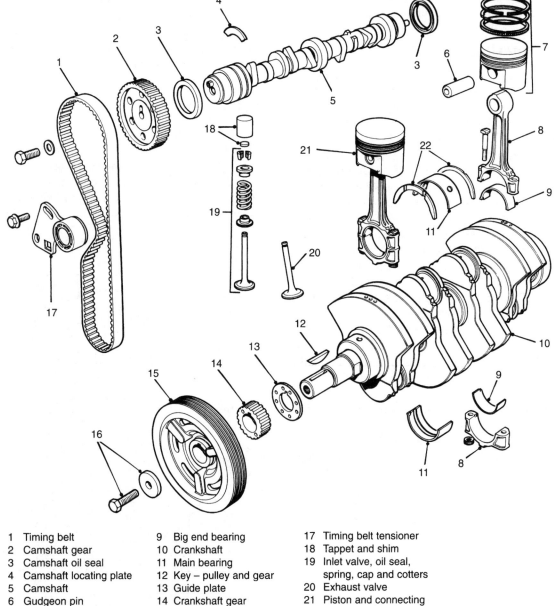

1 Timing belt	9 Big end bearing	17 Timing belt tensioner
2 Camshaft gear	10 Crankshaft	18 Tappet and shim
3 Camshaft oil seal	11 Main bearing	19 Inlet valve, oil seal,
4 Camshaft locating plate	12 Key – pulley and gear	spring, cap and cotters
5 Camshaft	13 Guide plate	20 Exhaust valve
6 Gudgeon pin	14 Crankshaft gear	21 Piston and connecting
7 Piston rings and piston	15 Crankshaft pulley	rod assembly
8 Connecting rod and cap	16 Pulley bolt and special	22 Crankshaft thrust and
	washer	washers

Fig. 1.33 A four-cylinder, five main bearing crankshaft

Crankshaft

The crankshaft (Fig. 1.33) represents the final link in converting the straight line movement of the piston to one of rotating movement at the flywheel. In the case of a multi-cylinder engine, the crankshaft also controls the relative movement of the pistons from TDC to BDC whilst at the same time receiving their power impulses.

Learning tasks

1. Take a look at an engine with cylinder liners and make a simple sketch to show how they are kept in place. Show one other way of doing this.
2. Give three reasons for fitting liners in a cylinder block.

The crankshaft rotates in plain bearing shells in the crankcase, held in position by bearing caps. Oil under pressure from the oil pump is forced into the bearings to lubricate the moving surfaces.

A one piece construction is most commonly used for the motor vehicle crankshaft. It extends the whole length of the engine and must therefore be fairly rigid. The drive for the camshaft is normally taken from the front end of the crankshaft, as is the pulley and belt drive for the engine auxiliaries such as the water pump and alternator.

The flywheel is fitted to the other end of the crankshaft as shown in Fig. 1.34. Its purpose is to store the energy given to it on the power stroke so that it can carry the rotating components over the induction, compression and exhaust strokes (these strokes do not produce any useable power but the engine cannot operate without them). It is also a convenient point from which to pass the drive to the clutch. The starter ring gear is commonly fitted to the outer circumference to enable the engine to be rotated during starting. Several methods of locating the flywheel to the crankshaft are used, such as dowel, key, flange and taper with slotted washer. The attachment must be positive and secure, and preferably in one position only. This provides for balance of both crankshaft and flywheel together.

Seals

Oil seals (Fig. 1.35) are placed at each end of the crankshaft and camshaft to prevent the loss of oil between the shafts and their housings. Where the surfaces do not move, gaskets are used to ensure a water-, oil- and gas-tight seal, for example between the cylinder head and the cylinder block.

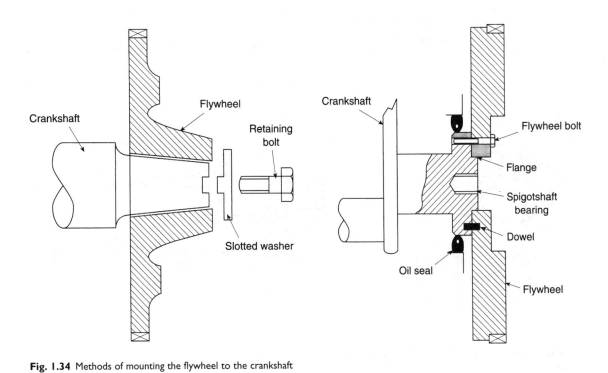

Fig. 1.34 Methods of mounting the flywheel to the crankshaft

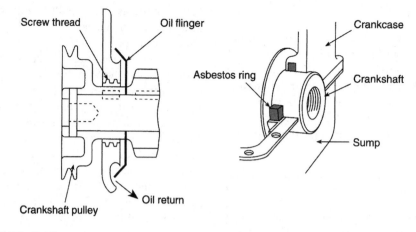

Fig. 1.35 Crankshaft oil seals

Sump

The bottom of the engine is enclosed by a sump which normally contains the engine oil. Commonly made from sheet steel it is often placed in the air flow under the car to assist in cooling the oil.

1.15 Lubrication

To understand how the oil does its work in the operation of the engine and other parts of the motor vehicle, we first need to understand what is meant by **friction**. The term friction is defined as a resistance to movement between any two surfaces in contact with each other.

Learning tasks

1. Why is it necessary to keep the components in the correct order when dismantling and removing them from an engine?
2. What safety precautions should be observed when dismantling an engine in the vehicle?
3. Draw up an operations schedule for removing an engine from a vehicle.
4. Identify four special tools and their uses when reassembling an engine block with its crankshaft and pistons.
5. The oil is contaminated with water. How would this fault be recognized? What would you suspect as the faulty component?
6. What equipment should be used to identify the above fault and how should it be used?

(a)

(b)

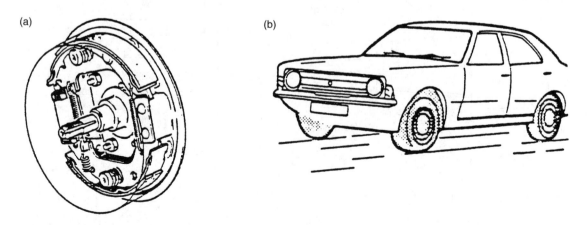

Fig. 1.36 Useful friction in the motor vehicle. (a) Friction between brake linings and brake drum, (b) Friction between tyres and road surface

7. State the method and equipment used to test for a low compression on one cylinder. Which other test could you do to identify which component is faulty?

In some cases friction on a vehicle is useful (Fig. 1.36). The type of friction which keeps our feet from slipping when we are walking also provides the frictional grip that is required between the tyres and the surface of the road, the brake pads and the brake disc, the drive belt and the pulleys of the fan and crankshaft.

If friction occurs in the engine it can cause serious problems as it destroys the effectiveness of the engine components due to the heat generated. This in turn causes wear and early failure of components such as bearings and their journals. It follows then that this type of friction must be reduced to a minimum to allow the engine to operate satisfactorily.

Many years ago it was found that considerable effort was required to drag or push a heavy stone along the ground. It was found to be much easier to roll the stone (Fig. 1.37). It was later discovered that when the stone was put onto a raft and floated on water it was easier still to transport the same stone (Fig. 1.38).

It was almost impossible to move the stone by sliding it along the ground because **dry sliding** friction creates a lot of resistance. This type of

Fig. 1.38 Examples of fluid friction

Ball/roller bearing

Fig. 1.37 Examples of rolling friction

friction is used in the brakes and so is useful (Fig. 1.39). When the stone was rolled it was found to be easier to move. **Rolling** friction creates a lot less resistance and therefore far less heat. This type of friction exists in the ball-and-roller-type wheel bearing (Fig. 1.37).

When the stone was placed on a raft and floated on the water it made the work lighter still. This is called **fluid** friction and exists in the sliding bearings under certain conditions as it does in the crankshaft bearings.

Note: The fact that the raft floats on the water is not the most important factor. When the water comes between the raft and the bed of the river the only force resisting the movement of the raft carrying the stone is the resistance caused by one particle of water sliding over another. This resistance is a lot less than the resistance of dry friction when an object is in direct contact with the solid ground (Fig. 1.39).

To summarize, less force is required to overcome rolling friction than sliding friction. However, when no lubricant is present, the same **wear, heat** or eventual **seizure** of the surfaces in contact will occur, but to a lesser degree in the case of rolling friction. From this we can see that friction can be classified into four basic types.

Clutch plate   Disc brake

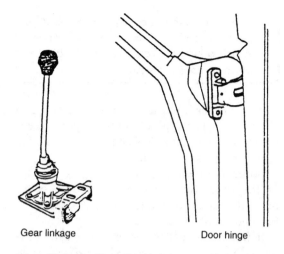

Gear linkage   Door hinge

Fig. 1.41 Areas of boundary friction

Fig. 1.39 Examples of dry sliding friction

- **Boundary friction** This is friction between two materials where very little lubrication is present (Fig. 1.41).
- **Fluid friction** (hydrodynamic friction) A film of fluid prevents contact between two materials, for example between the rotating crankshaft journal and the bearing. A film of oil under pressure ensures that the two metal surfaces slide over one another. The only friction that occurs is between the oil particles themselves. These have less resistance to sliding over one another than solid surfaces have. This reduces friction to a minimum resulting in minimum heat and therefore minimum wear (Fig. 1.42).
- **Mixed friction** This is friction between two materials with a good oil film which is not quite thick enough to prevent contact occurring between the metal surfaces. Here the metal itself can touch the surface of the

- **Dry friction** This is friction between two materials without any type of lubrication. It can generate a large amount of heat (Fig. 1.40).

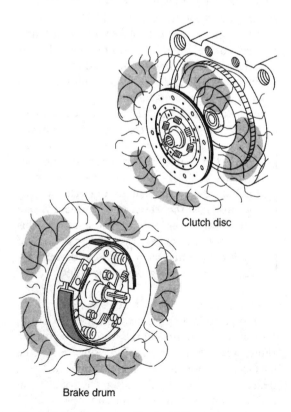

Clutch disc

Brake drum

Fig. 1.40 Heat is generated by dry friction

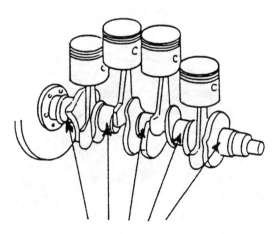

Fig. 1.42 Points of fluid friction

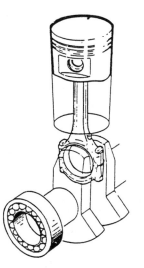

Fig. 1.43 Mixed friction occurs in this ball and roller bearing

second metal for a short time but this is not long enough to produce much heat; therefore wear is limited. This type of friction takes place in ball and roller bearings (Fig. 1.43). The more mixed friction moves towards boundary friction the more heat is produced, this causes more wear and eventually complete seizure.

Learning tasks

1. Name three areas on the motor vehicle where you will find the following types of friction:
   (a) dry friction
   (b) boundary friction
   (c) mixed friction
2. What can change fluid friction to mixed friction and mixed friction to boundary friction?

Engine oil – SAE viscosity classification

When selecting oil for an engine it is important that the one chosen is suitable for the engine itself and the conditions under which it will be used. Two important factors determining the choice of oil are:

- that the oil meets the quality requirements;
- that the oil has the right 'thickness' or **viscosity**.

The term viscosity refers to the relative thickness of a liquid. A thin free-flowing liquid has a low viscosity, and a thick, slow-flowing liquid

has a high viscosity. It will be seen that the viscosity of a liquid like oil changes as the temperature changes. At high temperatures the oil becomes thinner (giving a low viscosity) and at low temperatures the oil becomes thicker (giving a high viscosity). The correct oil viscosity is essential for the efficient operation of the engine because when the oil is too thick (high viscosity) it causes resistance and too much power is needed to turn the engine, making it difficult to start when cold.

Thick oil does not circulate freely enough during the starting period, causing insufficient lubrication of the bearings and thus increasing wear. Oil which is too thin (low viscosity) combined with a high temperature and a heavy load has the risk of oil being pressed out from between the bearings or other engine components. This would cause the oil film, which must be present to keep the components apart, to break down.

The most significant characteristic of lubricating oil is its viscosity. This can be measured in different ways. One way, shown in Fig. 1.44, is to check the quantity of oil which passes through a hole in a tube of standard size during a certain length of time at a given temperature.

A widely used system of grading oil, based upon viscosity, has been worked out by the American Authorities for Standardization (SAE – Society of Automotive Engineers). Various oils are grouped into viscosity grades marked with SAE numbers. These indicate the **viscosity index** of the oil (see Fig. 1.45).

For engine oils the **SAE grades** are numbered: 5 W, 10 W, 15 W, 20 W, 30, 40, 50. The lower numbers indicate thin oils and the higher numbers indicate thicker, or higher-viscosity, oils. For the SAE grades 20, 30, 40 and 50, the viscosity is measured at an oil temperature of 100 °C which is the normal oil temperature when the engine is running. The grading can be read on the right-hand side of the graph in Fig. 1.46.

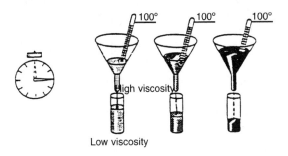

Fig. 1.44 Checking oil viscosity

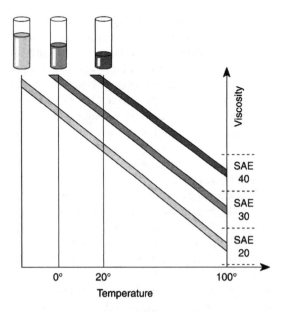

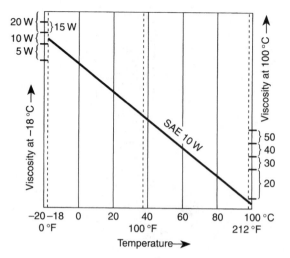

Fig. 1.47 SAE 10 W single grade oil

For the SAE grades 5 W, 10 W, 15 W and 20 W, the viscosity is measured at −18 °C (0 °F). This can be seen on the left-hand side of the graph in Fig. 1.47.

Now we have a very wide range of oil viscosity from very thin oil (SAE5W) up to thick oil (SAE50). Every SAE grade represents an oil suitable for use within a specified range of temperatures and for a certain type of engine, for example, oils within the viscosity range SAE 5 W, 10 W and 20 W are suitable for use in climates with temperatures ranging from very low to moderate. SAE 20 and 30 are suitable for use in moderate to hot climates. SAE 40 and 50 oils are mostly used in old engines designed for

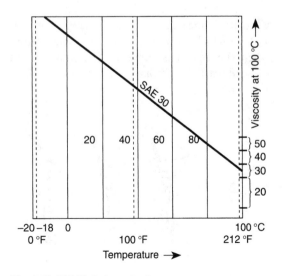

Fig. 1.46 SAE 30 single grade oil

rather thick oil or in badly worn engines with high oil consumption.

However, many engines operate in climates where the temperature varies considerably from season to season. In addition it is always preferable for any engine to run on low-viscosity oil during the starting period when the engine is cold, and on high-viscosity oil when the engine is hot and fully loaded. For practical reasons it is not possible to use a low-viscosity oil for starting and then change to a high-viscosity oil when the engine is hot. Therefore, we need an oil which is thin enough at low temperatures, but will also have a sufficiently high viscosity at high temperatures – that is a **multi-viscosity** or **multi-grade** oil.

We cannot change the fact that oil becomes thinner as it is heated, but it is possible, by the use of additives to the base oil, to reduce the extent to which this happens. For example, an oil having a viscosity of 10 W at −18 °C can be improved so that it also has a viscosity of SAE30 at 100 °C.

As the oil now has a viscosity equivalent to SAE10 W at −18 °C and a viscosity equivalent to SAE30 at 100 °C, it is marked with both numbers, that is SAE 10 W/30. Oils are available with a very wide viscosity range, e.g. 10 W/50 which may be used in any climate enables the cold engine to start on a thin oil (SAE10 W) and, when hot, to run on a sufficiently thick oil (SAE50).

Oils that meet the specification for more than one SAE grade are often referred to as multi-viscosity, **all season** or multi-grade oils. Remember that the SAE grades only tell you the viscosity of the oil and not the quality. The correct method for selecting the oil with the right SAE number corresponding to the temperature and running conditions would be to check in the specifications for the engine. Oil producers have worked out, together with the engine manufacturers, recommendations for the oil viscosity most suitable for given conditions.

It is important that only the correct grade of oil is used in a given vehicle and the information about that is given in the driver's handbook or the workshop manual. In most modern engines multi-grade oils are recommended regardless of temperature. An SAE10 W/40 oil, for instance, can be used in most climates thus simplifying the selection of oil. It is important to know that the SAE grades only give information on the viscosity of the oil and not the quality. This can be found by making reference to the API (American Petroleum Institute) classification.

Lubrication system

There are three main types of lubrication systems in common use on internal combustion engines. These are:

- **wet sump**
- **dry sump**
- **total loss**

The object of the lubrication system is to feed oil to all the moving parts of the engine to reduce friction and wear and to dissipate heat. Modern oils also clean the engine by keeping the products of combustion, dirt, etc. in suspension. This makes it essential that oil and filters are changed in accordance with the manufacturers' instructions. It can be seen from this that the oil performs four important functions.

- It keeps friction and wear on the moving parts to a minimum.
- It acts as a coolant and transfers the heat from the moving parts.
- It keeps the moving parts clean and carries the impurities to the oil filter.
- It reduces corrosion and noise in the engine. It also acts as a sealant around the piston and rings.

Learning tasks

1. What special measures should be observed when storing/using oil? How should waste oil be disposed of?

2. Write to an oil manufacturer and ask for details of their current products for the vehicles most commonly serviced in your workshop, e.g. types of oils recommended for the engine, gearbox, final drive, special additives used in the oils, types of grease available, etc.
3. Write to an oil manufacturer and ask for charts showing quantities of oil used in vehicles, viscosity and API ratings, etc. that could be displayed in the motor vehicle workshop.

Main components in the lubrication system

The lubrication system is mostly pressurized and consists of the following main components.

- **Oil pump** draws the oil from the sump and delivers it under pressure to the engine lubrication system.
- **Relief valve** limits the maximum pressure of the oil supplied by the pump to the system.
- **Sump** serves as a reservoir for the oil.
- **Oil galleries** are channels or drillings through which the oil passes to the different lubrication points in the engine.
- **Oil pressure indicator** shows whether the oil pressure is being kept within the manufacturers' limits.
- **Oil filter** filters the oil removing impurities to keep it clean.

A typical lubrication system works in the following way.

1. Oil is drawn from the sump by the oil pump.
2. The pump pressurizes the oil and passes it through the oil filter into the oil galleries and passages which lead to the crank-shaft and camshaft bearings and, in some engines, the rocker shaft and rocker arms.
3. The oil splashed from the crankshaft lubricates the pistons and other internal parts of the engine.
4. After lubricating the moving components the oil drips back down into the sump. Figure 1.48 shows the oil flow in an engine lubrication system.

Function of the oil

When looking at the working surfaces of, say, the crankshaft (A in Fig. 1.49) and the main bearings in which it runs (B) they

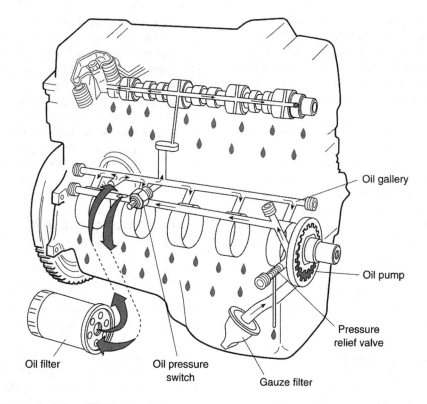

Fig. 1.48 Lubrication circuit for an OHC engine

appear to be blank and smooth. But when observed under a powerful microscope they will be seen to be uneven and rough. When oil is introduced between these surfaces, it fills up the slight irregularities and forms a thin layer, called an oil film. It is this oil film that separates the surfaces and, when the components are rotating, prevents metal-to-metal contact. When the engine is operating the oil must be strong enough to withstand the heavy loads imposed on all the moving parts. The oil is therefore delivered under pressure to the bearings and, to enable it to enter, a very small clearance between the shaft and bearing is necessary.

The clearance must be sufficient for the oil to enter, but small enough to resist the heavy loadings to which the bearings are subjected. This clearance is approximately 0.05 mm for a shaft of 60–70 mm diameter. When the shaft is not rotating but is resting on the bearings only a very thin film of residual oil separates the surfaces. As the engine starts the only lubrication for the first revolution is provided by this thin film of oil; as the revolutions increase, the oil pump starts to deliver the oil under pressure to the bearings. The oil is

drawn round by the rotating shaft which, together with the pressure, forms an oil wedge which lifts the shaft up from the bearing. The shaft then rotates freely, separated from the bearings by this thin film of oil. A correct bearing clearance is shown in Fig. 1.50.

It is important that the bearings have the correct clearance. Too much will cause the oil to escape from the bearing without being able to create the required **oil wedge**; too little will restrict the oil from entering the bearing, causing metal-to-metal contact. In both cases wear will increase.

The oil which is pressed out from the bearings is splashed around and forms an oil mist which lubricates the cylinder and piston (Fig. 1.51). In some cases a hole is drilled from the side of the connecting rod into the bearing shell to spray oil onto the thrust side of the cylinder wall.

Function of the sump

When the engine is filled with oil, it flows down through the engine into a container called the sump. This is attached to the bottom of the engine block with a series of small

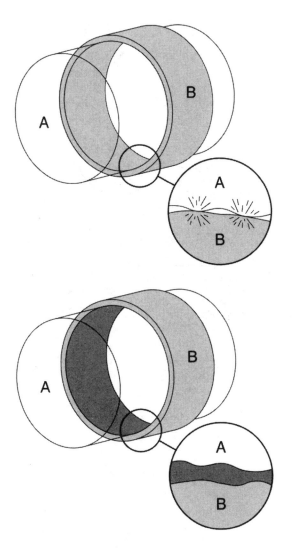

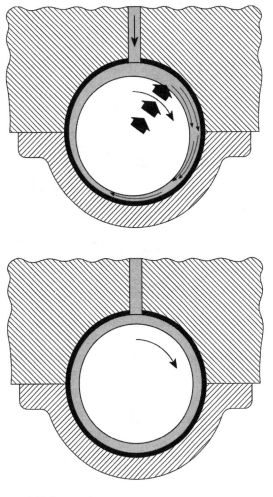

Fig. 1.50 Bearing clearance correct

Fig. 1.49 Bearing surface shown under a microscope. Oil under pressure keeps the surfaces apart

bolts, usually with a gasket between the block and the sump. It is commonly formed from sheet steel pressed to a shape that has one end slightly lower to form the oil reservoir. In the bottom of the sump is the drain plug. **Baffle plates** are fitted to prevent the oil from splashing around or surging when the vehicle is accelerating and braking or going round corners. If all the oil is allowed to move to the rear or to the side of the sump, the oil pick-up may become exposed causing air to be drawn into the lubrication. The sump also acts as an oil cooler because it extends into the air stream under the vehicle. To assist with the cooling process there may also be small fins formed on the outside to increase its surface area. Aluminium is sometimes used to give a more rigid structure to support the crankshaft and crankcase

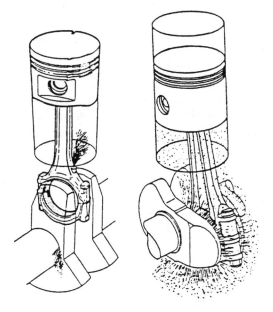

Fig. 1.51 Splash lubrication from the big-end bearing

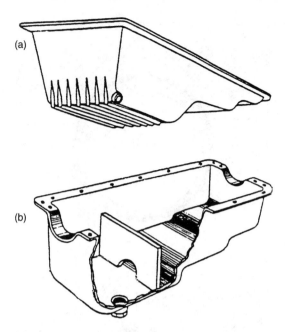

Fig. 1.52 Types of sump (a) aluminium cast, (b) pressed steel

of the engine. Two types of sump are shown in Fig. 1.52.

Oil level indicators

The level in the sump is checked by means of a dipstick on which the maximum and minimum oil levels are indicated. A number of vehicles fit indicators to show the driver the level of oil in the engine without having to lift the bonnet and remove the dipstick manually. One of the popular types used is the 'hot wire' dipstick where a resistance wire is fitted inside the hollow stick between the oil level marks. The current is only supplied to the wire for about 1.5 seconds at the instant the ignition is switched on. If the wire is not in the oil it overheats and an extra electrical resistance is created which signals the ECU (electronic control unit) to operate the driver's warning light. This is illustrated in Fig. 1.53.

Learning tasks

1. What would be the effects of running an engine with the oil level (a) above the maximum, and (b) below the minimum on the dipstick?
2. A vehicle is brought into the workshop with a suspected oil leak from the sump. Suggest one method of checking exactly where the

---

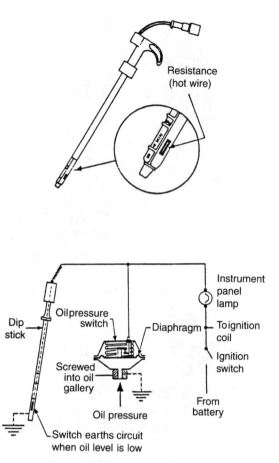

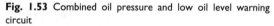

Fig. 1.53 Combined oil pressure and low oil level warning circuit

---

leak is coming from and describe the procedure for correcting the fault.
3. How should a resistance-type oil level indicator be checked for correct operation?

Oil pump

The oil enters the pump via a pipe with a strainer on the end which is immersed in the oil reservoir in the sump. This strainer prevents larger particles from being sucked into the lubrication system. The oil pump creates the required pressure that forces the oil to the various lubrication points. The quantity of oil delivered by the pump varies greatly from vehicle to vehicle and also depends on engine speed but will be approximately 120 litres when the speed of the vehicle is 100 km/h.

The most common types of pumps used in the motor vehicle engines are the **gear**, **rotary** or **vane**.

Gear pump

As shown in Fig. 1.54, gear pump consists of two gears in a compact housing with an inlet and outlet. The gears can be either **spur** or **helical** in shape (the helical being quieter in operation). The pump drive shaft is mounted in the housing and fixed to this is the driving gear. Oil is drawn via the inlet into the pump. It passes through the pump in the spaces between the gear teeth and pump casing and out through the outlet at a faster rate than is used by the system. In this way pressure is created in the system until the maximum pressure is reached at which time the pressure-relief valve will open and release the excess pressure into the sump.

Rotary pump

The main parts of this type of pump (shown in Fig. 1.55) are the **inner rotor**, the **outer rotor** and the **housing** containing the inlet and outlet

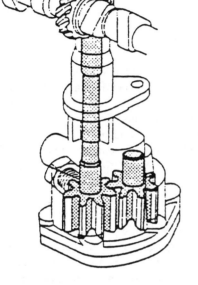

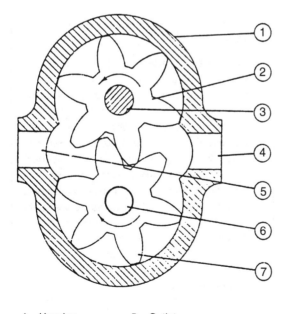

1 – Housing   5 – Outlet
2 – Driving gear   6 – Driven gear
3 – Driving shaft 7 – Gear lobe
4 – Inlet

Fig. 1.54 Gear-type pump

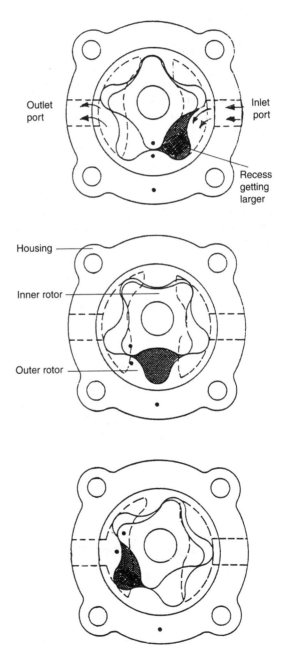

Fig. 1.55 Rotary pump

ports. The inner rotor, which has four lobes, is fixed to the end of a shaft; the shaft is mounted off-centre in the outer rotor which has five recesses corresponding to the lobes. When the inner rotor turns, its lobes slide over the corresponding recesses in the outer rotor turning it in the pump housing. At the inlet side the recess is small; as the rotor turns the recess increases in size drawing oil up from the sump into the pump. When the recess is at its largest the inlet port finishes, further movement of the rotor reveals the outlet port and the recess begins to decrease in size forcing the oil under pressure through the outlet port.

Vane-type pump

This pump, shown in Fig. 1.56, takes the form of a driven rotor that is eccentrically mounted (mounted offset) inside a circular housing. The rotor is slotted and the **eccentric vanes** are free to slide within the slots, a pair of thrust rings ensuring that the vanes maintain a close clearance with the housing. When in operation the vanes are pressurized outwards by the centrifugal action of the rotor rotating at high speed. As the pump rotates the volume between the vanes at the inlet increases, thus drawing oil from the sump into the pump; this volume decreases as the oil reaches the outlet, pressurizing the oil and delivering it to the oil gallery. This type has the advantage of giving a continuous oil flow rather than the pulsating flow that is rather characteristic of the gear-type pump.

Pressure-relief valve

As engine speed increases the oil pump produces a higher pressure than is required by the engine lubrication system. A **pressure-relief valve** (see Fig. 1.57) is therefore fitted in the system to take away

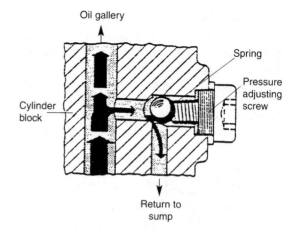

Fig. 1.57 Oil pressure-relief valve

the excess pressure and maintain it at a level appropriate for the bearings and seals used. It will be seen then that the relief valve performs two important functions: first, it acts as a pressure regulator; and second, it acts as a safety device in the lubrication system. The main types in use are the ball valve, the plate and the plunger or poppet valve. Each is held in the closed position by a spring. As the oil pressure in the oil gallery rises above the setting for the relief valve, the valve opens against spring pressure allowing the oil to bypass the system and return back to the sump via the return outlet. The force on the spring determines the oil pressure in the lubrication system.

Learning tasks

1. When next in the workshop, dismantle a vane-, rotor- and gear-type oil pump. Using the manufacturers' data measure and record the tolerances for each type and give your recommendations on serviceability.
2. Remove the oil pressure switch from the oil gallery of an engine and attach the appropriate adapter to measure the oil pressure. Record the results at the relevant speeds. Identify any major differences between the recorded results and the manufacturer's data and give your recommendations.
3. Give two reasons for low oil pressure together with recommendations for correcting the fault.

Oil filter

When the oil passes through the engine it becomes contaminated with carbon (the

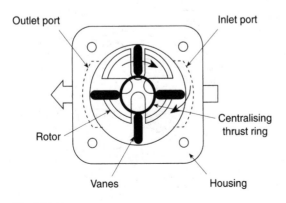

Fig. 1.56 Vane pump

byproduct of the combustion process), dust (drawn in from the atmosphere), small metal particles (from components rubbing together), water and sludge (a combination of all these impurities mixed together). All these will cause engine wear if they remain in the oil, so the engine must be equipped with a filtering system that will remove them and keep the oil as clean as possible. Most modern engines are equipped with a filtering system where all the oil is filtered before it reaches the bearings. This arrangement is called the **full-flow** system. There is another system also in use where only a portion of the oil passes through the filter, called the **bypass** filter system. The two systems are shown in Fig. 1.58.

The importance of filtering the oil is shown by the results of an investigation into the wear on the cylinder and piston, using the two filtering systems. It was found that maximum wear (100%) occurs in engines working without an oil filter. When a bypass filter is used, wear is reduced to about 43% on the cylinder and 73% on the piston, which means that the life of the piston and cylinder are almost doubled. Minimum wear occurs when a full-flow filter is used, wear is again reduced by a further 15% on the cylinder and 22% on the piston. This means that the life of the piston and cylinder is four to five times longer than in an engine working without a filter. A good oil filter must be capable of stopping the flow of very small particles without restricting the flow of oil through the filter. To meet this requirement different materials are used as the filtering medium. Resin-impregnated paper is widely used, the paper being folded in order to make a large surface area available for the oil to

flow through; particles are left on the paper and clean oil is passed to the lubrication system. In this way when the filter is changed the impurities are removed at the same time. In other types of oil filters different kinds of fibrous materials are used. The filtering material is enclosed in perforated cylinders, one outer and one inner to form a filter element (Fig. 1.59).

The oil enters through the perforations in the cylinder, passes through the filtering element and leaves through the central tube outlet. Many modern filters are now the cartridge-type which is removed complete. The advantages of this disposable type are that it cleans the oil very efficiently, it is relatively easy to change and it is less messy to remove. The filter element can also be located in a removable metal container. With the replaceable-element-type it is only the element itself that is changed, the container is thoroughly cleaned and the 'O' ring replaced.

Full-flow filter

Oil filter operation is shown in Fig. 1.60. The most widely used filtering system is the full-flow filter. The construction of the filter is very efficient because all the oil is passed through the filter before it flows to the bearings. After a certain length of time the element becomes dirty and less efficient and must therefore be changed. If the element is not changed regularly the impurities will accumulate and the element will become clogged, restricting or preventing the oil from passing through the filter. For this reason a relief valve is fitted into the filter which opens and allows the oil to bypass the clogged

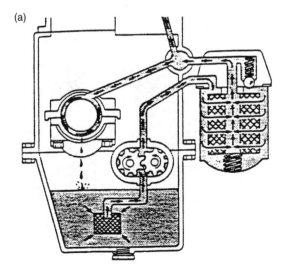
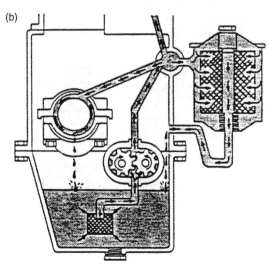

Fig. 1.58 Lubrication systems (a) full flow system, (b) by-pass system

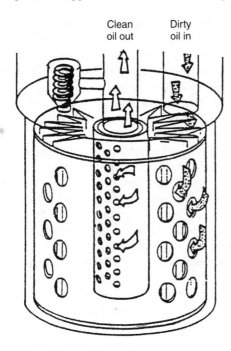
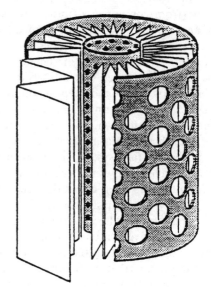

Fig. 1.59 The oil filter and pleated element

filter element and flow directly to the bearings unrestricted. If the condition is allowed to continue, unfiltered oil will carry abrasive particles to the bearings causing rapid wear.

Cartridge filter

In the cartridge filter the relief valve is in the filter (shown in the open position in Fig. 1.61). Many of the filters now contain a valve underneath the inlet hole which opens when oil pressure forces oil into the filter. When the engine stops and the oil flow ceases, the valve closes and the oil is kept within the filter. This prevents it from draining back into the sump. It also has the advantage of enabling the engine to develop the oil pressure more quickly when starting from cold. Correct operation is shown in Fig. 1.62.

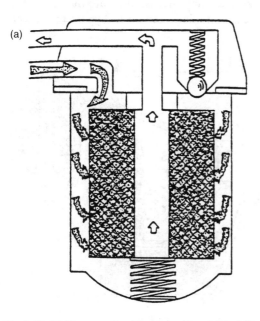
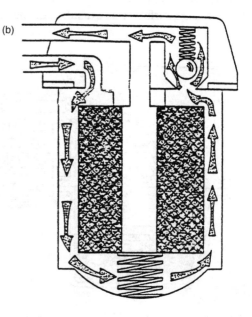

Fig. 1.60 Oil filter operation (a) oil being filtered, (b) oil filter blocked

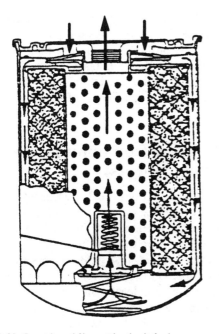

Fig. 1.61 Cartridge oil filter with oil relief valve open

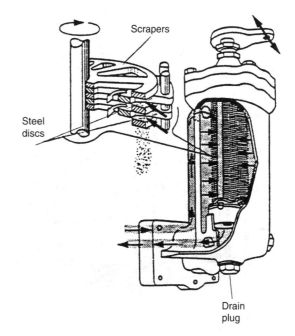

Fig. 1.63 Disc-type oil filter

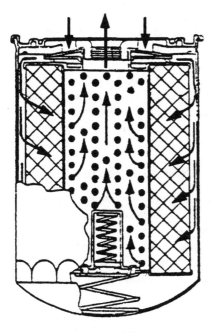

Fig. 1.62 Cartridge oil filter operating correctly

Disc filter

This type of full-flow filter (shown in Fig. 1.63) is used in large diesel engines. The oil is filtered by being forced through very narrow gaps (0.05 mm) between thin steel discs which form an assembly which can be rotated. The narrow gap between the discs prevents impurities in the oil from passing through. The deposits accumulate on the outside of the discs, which are kept clean by scrapers which scrape off the deposits as the disc assembly rotates. In most cases the assembly is connected to the clutch pedal; each time the pedal is operated the disc assembly is rotated a small amount. The filter must be drained as per manufacturers' recommendations, this being done by removing the drain plug allowing dirt and some oil to be flushed out.

Centrifugal filter

Again mainly found on larger engines, this consists of a housing with a shaft and rotor inside. The oil is forced through the inlet ports by the pump and fills the rotor through the inlet holes in the rotor shaft, passing down the pipes to the jets. Due to the force of the oil passing through the jets the rotor rotates at very high speed. Owing to the centrifugal force, the impurities (which are heavier than the oil) accumulate on the walls of the rotor. The filter must be periodically cleaned by dismantling the filter and washing with a suitable cleaning fluid.

> ### Learning tasks
>
> 1. Draw up a lubrication service schedule for a 12 000 mile major service.
> 2. Complete a lubrication service on a vehicle using the above service schedule. Identify any faults found, such as oil leaks, and complete the report on the workshop job card.

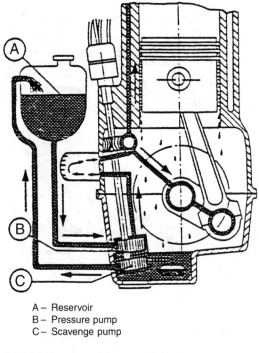

A – Reservoir
B – Pressure pump
C – Scavenge pump

Fig. 1.64 Operation of dry-sump lubrication system

Dry-sump lubrication

This type of system (Fig. 1.64) is fitted to vehicles where the engine is mounted on its side, or where greater ground clearance is required. It is also used for motor cycle engines, cross-country vehicles and racing engines where under certain conditions the pick-up pipe could be exposed for a period of time and therefore the oil supply to the engine lubrication system could be interrupted. To overcome this problem a dry sump system is often fitted.

The oil is stored in a separate oil tank instead of in the sump. The oil pump takes the oil from the tank and passes it to the lubrication system. The oil then drops down to the crankcase where a separate scavenge oil pump often running at a higher speed than the pressure pump returns it back to the oil tank. This means that the sump remains almost dry. The faster speed of the scavenge pump is due to the fact that it must be capable of pumping a mixture of air and oil (which has a larger volume than just oil) back to the tank.

Oil coolers (sometimes called heat exchangers)

Two types of oil coolers are fitted where the heat is removed from the oil: one is the oil-to-air where the heat is passed directly to the air; the other is the oil-to-water where the heat from the oil is passed to the water cooling system. Both types are shown in Fig. 1.65. On the water-cooled engines the oil cooler is normally located in front of, and sometimes combined with, the radiator. The advantage of the water-type heat exchanger is that the oil and water are operating at roughly the same temperature and each is maintained at its most efficient working temperature under most operating conditions. On air-cooled heat exchanger engines the cooler is usually located in the air stream of the cooling fan and is similar in construction to the cooling system radiator. An oil cooler bypass valve is fitted in the system which allows the oil to heat up more rapidly from cold by initially restricting its circulation to the engine only.

Total loss lubrication

There is one system commonly used that has so far not been mentioned; that is the total loss system. This is where the oil used to lubricate the piston, main and big-end bearings, is burnt during the combustion stroke and therefore lost through the exhaust system to the atmosphere.

(a)

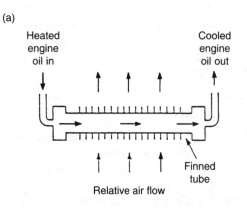

(b)

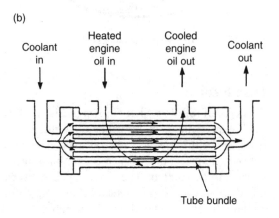

Fig. 1.65 Basic principles of oil coolers (a) oil-to-air, (b) oil-to-water

One example of this is the two-stroke petrol engine used in the motor cycle.

Learning tasks

1. Remove the pump from a motor cycle engine and identify the type of pump fitted and method of operation. After reassembling immerse the pump in clean engine oil and test the pumping side for pressure and the scavenge side for suction. Record the results.
2. Give two methods used in the total loss lubrication system of introducing oil into the engine.
3. How should engine oil be removed from the workshop floor?
4. List the main safety precautions that should be observed when working on the lubrication system.

Practical assignment – removing and refitting the cylinder heads

Objective

To visually check the condition of:

- the valves and valve seats
- the cylinder head for cracks and distortion
- the piston crowns for build-up of carbon
- the cylinder bores for scoring and excessive wear

Vehicle/engines used

This exercise should be carried out on both OHC and push-rod operated engines.

Equipment required

- Specialist test equipment
- Small hand tools, e.g. spanners, sockets, etc.
- Valve spring compressors
- Torque spanner
- Relevant repair manual
- Drain trays

Task

Before removing the cylinder head the following checks/tests should be carried out:

- compression test
- cylinder leakage test
- cylinder balance test
- pressure check the cooling system
- check the anti-freeze content of water
- check condition of engine oil
- check condition of water

Activity

1. Drain both engine oil and water.
2. Remove ancillary components, e.g.
    (a) all electrical connections
    (b) accelerator and choke cables
    (c) water and air intake hoses/connections
    (d) cam belt OHC (having set camshaft timing) inlet and exhaust manifolds
3. Remove camshaft (OHC) or rocker shaft and push rods.
4. Remove cylinder head (ensure bolts/nuts are released in the correct order).
5. Remove all the inlet and exhaust valves (ensure these are kept in the correct order).
6. Thoroughly clean all the components.
7. Visually inspect valve seats for pitting on seats and measure for wear on stems.
8. Visually check cylinder head for cracks and distortion.
9. Visually check valve seats in cylinder head for pitting and burning.
10. Reface valves and seats or replace as necessary.
11. Regrind valves into cylinder head.
12. Reassemble valves into cylinder head using new valve stem seals. Adjust clearances as necessary OHC.
13. Refit cylinder head to engine using new gaskets and bolts where required.
14. Tighten cylinder head bolts to the correct torque using recommended procedure.
15. Replace all ancillary equipment/components.
16. Fill with fresh oil and where appropriate water/anti-freeze.
17. Check operation by starting and running engine.
18. Check for water and oil leaks, check levels and adjust settings as necessary.

Typical assessment sheet

This should be completed by the assessor during the assignment. Student will also produce a report to go with this assessment sheet.

Self assessment questions

1. If a worn engine oil pump causes low oil pressure:
   (a) the oil pressure can be increased by increasing the spring pressure of the oil pressure relief valve
   (b) increasing the oil pressure relief valve spring pressure will make no difference to the oil pressure
   (c) the engine oil should be changed for one with a higher viscosity index
   (d) the by-pass valve in oil filter should be adjusted.

2. When checking valve clearances on a 4 cylinder in-line engine:
   (a) the clearance on number 1 valve may be checked when number 8 valve is fully open
   (b) the valves on number 4 cylinder should be 'rocking' when the valve clearances on number 1 cylinder are being checked
   (c) the engine must be hot
   (d) the clearance should be checked by means of an engineer's rule.

3. Piston rings are gapped:
   (a) so that they can be fitted to the piston
   (b) to allow a path for oil to reach the combustion chamber
   (c) to allow for thermal expansion
   (d) to prevent ring flutter.

4. Big end bearings are lubricated:
   (a) by oil mist that rises from agitation of oil in the sump
   (b) by splash lubrication piston
   (c) by oil that is fed to them through oilways in the crankshaft
   (d) by oil that is fed to them through holes drilled in the connecting rod.

5. The diameter at the top land of an aluminium alloy piston is:
   (a) smaller than the diameter at the piston skirt to allow for expansion
   (b) smaller than the diameter at the piston skirt to prevent the piston hitting the wear ridge
   (c) larger than the diameter on the skirt to provide a good gas seal
   (d) the same size as the remainder of the piston.

6. When the piston of number 1 cylinder of a 4 cylinder, 4 stroke, in-line engine is at TDC at the beginning of the firing stroke:
   (a) number 4 piston is at the beginning of exhaust stroke
   (b) number 4 piston is at the beginning of the induction stroke
   (c) the valves on number 1 cylinder are open
   (d) combustion will be taking place in cylinder number 3.

7. When replacing a cylinder head:
   (a) the bolts should be tightened by starting at the front of the cylinder head and working towards the back
   (b) the bolts should be tightened by starting at the middle of the cylinder head and working outwards in accordance with the order shown in the instruction manual
   (c) the bolts should be torqued down to the final setting in a single operation
   (d) the bolts should be 'run down' finger tight and then fully tightened after the engine has been restarted and warmed up to normal operating temperature.

8. End thrust on the crankshaft shown in Fig. 1.32 is taken by the component marked:
   (a) 13
   (b) 16
   (c) 3
   (d) 22.

9. A full flow oil filter is fitted with a by-pass valve:
   (a) to prevent excessive oil consumption
   (b) to allow oil to reach the oil galleries in the event of filter blockage
   (c) as a substitute for an oil pressure relief valve
   (d) to allow excess oil from the gallery to leak back through the oil pump.

10. In a certain engine , the inlet valve opens 5 degrees before TDC and closes 40 degrees after BDC. The inlet valve period is:
    (a) 215 degrees
    (b) 45 degrees
    (c) 225 degrees
    (d) 135 degrees.

Name								Date	from
									to

Practical assignment								**Tutor/assessor**
Tasks	Chooses correct tools and operates equipment properly	General cleanliness of work/task	Uses safe working practices	Identifies components correctly	Locates and uses technical information	Completes task successfully	Identifies any problems	Compiles report on completed task	General comments
Test equipment									Comments on specific tasks
Removing ancillaries									
Removing/ refitting valves									
Refitting cylinder head									
Checking and running engine									

2
Cooling systems

Topics covered in this chapter

Cooling system terminology
Liquid (water) cooled system
Air-cooled system
Anti-freeze
Specific heat capacity
Fans and pumps

During combustion, when the engine is operating at full throttle, the maximum temperature reached by the burning gases may be as high as 1500–2000 °C. The expansion of the gases during the power stroke lowers their temperature considerably, but during the exhaust stroke the gas temperature may still be approximately 800 °C. All the engine components with which these hot gases come into contact will absorb heat from them in proportion to:

● the gas temperature;
● the area of surface exposed to the gas;
● the duration of the exposure.

Engine operating temperatures are shown in Fig. 2.1.

2.1 Over-heating

For all these reasons the heat will raise the temperature of the engine components. If the temperature of the exhaust gas is above red heat it will be above the melting point of metals such as aluminium from which the pistons are made. Unless steps are taken to reduce these temperatures a number of serious problems could arise.

● The combustion chamber walls, piston crown, the upper end of the cylinder and the region of the exhaust port are exposed to the hottest gases and will therefore reach the highest temperatures. This will create distortion causing a leakage of gas, water or oil. It may even cause the valve to burn or the cylinder head to crack and as a consequence there will be a loss of power output.
● The oil film will be burnt causing excessive carbon to form. The loss of lubrication of the piston and rings will cause excessive wear or the piston to seize in the cylinder.
● Power output will be reduced because the incoming mixture will become heated so reducing its density. It may also cause **detonation** (this is an uncontrolled explosion in the cylinder) making it necessary to reduce the compression ratio.
● Some part of the surface of the combustion chamber could become hot enough to ignite the incoming charge before the spark occurs (called **pre-ignition**) which could cause serious damage to the engine if allowed to continue.

For these reasons the engine must be provided with a system of cooling, so that it can be maintained at its most efficient practicable operating temperature. This means that the average temperature of the cylinder walls should not exceed about 250 °C, whereas the actual temperature of the gases in the cylinder during combustion may reach ten times this figure. One of the other things to remember is that the engine should not be run too cool as this would reduce **thermal efficiency** (this is how good the engine is at converting heat into mechanical power), increase fuel consumption and oil dilution and cause wear and corrosion of the engine.

2.2 Heat transfer

The cooling system works on the principles of **heat transfer**. Heat will always travel from

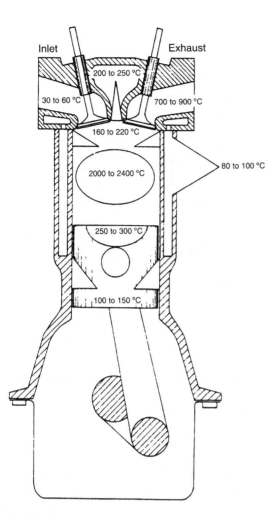

Fig. 2.1 Engine operating temperature ranges

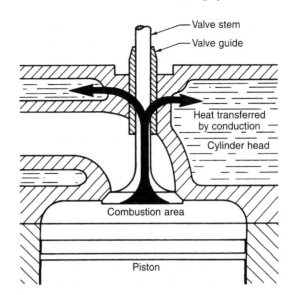

Fig. 2.2 Heat is transferred by conduction from the valve stem to the valve guide. Both objects are solid

emitted by all substances and may be reflected or absorbed by others. This ability will depend upon the colour and nature of the surface of the objects, for example, black rough ones are best for absorption of heat and light polished ones best for reflection of heat.

The cooling system relies on all three of these principles to remove excess heat from the engine.

hot to cold (e.g. from a hot object to a cold object, this would be by conduction). This transfer occurs in three different ways:

* conduction
* convection
* radiation

Conduction is defined as the transfer of heat between two solid objects, e.g. valve stem to valve guide as shown in Fig. 2.2. Since both objects are solid, heat is transferred from the hot valve stem to the cool valve guide by conduction and also from the guide to the cylinder head.

Convection is the transfer of heat by the circulation of heated parts of a liquid or gas. When the hot cylinder block transfers heat to the coolant it produces a change in its density and causes the warmer less dense water to rise, thus setting up convection currents in the cooling system.

Radiation is defined as the transfer of heat by converting it to radiant energy. Radiant heat is

---

Learning tasks

1. Take a look at three different types of radiators and check the colour and texture of the surface finish. Why are they finished like this?
2. How is the heat taken away from the top of the piston and spark plug? Which of the three methods named above are used?

2.3 Over-cooling

As we have seen, various problems can occur if the engine temperature gets too high but if the temperature becomes too low then another set of problems can occur.

* Fewer miles per gallon as the combustion process will be less efficient.
* There will be an increase in the build-up of carbon (as the fuel enters the cylinder it will condense and cause excessive build-up of carbon on the inlet valves).

- There will be an increase in the varnish and sludges formed within the lubrication system. Cooler engines make it easier for these to form.
- A loss of power, because if the combustion process is less efficient the power output will be reduced.
- The fuel not being burned completely which will cause fuel to dilute the oil and cause excessive engine wear.

The purposes of the cooling system can be summarized as follows:

- to maintain the highest and most efficient operating temperature within the engine;
- to remove excess heat from the engine;
- to bring the engine up to operating temperature as quickly as possible – in heavy duty driving, an engine could theoretically produce enough heat to melt an average 100 kg engine block in 20 minutes.

2.4 Types of cooling systems

There are two main types of cooling systems in common use, air and water. Both dissipate (radiate) heat removed from the cylinder into the surrounding air. Air cooling is described below and water cooling in Section 2.6.

Air cooling

In this system heat is radiated from the cylinder and head directly into the surrounding air. The rate at which heat is radiated from an object is dependent on:

1. the difference in temperature between the object and the surrounding air;
2. the surface area from which the heat is radiated (since (1) must be limited, the surface area of the cylinder and head exposed to the air must be increased, by forming fins on their external surfaces) (Fig. 2.3);
3. the nature of the surface.

It is also necessary to remove the heated air from around the cylinder and deliver a constant supply of cool air around and between the fins. This means that the cylinders must be sufficiently widely spaced to permit a suitable depth of finning all around them, and the engine must be placed where the movement of the vehicle can provide the necessary supply of cool air.

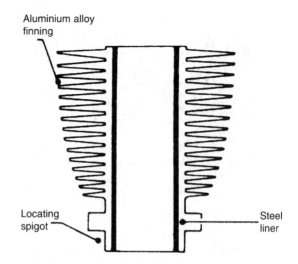

Fig. 2.3 Air-cooled finned cylinder wall method of heat transfer

A large **fan** is often used and the engine is surrounded by large **cowls** to direct air to where it is required, e.g. around the cylinder head and valve area.

The choice of air or liquid cooling has always been controversial. Air is cheaper, lighter and more readily obtainable than water – though to remove a given quantity of heat demands four times the weight and 4000 times the volume of air than it does water. It also gives less control of the engine temperature and air-cooled engines tend to be noisier. But air can be collected and rejected, whereas water must be carried on the car and the jacketing, hoses, pump and radiator of a water-cooled engine will probably weigh more than the substantial fins of an air-cooled engine.

However, the fins force the cylinders of an air-cooled engine to be more widely separated than those of a water-cooled engine, so the crankshaft and crankcase must be longer and therefore heavier. For an engine of many cylinders, this is one of the greatest objections to air cooling.

2.5 Engine air-cooling system

Circulation of cooling air

With air cooling the engine structure is directly cooled by forcing air over its high-temperature surfaces. These are finned to present a greater cooling surface area to the air, which in nonmotor-cycle applications is forced to circulate over them by means of a powerful fan.

The engine structure is almost entirely enclosed by sheet metal ducting (called a cowl), which incorporates a system of partitions (called **baffles**); these ensure that the air flow is properly directed over the cylinders and cylinder heads. To obtain uniform temperatures the air is forced to circulate around the entire circumference of each cylinder and its cylinder head. The direction

of air flow will be along the cooling fins, the greatest number of which are found towards the top of the cylinder (around the exhaust valve) as this is the hottest part of the engine.

The complete system forms what is known as the **plenum chamber** in which the internal air pressure is higher than that of the atmosphere. The heated air is discharged from the plenum chamber to the atmosphere, or redirected to heat the car interior. Figure 2.4 shows a

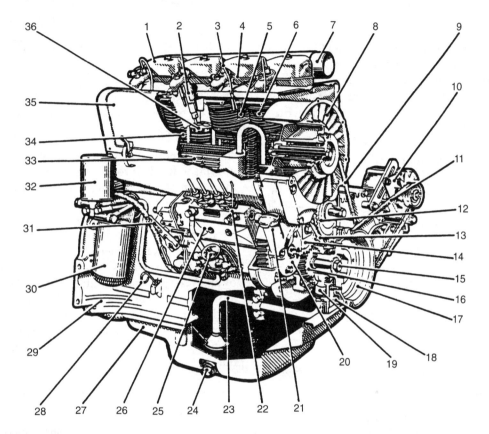

1 – Rocker chamber cover
2 – Injector
3 – Injection line to no.3 cylinder
4 – Back-leakage line
5 – Cylinder head anti-fatigue bolt (four bolts securing each cylinder head with cylinder to crankcase)
6 – Cylinder head (light alloy)
7 – Air intake manifold
8 – Cooling blower (V-belt driven)
9 – Cooling blower V-belt
10 – Generator (dynamo or alternator)
11 – Generator V-belt
12 – Camshaft gear
13 – Oil gallery
14 – Idler gear (driving injection pump and camshaft)
15 – Anti-fatigue bolt (securing V-belt pulley to crankshaft)
16 – Crankshaft gear
17 – V-belt pulley
18 – Vibration damper
19 – Oil pump
20 – Injection pump drive gear with advance/retard unit
21 – Oil filler neck
22 – Overflow line
23 – Oil suction pipe
24 – Oil drain plug
25 – Fuel feed pump
26 – Bosch in-line injection pump with mechanical centrifugal governor
27 – Oil sump (sheet metal or cast iron)
28 – Oil dipstick
29 – Crankcase (cast iron)
30 – Oil filter
31 – Speed control lever
32 – Fuel filter
33 – Integral oil cooler
34 – Finned cylinder (grey cast iron), separately removable
35 – Removable air cowling
36 – Piston

Fig. 2.4 Cut-away view of modern air-cooled diesel engine

modern, air-cooled diesel engine. In order to provide the necessary air flow around the cylinders of an enclosed engine, a powerful fan is essential.

Types of air-cooling fans

Axial flow

This is a simple curved blade type in which the direction of air flow is parallel to the axis of the fan spindle (Fig. 2.5).

Radial flow

Often called a **centrifugal** this type (Fig. 2.6) is more commonly used because it is more effective and a fan of smaller diameter can be used for a given air flow. This type of fan has a number of curved radial vanes mounted between two discs, one or both having a large central hole. When the fan is rotated, air between the vanes rotates with it and is thrown outwards by centrifugal force.

Figure 2.7 shows a simple air-cooled system for a four-cylinder in-line engine. A centrifugal fan, driven at approximately twice crankshaft

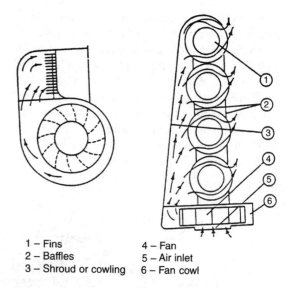

1 – Fins   4 – Fan
2 – Baffles   5 – Air inlet
3 – Shroud or cowling   6 – Fan cowl

Fig. 2.7 Simple in-line air-cooled engine using a radial type fan

speed, is mounted at the front of the engine and takes in air through a central opening (5) in the fan casing. This air is delivered into the cowl (3) where it is directed over the fins of the cylinders (1). Baffles (2) ensure that the air passes between the fins where it picks up heat, thus cooling the cylinders.

The in-line engine is the most difficult to cool by air. Vee-type or horizontally opposed engines are easier to cool as the cylinders are spaced further apart to leave room for the crankshaft bearings and this allows more room between the cylinders for a good air flow whilst at the same time keeping the total engine length fairly short.

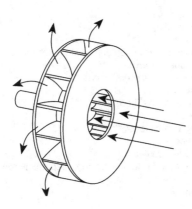

Fig. 2.5 Axial air flow fan

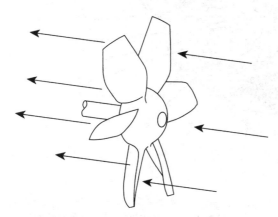

Fig. 2.6 Radial air flow fan

Learning tasks

1. What safety precautions should be observed when working on air-cooled systems?
2. What are the main difficulties with working on air-cooled engines?
3. Remove the cowling from an air-cooled engine, clean any excessive dirt, etc. from between the fins of the cylinders. Reassemble the cowling and test engine for correct operating temperature.

2.6 Liquid cooling

In this arrangement the outer surfaces of the cylinder and head are enclosed in a jacket, leaving a space between the cylinder and the jacket through which a suitable liquid is circulated.

The liquid generally used is water, which is in many ways the most suitable for this purpose, even though it has a number of drawbacks. Whilst passing through the jacket the water absorbs heat from the cylinder and head, and it is cooled by being passed through a radiator before being returned to the jacket.

Thermo-syphon system

When heated, the water becomes less dense and therefore lighter than cold water. Thermosyphon is the action of the water being heated, rising and setting up convection currents in the water. The thermo-syphon system is no longer used in the modern motor vehicle as it has a number of disadvantages.

- To ensure sufficient circulation the radiator must be arranged higher than the engine to ensure that the heated coolant will rise into the top of the radiator header tank and the cooled water in the radiator will flow into the bottom of the engine.
- Water circulation will be slow, so a relatively large amount of water must be carried.
- Large water passages must be used to allow an unrestricted flow of water around the system.

This system is usually now confined to small stationary engines such as those used to power narrow boats, small generators, water pumps, etc.

Pump-assisted circulation

Most modern engines use a pump to provide a positive circulation of the coolant. This is shown in Fig. 2.8 and gives the following advantages:

1. a smaller radiator can be used than in the thermo-syphon system;
2. less coolant is carried as the water is circulated faster and therefore the heat is removed more quickly;
3. smaller passages and hoses are used because of (2) above;
4. the radiator does not need to be above the level of the engine, giving a lower bonnet line; this also has the advantage of less wind resistance giving a better fuel consumption;
5. because the water flow is given positive direction the engine will operate at a more even temperature.

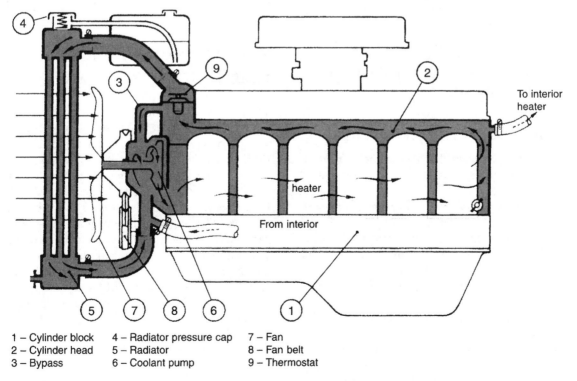

1 – Cylinder block	4 – Radiator pressure cap	7 – Fan
2 – Cylinder head	5 – Radiator	8 – Fan belt
3 – Bypass	6 – Coolant pump	9 – Thermostat

Fig. 2.8 Coolant is pumped from the water pump, through the cylinder block and heads, through the thermostat into the radiator, and back to the water pump

2.7 Comparison of air- and water-cooled systems

Advantages of air cooling

- An air-cooled engine is generally lighter than an equivalent water-cooled engine.
- It warms up to its normal running temperature very quickly.
- The engine can operate at a higher temperature than a water-cooled engine.
- The system is free from coolant leakage problems and requires no maintenance.
- There is no risk of damage due to freezing of the coolant in cold weather.

Disadvantages of air cooling

- A fan and suitable cowls are necessary to provide and direct the air flow. The fan can be noisy and absorbs a large amount of engine power. The cowl makes it difficult to get at various parts of the engine when servicing is required.
- The engine is more liable to over-heating under difficult conditions than a water-cooled engine.
- Mechanical engine noises tend to be amplified by the fins.
- The cylinders usually have to be made separately to ensure proper formation of the fins. This makes the engine more costly to manufacture.
- Cylinders must be spaced well apart to allow sufficient depth of fins.
- It is more difficult to arrange a satisfactory car-heating system.

Advantages of water cooling

- The temperatures throughout the engine are more uniform, thus keeping distortion to a minimum.
- Cylinders can be placed closer together making the engine more compact.
- Although a fan is usually fitted to force air through the radiator, it is much smaller than the type required for an air-cooled engine. It therefore absorbs less power and is quieter in operation.
- There is no cowl to obstruct access to the engine.
- The water jacket absorbs some of the mechanical noise making the running engine quieter.
- The engine is better able to operate under difficult conditions without over-heating.

Disadvantages of water cooling

- Weight – not only of the radiator and connections but also of the water; the whole engine installation is likely to be heavier than an equivalent air-cooled engine.
- Because the water has to be heated, it takes longer to warm up after starting from cold.
- If water is used, the maximum temperature is limited to about 85–90 °C to avoid the risk of boiling away the water. However, modern cooling systems are pressurized and this permits higher temperatures and better efficiency.
- If the engine is left standing in very cold weather, precautions must be taken to prevent the water freezing in the cylinder jackets and cracking them.
- There is a constant risk of a coolant leakage developing.
- A certain amount of maintenance is necessary, for example, checking water level, anti-frost precautions, cleaning out deposits, etc.

Learning task

Using the above information write a short paragraph to explain the main components and their purpose in the water cooling system.

2.8 Radiator and heater matrix

The purpose of the radiator is to provide a cooling area for the water and to expose it to the air stream. A reservoir for the water is included in the construction of the radiator. This is known as the header tank and is made of thin steel or brass sheet and is connected to the bottom tank by brass or copper tubes; these are surrounded by 'fins'. This assembly is known as the **matrix, core, block** or **stack**. The more modern radiator uses plastic for the tanks and aluminium for the matrix. Shown in Fig. 2.9(a) is a conventional type of radiator. Figure 2.9(b) shows a typical type of cross-flow radiator which has an integral oil cooler fitted.

The function of the radiator as we have said is to transfer heat from the coolant to the air stream. It is designed with a very large surface area combined with a relatively small frontal area, and it forms a container for some of the

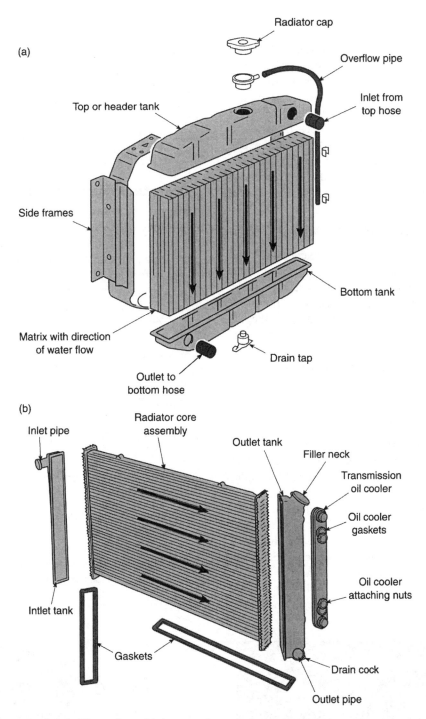

(a)

Radiator cap

Overflow pipe

Inlet from top hose

Top or header tank

Side frames

Bottom tank

Matrix with direction of water flow

Outlet to bottom hose

Drain tap

(b)

Radiator core assembly

Inlet pipe

Outlet tank

Filler neck

Transmission oil cooler

Oil cooler gaskets

Oil cooler attaching nuts

Intlet tank

Gaskets

Drain cock

Outlet pipe

Fig. 2.9 (a) Conventional vertical flow radiator, (b) the parts of a standard cross-flow radiator

coolant in the system. The radiators usually have mounting feet or brackets, a filler cap, an overflow pipe and sometimes a drain tap is fitted to the lower tank.

A large number of different types of radiator matrix are in common use depending on application size of engine, etc. Details of the main types in common use are given below.

Film core

The tubes are the full width of the core and are bent to form square spaces through which air can pass. They are sometimes crinkled to extend their length. The top and bottom tanks are secured to side frames with the core located between them and a fan cowl often completes the unit.

Tube and fin

This consists of copper or brass tubes of round, oval or rectangular cross-section. The tubes pass through a series of thin copper fins with the top and bottom tanks attached to the upper and lower fins, respectively. The fins secure the tubes and increase the surface area from which the heat can be dissipated. The tubes are placed edge on to the air flow for minimum air resistance and they are now produced from strip lock-seaming.

Tube and corrugated film

Sometimes used as an alternative to the tube and fin, the corrugated separator filming is made from copper and laid between the tubes to provide an airway. Each face of the filming is louvred to increase air turbulence as the air passes through. This improves the cooling efficiency of this design. A commonly used form of radiator matrix (core) construction is shown in Fig. 2.10.

Separate tubes

Radiators with separate coolant tubes are occasionally used. They provide a stronger core than the other types but they are more costly to build, heavy, time-consuming to repair and because of this are mainly confined to commercial vehicle applications. The tanks and side frames are usually bolted together and locate the thick walled tubes of rectangular or circular cross-section. The tubes are made watertight in the upper and lower tanks with rubber and metal seals, and they have bonded copper fins or a spiral copper wire wound over their complete length to increase their ability to dissipate the heat. Tube removal and refitting may be done by two methods depending on their construction. One is to remove top and bottom tanks from their side frames, the other is to spring the tubes in and out, which is only possible because of their flexibility.

Separate expansion or header tanks

Separate expansion or header tanks are now commonly used. These allow the radiator to be fitted lower than the engine and, on a commercial vehicle, to be fitted in a more accessible position for checking and refilling the coolant. The tank is also used to reduce the risk of **aeration** (this is air bubbles forming in the water) of the coolant when the engine is running. The early types of radiators were all of the conventional type, for example, the coolant flows from the top of the radiator to the bottom (vertically). Many vehicles now use a cross-flow type in which the coolant flows horizontally through the core from the top of one side tank across to the bottom of the other side tank as shown in Fig. 2.11.

> *Learning tasks*
>
> 1. What are the advantages of a cross-flow radiator?
> 2. How should the radiator be tested for leaks when removed from the vehicle?
> 3. Why is the radiator painted matt black?

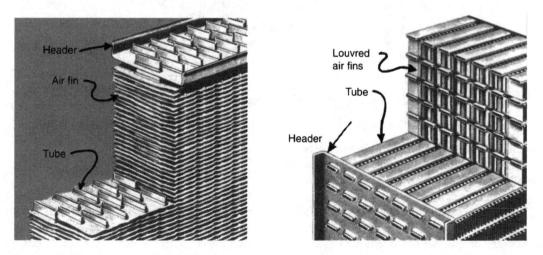

Fig. 2.10 Inside the radiator there are small tubes through which the coolant flows, helping to remove the heat into the air

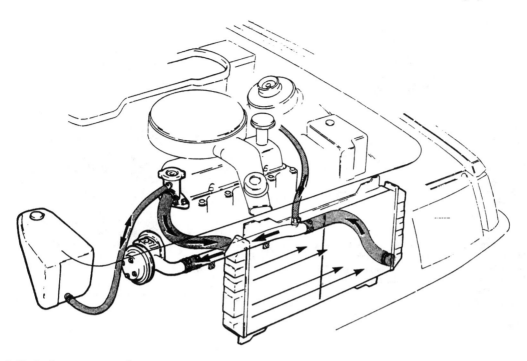

Fig. 2.11 Cooling system water flow

2.9 Water pump

The purpose of the water pump is to provide a positive means of circulating the water (it gives a direction to its flow, it does not pressurize the system).

The body of the pump is usually made of cast iron or aluminium. In most cases it is bolted to the front of the engine block and draws cool water from the bottom of the radiator via the bottom hose. This cooler water is directed into the water jacket and over the hottest part of the engine such as the exhaust valve seats. Two types of pump are used, **axial flow** and **radial flow**.

Axial flow

With the axial flow, when the pump is full of water, the rotating impeller carries with it the water contained in the spaces between the impeller blades and the casing. This water is subjected to centrifugal force which causes it to flow outwards from inlet to outlet. A carbon ring bonded to a rubber sleeve is fitted into the housing and pressed into light contact with a machined face on the impeller by a light spring.

This provides a water-tight seal along the shaft. The pump is driven by a belt from the crankshaft via a pulley which is a press fit on the end of the pump spindle. The fan is then bolted to the pulley which draws air through the radiator.

Radial flow

The radial flow centrifugal pump operates with a slightly higher flow and therefore the circulation is slightly faster. This type is fitted to commercial vehicles. The construction of such a water pump is shown in Fig. 2.12.

Learning tasks

1. What are the symptoms of the water pump not working? What could be the cause of this fault and how should it be repaired?
2. How should the drive belt to the pump be checked? Remove the belt, check for signs of wear or cracking. Replace and correctly tension the belt.
3. Remove one type of water pump from an engine, check for premature bearing failure, signs of water leaks and general serviceability. Replace the pump, refill with coolant and pressure test the system.

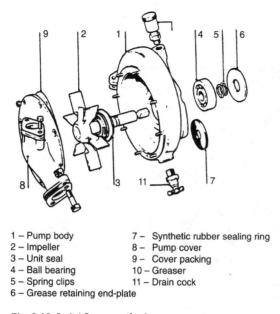

1 – Pump body
2 – Impeller
3 – Unit seal
4 – Ball bearing
5 – Spring clips
6 – Grease retaining end-plate
7 – Synthetic rubber sealing ring
8 – Pump cover
9 – Cover packing
10 – Greaser
11 – Drain cock

Fig. 2.12 Radial flow centrifugal pump

2.10 Thermostat

This is a temperature sensitive valve that controls the water flow to the radiator. There are two main reasons for its use:

- to enable the engine to warm up quickly from cold;
- to control the rate of flow and so maintain a constant temperature in the engine.

Two types are in common use, the **bellows** type and the **wax** type.

Wax type

This is used in the pressurized system as it is not sensitive to pressure like the bellows type. A special wax is used, contained in a strong steel cylinder. The reaction pin is surrounded by a rubber sleeve and is positioned inside the cylinder.

As the temperature increases the wax begins to melt, changing from a solid to a liquid, and at the same time it expands. This forces the rubber against the fixed reaction pin, opening the valve against spring pressure thus allowing the water to circulate through the radiator. There is a small hole in the valve disc to assist in bleeding the system as filling takes place. The 'jiggle pin' closes the hole during engine warm-up. This thermostat is shown in Fig. 2.13.

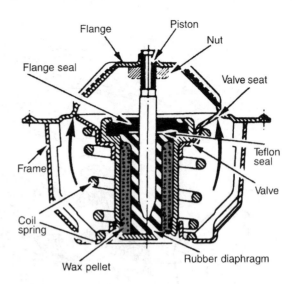

Fig. 2.13 Wax pellet-type thermostat

Bellows type

As shown in Fig. 2.14. Bellows type consists of a flexible metal bellows which is partly filled with a liquid which has a boiling point that is lower than that of water (e.g. alcohol, ether or acetone). Air is removed from the bellows, leaving only the liquid and its vapour. The pressure in the bellows is then only due to the vapour pressure of the liquid. This varies with temperature, and is equal to atmospheric pressure at the boiling temperature of the liquid, less at lower temperatures and more at higher temperatures. As the temperature of the water increases, the liquid in the bellows begins to turn to a vapour and increase in pressure; this expands the bellows and opens the valve allowing water to pass to the radiator.

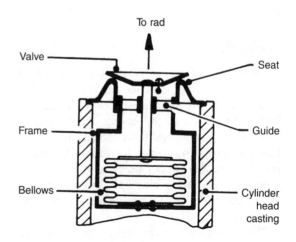

Fig. 2.14 Bellows-type thermostat

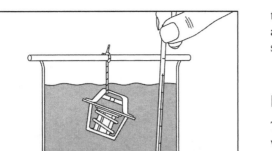

Fig. 2.15 Thermostat suspended in water container

> *Note*: This type is not suitable for pressurized systems as the valves are pressure-sensitive. The wax element type does not have this disadvantage.

Testing thermostats

The thermostat cannot be repaired and so must be replaced if found to be faulty. It can be tested by placing the thermostat in a beaker of water and gradually heating it. A thermometer is used to check the temperature of the water (Fig. 2.15). The thermostat should begin to open when the temperature marked on the valve is reached. An increase of approximately 10–20 °C will elapse before the valve is fully open.

> *Learning tasks*
>
> 1. Write a schedule for removing, testing and replacing a thermostat that is suspected of being faulty.
> 2. What is the symptom, fault, cause and remedial action that should be taken when the driver complains of the heater blowing hot and cold?
> 3. Remove, test and replace a thermostat according to your work schedule drawn up at (I) above. State the type of unit fitted, the temperature at which it opened/closed and check with the manufacturer's data.

2.11 Pressurized cooling systems

Pressurised cooling systems are used because they allow the engine to operate at a higher temperature. Figure 2.16 shows the layout and main components of a modern pressurized system.

Pressure cap

The cap contains two valves; one is the **pressure valve** the other is the **vacuum valve**. As the temperature of the water increases it expands and in a sealed system this expansion increases the pressure until it reaches the relief pressure of the cap. As the system cools down it contracts and opens the vacuum valve drawing in air. If no vacuum valve was fitted the depression in the system, caused by the contracting effect of the water as it cools, could cause the rubber hoses in the system to collapse. Most pressure caps operate at 28–100 kN/m. The pressure is usually stamped on the cap indicating the maximum relief pressure for the system to which it is fitted. Figure 2.17(c) shows a pressure cap in the open position.

> *Safety note*
>
> Never remove the cap when the coolant temperature is above 100 °C as this will allow the water to boil violently: the resulting jet of steam and water from the open filler can cause very serious scalds. The system should be allowed to cool down and the cap removed slowly. It is designed so that the spring disc remains seated on the top of the filler neck until after the seal has lifted. This allows the pressure to escape through the vent pipe before it can escape from the main opening.

The temperature at which a liquid boils rises as the pressure acting on it rises. This is shown in Fig. 2.18. The cooling system's pressure is maintained by the use of a pressure cap fitted to the top of the radiator or expansion bottle. This closes the system off from the atmosphere creating a sealed system.

The advantages of using a pressurized system are:

- elimination of coolant loss by surging of the coolant during heavy braking;
- prevention of boiling during long hill climbs, particularly in regions much above sea level;
- raising of the working temperature improving engine efficiency;
- allowing the use of a smaller radiator to dissipate the same amount of heat as a larger one operating at a lower temperature.

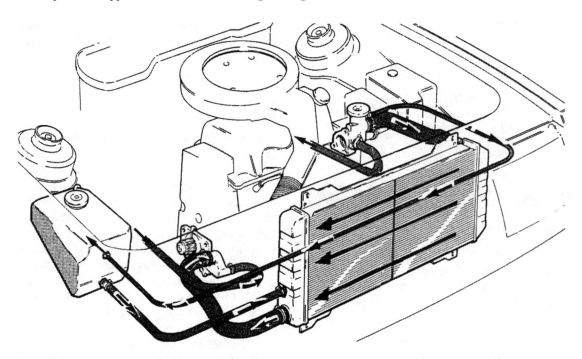

Fig. 2.16 'Degas' system with coolant at normal operating pressure and temperature

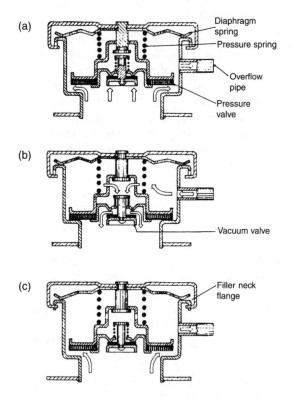

Fig. 2.17 Open-type pressure cap used in the semi-sealed system. (a) Pressure-relief valve open, (b) vacuum-relief valve open and (c) cap-removal precaution action

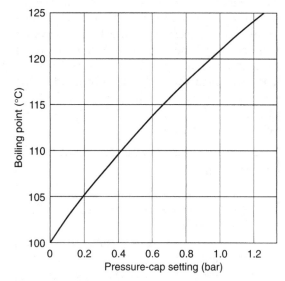

Fig. 2.18 Variation of boiling point of water with pressure

Semi-sealed system

The advantage of the modern sealed cooling system is that it reduces the need for frequent inspection of the coolant level and the risk of weakening the anti-freeze solution by topping up. Although the pressure cap provided a semi-sealed system the fully sealed system has in effect a means of recovering the coolant that is lost when the engine is at its operating temperature and the pressure cap is lifted. It consists of an **expansion tank** that is mounted independently from the radiator and vented to the atmosphere. A flexible hose connects the overflow pipe of the radiator to a dip tube in

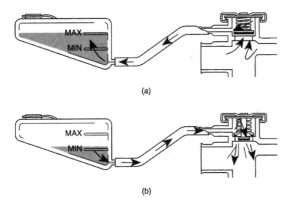

(a)

(b)

Fig. 2.19 Operating function of the overflow bottle. (a) Water heating up, (b) water cooling down

the overflow tank. As the coolant heats up the pressure valve in the cap opens and excess coolant passes to the expansion tank. When the system cools down the coolant is drawn back into the radiator again through the vacuum valve in the cap. This is shown in Fig. 2.19.

Fully sealed system

In the fully sealed system, the pressure cap is fitted to the expansion tank and a simple cap is used on the radiator. The operation of the system is much the same as in the semi-sealed system. The reservoir is vented to the atmosphere and should be approximately two-thirds full. If the coolant falls below the minimum level indicated on the tank air may be drawn into the radiator.

Learning tasks

1. What is the correct method for testing the pressure and vacuum valves in the cap? What would be the effect of fitting a cap with: (a) a stronger spring? (b) a weaker spring?
2. How would you test for water/oil contamination and which component would you suspect to be faulty?

2.12 Fans and their operation

Fans and temperature control

An important aspect of the cooling capabilities of the radiator is the volume of air, in unit time, which can be caused to flow through the matrix. Hence, the purpose of the cooling fan is to maintain adequate air flow through the matrix, at low and engine idling speeds. The speed of the fan ranges from slightly less to rather more than that of the engine. Excessive fan speeds are avoided because of noise and the power required to drive it.

Electrically driven fans

With this type of fan arrangement an electric motor is used together with a fan and a temperature-sensitive control unit. The advantages of this arrangement are:

- the fan only operates when the engine reaches its predetermined temperature;
- the engine will be more efficient as the fan is not being driven all the time;
- the radiator and fan can now be fitted in any convenient position, ideal for transversely mounted engines;
- the fan assembly can be mounted in front of or behind the radiator;
- engine temperature is more closely controlled as the temperature sensor will automatically switch the fan on and off within very close limits as required.

A typical layout is shown in Fig. 2.20.

To improve the efficiency of the fan a cowl is often fitted. Its function is to prevent heated air from reversing its flow past the fan and recirculating through the matrix which could lead to engine overheating. The fan can be formed from some of the following materials:

- one piece steel pressing
- aluminium casting
- plastic moulding with a metal insert

An uneven number of irregularly spaced blades (Fig. 2.21) are often employed to minimize fan noise. In all applications the fan assembly must be accurately balanced and the blades correctly aligned to avoid vibration.

Flexible fan blades

Used to reduce frictional power loss from the fan, these are made from fibreglass, metal and moulded plastic such as polypropylene. The plastic fans are lighter, easier to balance, look better, are more efficient aerodynamically, have

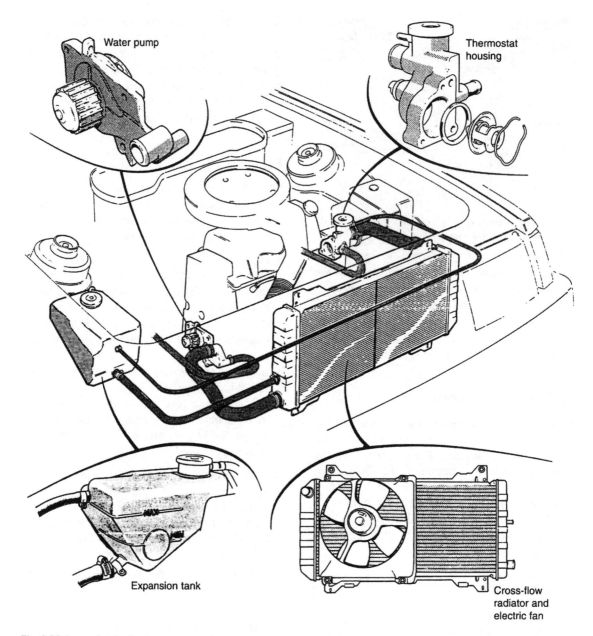

Fig. 2.20 Layout for the Ford transverse engine

a reduced noise level, a reduction in vibration and offer less risk of serious injury. They are also cheaper to manufacture. For these reasons the plastic fan is the most common of this type fitted to light vehicles.

A number of vehicles have a cover surrounding the fan (called a **shroud**) fitted to make sure the fan pulls air through the *entire* radiator. Figures 2.20 and 2.22 show arrangements commonly used. When no shroud is used the fan will only pull air through the radiator directly in front of the blades. There is very little air moving through the corners of the matrix.

With flexible fan blades, as the speed of the engine increases the blades flatten and therefore move less air, this give a reduction in the power lost to friction. An example of a flexible fan with a shroud is that used in the Mini; the air is forced through the radiator, not drawn through as in most vehicles.

Cooling fan with swept tips

The flexible tips of the fan tend to straighten out as speed increases and this has the effect of reducing the amount of air drawn through the fan.

Fig. 2.21 Unevenly spaced fan blades

Cooling fan surrounded by cowl or shroud attached to the radiator

The shroud is a close fit around the radiator and fan to prevent the air from just circulating around the fan and not passing through the radiator matrix and cooling the water.

Directly driven fans

Directly driven fans are not normally used in light vehicles because they have the following disadvantages:

- rising noise level
- increase in power used
- tendency to over-cooling at higher speeds

Some engine manufacturers still use directly driven fans mainly because of the type of engine fitted or because space is limited. The opposed piston engine often has the fan blades bolted directly on the end of the crankshaft, usually with some means of limiting the maximum speed of the fan. LGVs may also use the same method of driving the fan.

Fan drives

The drive for the fan is normally located on the water pump shaft. Its purpose is to draw sufficient air through the matrix for cooling purposes in heavy traffic on a hot day and to ensure adequate cooling at engine idling speeds.

At moderate speeds, say, driving down the motorway, the engine does not require the same amount of assistance from the fan, as the natural flow of air passing through the radiator will often be sufficient to cool the water. Power would be lost unnecessarily to the fan under these conditions. A number of different methods of controlling fan operation and speed are becoming more popular.

Automatically-controlled fans are generally classified into four types:

- **free-wheeling**
- **variable-speed**
- **torque-limiting**
- **variable-pitch**

Free-wheeling

The free-wheeling-type is mounted on the coolant pump shaft and may form part of an **electromagnetic clutch** with the drive pulley. Connected into the electrical supply circuit to the clutch assembly is a thermostatically controlled switch. When the fan is no longer required, the supply to the electromagnet is

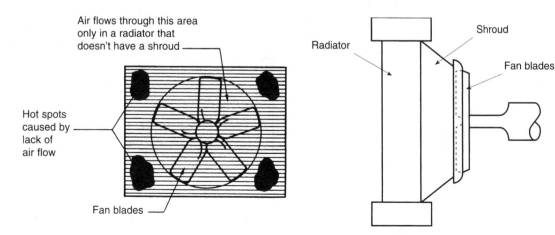

Fig. 2.22 Fan shrouds

interrupted, so that it disengages and allows the fan to free-wheel on the pump shaft.

Variable-speed

The variable-speed-type generally has a **viscous coupling**, which permits motion or **slip** to occur between the driving and the driven members. The driving member consists of a disc mounted on the pump shaft. Attached to the fan is a sealed cylindrical casing or chamber, freely mounted on the pump shaft. The driving disc revolves within the coupling chamber, which is partially filled with a highly stable **silicone fluid** (this means it retains a nearly constant viscosity with varying operating temperatures). The disc initially revolves in the stationary mass of fluid and sets up a **drag**, so that the fluid begins to circulate in the chamber. As a result of centrifugal force it moves outwards to fill the gaps between the drive faces of the coupling members. Since these faces are very close to one another, drive is transmitted between them by the viscous drag of the silicone fluid. This drag is maintained with increasing fan speed, until the resisting torque imposed upon the fan by the air flow rises to an extent that causes shear breakdown of the fluid film, and the coupling slips. Fan speed is thus controlled by the slipping action of the coupling, which is predetermined by the viscosity of the fluid and the degree to which the coupling is filled. Where no provision is made for the quantity of oil, the coupling is termed a torque-limiting type. The layout of a viscous fan drive is shown in Fig. 2.23.

Torque-limiting

The torque-limiting-type has the disadvantage of operating within a fixed speed range, regardless of cooling system demands. Thus air circulation may be inadequate when car speed is low and engine speed is high. To overcome these difficulties, a temperature sensitive fan control is used in some installations.

Variable-pitch

In the variable-pitch fan the volume of air displaced is controlled by twisting its blades. The term pitch is considered as the distance the blade can twist, if it were turning one revolution in a solid substance (that is with no slip taking place). When the least amount of cooling is required, the blades are automatically adjusted to a low pitch, thereby reducing airflow to a

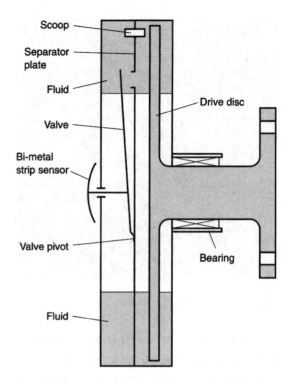

Fig. 2.23 Simplified diagram showing internal details of a viscous fan drive

minimum. The variable-pitch fan may be actuated by one of the following methods:

- centrifugal
- torque-limiting
- thermostatic means

An alternative to the engine driven fan is a separate electrically driven fan, which can be controlled automatically by the thermostat that responds to changes in engine coolant temperature. Although fan installations of this type absorb no engine power, they do impose an electrical load upon the car battery. It has a fixed higher operating speed and it may be noisier than a conventional driven fan, but it has the advantages of greater convenience of mounting position.

Learning tasks

1. Give reasons for the use of an electrically driven fan over other methods of operation.
2. Draw up in logical sequence the way an electric fan should be tested, apart from running the engine up to its normal operating temperature.
3. How should a viscous coupling be tested/repaired?

Radiator blinds and shutters

A very simple arrangement of controlling the air flow through the radiator is shown in Fig. 2.24. A spring-loaded **roller blind** is carried at the lower end of a rectangular channel-section frame attached to the front of the radiator. A simple cable control raises the blind to close off as much of the radiator as may be necessary to maintain the correct running temperature of the engine. This is normally operated by the driver from inside the cab.

Another method of controlling the air flow is by the use of a **shutter**. This operates in a similar way to a venetian window blind, each shutter rotates on a separate shaft. Mounted in front of the radiator it can be manually or thermostatically operated.

Fan belts

Most fan belts are of the V-type in construction and use the friction produced between the sides of the belt and pulley to provide the drive. The **vee** belt has a larger area of contact and therefore provides a more positive drive between the belt and the pulley. Another popular type is the flat **serpentine** belt; about 35 mm wide it has several grooves on one side. A larger area is provided by the grooves to give a more positive grip. This type is also used to transmit a drive around a smaller pulley than a conventional vee belt.

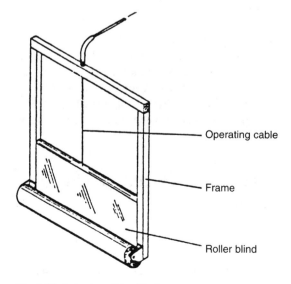

— Operating cable

— Frame

— Roller blind

Fig. 2.24 A simple radiator blind

2.13 Corrosion in the cooling system

This can be very damaging to the engine. It can be caused in several different ways.

Direct attack

This means the water in the coolant is mixed with oxygen from the air. This process produces rust particles which can damage water pump seals and cause increased leakage.

Electromechanical attack

This is a result of using different metals in the construction of the engine. In the presence of the coolant, different metals may set up an electrical current in the coolant. If this occurs, one metal may deteriorate and deposit itself on the other metal. For example, a core plug may deteriorate to the point of causing a leak.

Cavitation

Cavitation is defined as high shock pressure developed by collapsing vapour bubbles in the coolant. These bubbles are produced by the rapid spinning of the water pump impeller. The shock waves cause small pin holes to form in nearby metal surfaces such as the pump impeller or the walls of the wet-type cylinder liner which could reach all the way through into the cylinder.

Mineral deposits

Calcium and silicate deposits are produced when a hard water is used in the cooling system. Both deposits restrict the conduction of heat out of the cooling system. As can be seen from Fig. 2.25 the deposits cover the internal passages causing uneven heat transfer.

2.14 Anti-freeze

Function

When an **ethylene-glycol** anti-freeze solution is added to the coolant many corrosion problems are overcome. Chemicals are added to the anti-freeze to reduce corrosion. It is not therefore

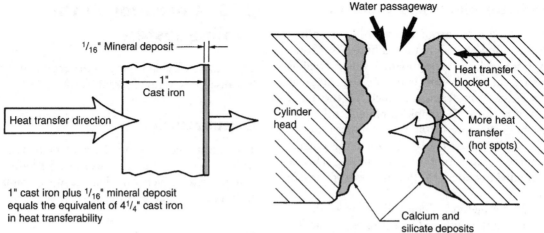

Fig. 2.25 Build-up of deposits in the cooling system

necessary to add a corrosion inhibitor to the coolant when using an anti-freeze solution. In some cases, mixing different corrosion inhibitors or anti-freezes produces unwanted sludges within the cooling system.

Fortunately the freezing point of water can be lowered quite considerably by the simple addition of certain liquids such as ethylene-glycol. The recommended mixture varies with each manufacturer but is usually 33–50% of anti-freeze in the cooling water.

Checking the level

The proportion of ethylene-glycol-based anti-freeze present in a cooling system can be determined by checking the **specific gravity** of the coolant and by reference to its temperature. The percentage of anti-freeze is measured by the use of a hydrometer and a thermometer.

The graph in Fig. 2.26 shows what happens to water when an anti-freeze solution is added. The water changes from a liquid to a mush before becoming solid ice. This ability to form mush before becoming ice gives some warning of freezing and consequently of the danger of damage to the engine.

Methanol-based anti-freeze is also used but has the disadvantage of losing its anti-freeze effect due to evaporation. It is also inflammable. Both types are toxic, and if spilt on the paint work of the vehicle would damage it.

Figure 2.27 shows how to check the anti-freeze content using the 'Bluecol' hydrometer. The coolant is drawn into the hydrometer to a level between the two lines. Note the letter on

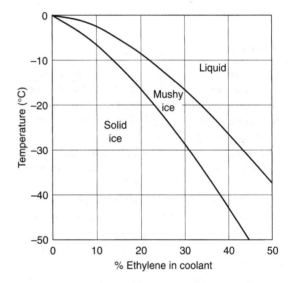

Fig. 2.26 Variation of freezing range with different strengths of anti-freeze

the float at the water line and the temperature of the coolant on the thermometer. Using the slide rule, line up the two readings of temperature and letter. The true percentage content of anti-freeze can be identified at the 'read off' point.

Faults and their possible courses are shown in Table 2.1.

2.15 Engine temperature gauge

The engine temperature gauge permits the driver to observe engine operating temperature. In the temperature indicator circuit shown in Fig. 2.28 the main features are the thermal

(a)

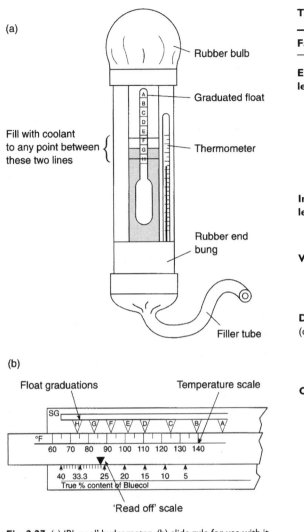

Rubber bulb

Graduated float

Fill with coolant to any point between these two lines

Thermometer

Rubber end bung

Filler tube

(b)

Float graduations

Temperature scale

'Read off' scale

Fig. 2.27 (a) 'Bluecol' hydrometer, (b) slide rule for use with it

Table 2.1 Liquid cooling systems fault chart

Fault	Possible cause
External leakage	1. Loose hose clips or split rubber hose 2. Damaged radiator (cracked joints or corroded core) 3. Water pump leaking (bearing worn) 4. Corroded core plugs 5. Damaged gaskets 6. Interior heater, hoses, valves 7. Temperature sensor connection leaking
Internal leakage	1. Defective cylinder head gasket 2. Cylinder head not correctly tightened 3. Cracked water jacket internal wall 4. Defective cylinder liner seals
Water loss	1. Boiling 2. Leaks – internal and external 3. Restriction in radiator 4. Airways in radiator matrix blocked
Dirty coolant (corrosion)	1. Excessive impurity in coolant water 2. Infrequent draining and flushing of system (where required) 3. Incorrect anti-freeze mixtures 4. Lack of inhibitor
Over-heating	1. Loose, broken, worn or incorrect fan belt tension 2. Defective thermostat 3. Water pump impeller loose on shaft 4. Restricted circulation – through radiator, hoses, etc. 5. Radiator airways choked 6. Incorrect ignition timing 7. Incorrect valve timing 8. Tight engine 9. Low oil level 10. Insufficient coolant in system
Over-cooling	1. Defective thermostat 2. Temperature gauge incorrect 3. Electric fan operating continuously

type gauge, the negative temperature coefficient sensor (thermistor), the voltage stabiliser and the interconnecting circuit.

The sensor is normally situated in the cylinder head of the engine and the sensing element is surrounded by coolant. In a typical cooling system sensor, the resistance of the sensing element varies from approximately 230 ohms at 50 °C to approximately 20 ohms at 110 °C. This variation of resistance causes the current flowing through the bi-metallic element of the gauge to vary. The variation of current causes the temperature of the bi-metallic element to change. The change of temperature causes the bi-metallic element to bend and so causes the gauge pointer to move to indicate coolant temperature. Because the operation of the gauge is dependent on a steady voltage being applied, the circuit includes a vibrating contact voltage stabiliser which also relies on a bi-metallic strip for its operation.

> *Learning tasks*
>
> 1. Explain in your own words what is meant by semi-conductor.
> 2. How should the temperature sensor/transmitter be checked for correct operation? Name any special equipment that might be required.
> 3. Remove a temperature sensor and test for electrical resistance, both when cold and when hot, note the readings and check your results with the manufacturer's data. Make recommendations on serviceability.
> 4. Where in the cooling system would a temperature sensor be located? Why would it be fitted in this position?

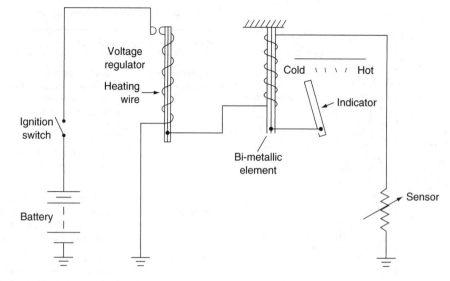

Fig. 2.28 A thermal type gauge and voltage regulator

2.16 Engine core plugs

These are fitted for the following reasons:

- they may blank off the holes left by the jacket cores during casting or machining;
- they may be removed for cleaning out corrosive deposits from the jacket.

The plugs may be of the **welsh plug**, drawn **steel cup** or, less commonly, the **screwed plug** type. The first two are expanded and pressed into core holes that have been machined to size. All three are shown in Fig. 2.29.

2.17 Heating and ventilation

The most common method of providing a comfortable atmosphere in the car is through the heating and ventilation system. In some countries it could mean that a full air-conditioning unit is required where refrigeration cooling is fitted to the vehicle. Fresh air ventilation comes under two headings, **direct** and **indirect**, the heat source being the hot water from the engine. But it may also be gained from a number of other sources such as the exhaust system, as in the air-cooled engine, or in a separate heater where fuel is burnt to heat a chamber over which air is passed, or electrically, in which an element is heated and air passed over the element. The water type are normally fitted to the bulkhead or behind the facia panels and are the most common arrangement in light vehicles.

Direct ventilation

This can be achieved by simply opening one or more of the windows, but this could cause draughts, noise and difficulties in sealing against rain when closed. A number of different methods have been tried to overcome these problems, such as the swivelling quarter light window, but this created problems with water leaking in when the window was shut.

Indirect ventilation

This is routed through a **plenum chamber** located at the base of the windscreen. The internal pressure in the chamber is higher than that of the surrounding atmosphere. It is important therefore that the position of the plenum

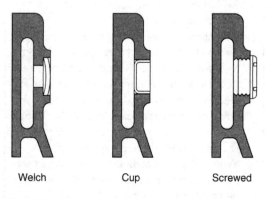

Welch   Cup   Screwed

Fig. 2.29 Types of core plugs fitted to the engine block

chamber and its entrance is chosen to coincide with a high-pressure zone of air flow over the car body, and also where it is free from engine fumes. A heating and ventilation system is shown in Fig. 2.30.

The air flow through the interior of the car can be derived simply from the ram effect of air passing over the car and spilling into the plenum chamber or, at low speeds, it may be boosted by the use of an electrically driven fan connected into the intake of the plenum chamber. Directional control for the air flow is adjustable by deflectors fitted to the outlets on the facia.

Provision for the extraction of the stale air is provided for in grills incorporated in the rear quarter panels of the body, their siting coinciding with **neutral pressure zones** in the air flow over the car.

Learning tasks

1. The volume of air delivered by the interior heater is controlled by the speed of the fan. Identify two methods that are used to control the temperature of the air delivered by the heater.

2. How would you check for the correct operation of the three-speed switch which operated the fan at only two speeds?
3. What symptoms would indicate that the cooling system needs bleeding?
4. How would you identify a suspected problem with a noisy heater fan? Suggest methods you would use to rectify the fault?

2.18 Routine maintenance

The reasons for carrying out routine maintenance are:

- to maintain engine efficiency;
- to extend the life of the engine;
- to reduce the risk of failure in the cooling system;
- to reduce the time the vehicle is off the road when a failure does occur.

There are a number of simple checks that should be made when a routine service is carried out on the cooling system. Check the water level, the condition of hoses and clips

A fresh air into plenum chamber
B warm air into car interior
C stale air out

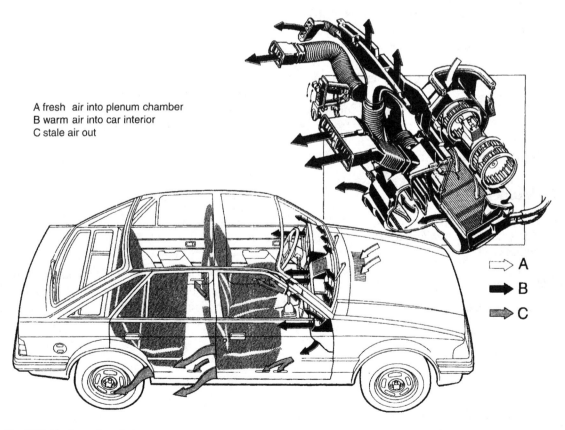

⇨ A
➡ B
⇨ C

Fig. 2.30 Heating and ventilation system

and the heater/radiator matrix for blockages. Drain and flush the system and refill with fresh anti-freeze mixture. Check the anti-freeze content, adjust the belt tension, pressure-test the system, test the thermostat and flow-test the radiator. Bleed the heater system and adjust the control, check the fan cowling and thermostatically controlled components for correct operation.

2.19 Heat losses

These occur when heat transfers from a hot body to a cold body. This process of transfer continues until all the parts are at the same temperature. Heat is measured in **joules** and can be directly converted or expressed as mechanical energy, for example:

1 newton metre (N m) = 1 joule (J)

When a mass of 1 kg is used as a standard the heating value is called the **specific heat capacity** (SHC). This is the amount of heat required to raise the temperature of a substance through 1 °C. The SI unit of heat capacity is the **joule per degree C**. As an example the heat required to raise 1 kg of water through 1 °C is 4.18 kJ. Therefore the SHC of water is 4.18 kJ/kg °C.

Different materials accept or lose heat at different rates. Therefore for a similar mass they will increase or decrease their temperature at different rates.

Table 2.2 shows the SHC of various substances.

The quantity of heat transferred from one substance to another can be calculated from mass, SHC and temperature change.

Calculations

One litre of water has a mass of 1 kg. The formula used in calculating the heat lost or gained by a substance is:

$Q = m \times c \times t$

where : Q = heat loss

m = mass

c = SHC

t = temperature change

1. A pump circulates 150 litres of water through a cooling system in two minutes. The

Table 2.2 The specific heat capacity of substances used in motor vehicles

Substance	SHC kJ/kg °C	Substance	SHC kJ/kg °C
Water	4.18	Steel	0.48
Lub. Oil	1.70	Brass	0.40
Aluminium	0.88	Lead	0.13

temperature at the top of the radiator is 90 °C, and at the bottom 70 °C. Calculate the heat energy radiated per second.

$Q = \dfrac{m \times c \times t}{\text{time}}$

$= \dfrac{150 \times 4.18 \times (90 - 70)}{60 \times 2}$

$= 1.25 \times 4.18 \times 20$

$= 104.5 \, \text{kJ/sec}$

2. An impeller unit circulates two litres of coolant to the radiator per second. Calculate the heat lost to air per second, when the temperature difference between top and bottom tank is 25 °C.

$Q = m \times c \times t$

$= 2 \times 4.18 \times 25$

$= 209 \, \text{kJ/sec}$

3. a) A cooling system contains 15 kg of water. Calculate the quantity of heat gained by the water if its temperature rises from 12 °C to 88 °C on starting.

$Q = m \times c \times t$

$= 15 \times 4.18 \times (88 - 12)$

$= 15 \times 4.18 \times 76$

$= 4765.2 \, \text{kJ}$

b) What heat is lost during cooling if the flow rate is two litres per second and the temperature at the bottom of the radiator is 53 °C?

$t = 88 - 53 = 35 \, °C$

$Q = m \times c \times t$

$= 2 \times 4.18 \times 35$

$= 292.6 \, \text{kJ/sec}$

Practical assignment – cooling systems

Introduction

At the end of this assignment you will be able to:

- recognize water cooling and heater systems
- test the system for leaks
- test the system for anti-freeze content
- remove and test the thermostat
- test the pressure and vacuum valves
- check the operation of the temperature sensor and electric fan
- remove and check the drive belts for serviceability
- refit and set drive belt tension
- remove the radiator and test the flow rate
- bleed the cooling and heater system
- remove the water pump and state the type of impeller fitted
- flush the cooling and heater system

Tools and equipment

- A water-cooled engine
- Suitable drain trays
- Selection of tools and spanners
- Cooling system pressure tester
- Equipment for testing thermostat
- Hydrometer for testing anti-freeze content
- Relevant workshop manual or data book

Objective

- To check the correct operation of each of the components in the cooling system
- To prevent loss of water and over-heating which could cause damage to the engine
- To prevent freezing of the water during very cold weather

Activity

1. Before starting the engine, remove the radiator pressure cap and pressure test the cooling system.
2. Test the pressure and vacuum valves that are situated in the cap. Note any leaks which occur in tests 1 and 2 on the report sheet.
3. Remove the thermostat and check for correct operation using the equipment provided.
4. Remove the drive belts and check for cracks, splits, wear and general serviceability.
5. Remove the radiator, reverse flush it and where necessary complete a flow test. Remove any dirt, leaves, etc. from between fins of the matrix.
6. Remove the water pump, check the bearings for play and the seal for signs of leakage.
7. Refit the thermostat, water pump (using new gaskets), radiator and drive belts, setting the correct torque on the bolts and drive belt tension.
8. Flush the heater system, refit all the hoses and refill with water/anti-freeze mixture, bleeding the system as necessary.
9. Check the operation of the temperature sensor by running the engine up to its normal operating temperature.
10. Remove the fan. If it is an electric fan check for any play or tightness in the bearings, undue noise or loose mountings. Check for damage to the fan blades. Where necessary lubricate the bearings and refit the fan.
11. Complete the report sheet on the cooling system, identify any faults and report on serviceability.
12. Answer the following questions:

    (a) State the types of cooling systems used on motor vehicles and give two advantages for each.
    (b) State the purpose of:
        (i)   the pressure valve
        (ii) the vacuum valve
        (iii) the thermostat
        (iv) the temperature sensor
        (v)   the water pump

(c) What is the purpose of pressurizing the cooling system?

(d) Give two advantages of fitting an electric fan compared with belt driven fans.

(e) Describe a possible cause and the corrective action to be taken for the following faults:
   (i) Squealing noise from the front of the engine
   (ii) External leak from the bottom of the radiator
   (iii) Internal leak in the engine
   (iv) Over-heating with no loss of water
   (v) Heater does not get warm enough
   (vi) Coolant is very dirty.

Checklist

Vehicle

Visual checks on

Type of radiator fitted
Type of fan fitted
Type of cooling system

Practical tests on the cooling system

Anti-freeze content
Pressure test of cap
Pressure test of system
Fan belt tension and condition
Operation of heater
Operation of fan
Operation of thermostat
Radiator flow test
Manufacturer's technical data

Comments on serviceability

Student's signature

Supervisor's signature

Practical assignment – liquid cooling systems

Objective

To carry out a number of tests on the cooling system to ascertain its serviceability.

- Thermostat setting
- System pressure testing
- Radiator flow test.

Tools and equipment

- Running engine/vehicle
- Assorted hand tools
- Thermostat testing equipment
- Pressure tester
- Header tank, drain tank and fittings
- Stopwatch.

Safety aspects

- Cables, hands, hair and loose clothing must be kept clear of fans and other rotating parts. Stationary engines require guards to be fitted to the fan drive.
- Keep hands clear of hot parts of the engine.
- If a vehicle is used, the gear lever must be in the neutral position, the hand brake applied and the wheels chocked before operating the starter motor.
- Arrangements must be made for the exhaust gases to pass directly out of any enclosed space.
- Remove radiator cap *slowly* and cover with a cloth if the engine is hot.

Safety questions

1. Why must the exhaust gases not be discharged into the garage?
2. What are the dangers from unguarded fan blades and fan belts?
3. What is likely to happen if the radiator cap is removed quickly when the engine is hot?

Activity

1. (a) Disconnect hoses and housing and remove the thermostat from the engine.
   (b) Place the thermostat in the tester and heat the water to the opening temperature of the thermostat, watch for commencement of opening and fully open point, note the temperatures and record the results.

   Note: It may be necessary to first subject the thermostat to boiling water so that the fully open position can be assessed.

   (c) The recommended opening temperatures should be obtained from the data book.
   (d) Replace the thermostat in the correct position, refit the housing using a new gasket as required, fit hoses and tighten hose clips, run engine and check for water leaks.

2. (a) Remove radiator cap. (Observe safety instructions.)
   (b) Fit pressure testing equipment in place of the radiator cap, operate pump to pressurize system to recommended pressure.
   (c) Carry out visual inspection of components in the cooling system and joints for coolant leakage.

Note: Look for a steady reading of the pressure tester gauge; if the pressure falls steadily and there is no coolant leakage then an internal leakage of the coolant can be suspected. (Check that the tester unit is fully sealed on the radiator neck.)

   (d) To identify internal leakage there are products on the market which can be mixed with the coolant, the system is closed and the engine is run until it reaches its normal operating temperature. After the system cools the radiator cap is removed and the colour of the coolant is inspected. The chemical changes colour when it comes into contact with oxygen. Therefore if the colour of the coolant changes it can be assumed that air is entering the system, most probably through the cylinder head gasket.
   (e) The radiator cap can be checked for correct operation by fitting the correct adaptor to the pump and fitting the cap in place on the tester. The pressure is then raised by operating the pump and recorded when the stage is reached for the seal and the spring in the cap to lift and so relieve the pressure.
3. To flow test the radiator.
   (a) Remove the radiator and fit it to the test rig.
   (b) Fill the header tank of the rig with a known quantity of water.
   (c) Open the tap to discharge the water through the radiator and measure the draining time with a stop watch. Compare this with the manufacturing data (20 litres takes approximately 20 seconds for a car radiator with a water head of 0.7 metres).
   (d) The mineral deposits in the coolant tend to block the water ways, so an approximation of the flow rate can be ascertained by comparing the mass of the radiator under test with the mass of a new radiator of the same type. The deposits are heavy so a 25% increase in weight will give an indication of several mineral deposits.

Questions

1. What are the effects of pressurizing the cooling system?
2. Explain the function of the two valves in the pressure cap.
3. List five reasons for an engine becoming overheated.

Self assessment questions

1. A thermostat is fitted to cooling systems:
   (a) to provide a variable current in the temperature gauge circuit
   (b) to control circulation of coolant to help the system to warm up quickly and to maintain a constant operating temperature
   (c) to act as a control to switch the cooling fan on and off
   (d) to operate the compressor on an air conditioning system.
2. The lubrication system of a certain engine contains 0.8 kg of oil that has a specific heat capacity of 1.7 kJ/kgC. During the warm-up period the temperature of the oil rises by 30 °C. The amount of energy transferred to the oil during this warm up period is:
   (a) 32.5 J
   (b) 325 MJ
   (c) 40.8 kJ
   (d) 408 J
3. Inside a pressurised cooling system:
   (a) the coolant boiling temperature is above 100 °C
   (b) the increased pressure slows down the circulation of coolant
   (c) the boiling point of the coolant is lowered
   (d) the convection currents cease to operate
4. Ethylene glycol based anti freeze when added to the coolant causes:
   (a) less evaporation of the coolant
   (b) the freezing temperature of the coolant to be raised
   (c) the freezing temperature of the coolant to be lowered
   (d) the boiling point temperature of the coolant to be lowered
5. The purpose of the water pump in a cooling system is to:
   (a) pressurise the cooling system
   (b) circulate the coolant
   (c) eliminate the need for a radiator
   (d) reduce heat loss

6. After combustion the temperature inside the engine cylinder is:
   (a) approximately 200 °C
   (b) approximately 12 000 °C
   (c) approximately 1400 °C to 2000 °C
   (d) 250 °C
7. Cooling systems rely on heat transfer by:
   (a) pressure differentiation
   (b) electrical conductivity
   (c) Archimedes principle
   (d) conduction, convection and radiation
8. Radiator surfaces are often finished in matt black because:
   (a) this surface is effective against corrosion
   (b) this surface is most effective in radiating heat
   (c) this type of finish reflects most heat
   (d) this surface finish prevents heat escaping into the atmosphere

3
Fuel systems

Topics covered in this chapter

Petrol fuel system
Petrol injection
Carburettors
Diesel fuel systems
Direct injection – indirect injection
Common rail fuel system

3.1 Petrol fuel

Petrol is a colourless liquid and is one of the fuels most commonly used in the motor vehicle engine. This is because it is a clean liquid, is easily stored and flows freely. It gives off an inflammable vapour even at very low temperatures and when burnt gives off a large amount of heat.

Before petrol can be burnt it must be **atomized** (that is it has to be broken down into very small droplets like a mist) so that it can be mixed with a suitable quantity of air. This is usually done by the carburettor. The **combustion** process (the burning of the air/fuel mixture) involves the chemical combination of a fuel with oxyge; during this process heat is given off. This heat given off by the complete combustion of a unit mass of a fuel is called the **calorific value** of a fuel, e.g. an average sample of petrol has a calorific value of 44 MJ/kg. Crude petroleum, from which petrol is refined, is a mixture of various compounds of hydrogen and carbon (called **hydrocarbons**).

The mass of air per kilogram of fuel in a mixture of air and fuel gives the air/fuel ratio. For complete combustion the chemically correct mixture is approximately 14.7 parts of air to 1 part of fuel (**14.7:1**). Ratios of less than about 8:1 (rich) or more than 22:1 (weak) cannot normally be ignited in petrol engine cylinders. The air/fuel ratio has a considerable effect on the engine's performance and power output. The chemically correct ratio does not always give the best results: for instance, under cold starting conditions, an air/fuel ratio as low as 2:1 may be required; for acceleration a ratio of say 12:1; and for the most economical running 17 or 18:1 may be required.

Any change in the air/fuel ratio also changes the composition of the exhaust gases. Under normal operating conditions a ratio of approximately 14.7:1 will give the least toxic exhaust gas. With a richer mixture the exhaust gas will contain more **carbon monoxide**, with a weaker mixture **oxides of nitrogen** will be present. Both these gases are harmful to the human body, especially in confined spaces.

Learning tasks

1. What are the main dangers associated with petrol? What safety regulations must be observed in the workshop when working on the fuel system?
2. How may fuel be used or stored in the garage environment?
3. List the dangers of running a petrol engine in the workshop and any safety precautions that should be taken.
4. Complete a survey of the local service stations and garages in your area. Identify the types of fuel and services offered, stating price and grade of fuel, e.g. diesel, petrol, leaded, unleaded, super, etc. From your survey draw up a graph to show price and variety of services offered.
5. Identify the main differences between leaded and unleaded fuels, and give the advantages of each. What effect (if any) does each have on the running of the vehicle, e.g. power output, MPG, service intervals, engine wear, etc.?

3.2 Layout of the petrol fuel system

A complete fuel system consists of the following components.

- **Fuel tank** in which to store the fuel. This also contains a **sensor unit** to indicate to the driver how much fuel the tank contains.
- **Pipelines** that connect the tank to the lift pump and carburettor and return excess fuel back to the tank. This helps to reduce the formation of vapour locks in the system, which in warm weather may stop the engine from running.
- **Fuel filter** to remove unwanted sediment (particles of dirt and water) from the fuel before it reaches the carburettor.
- **Fuel lift pump** to transfer the fuel from the tank to the carburettor or fuel injection unit.
- **Carburettor or fuel injection unit** which meters the fuel and mixes it with the air in the correct proportions to suit the engine needs.
- **Inlet manifold** which directs the mixture to the inlet ports in the cylinder head.
- **Air filter** to remove small particles from the air as it is drawn into the inlet manifold via the carburettor.

A typical carburettor-type fuel system is shown in Fig. 3.1.

Learning task

1. Take a look at one of the vehicles in the workshop and identify the main units. Draw and label a simple block diagram to show the layout of the system and how the points are connected together. Show on the diagram the direction of the fuel flow.
2. What personal safety precautions should the mechanic take when working on the fuel system?

3.3 The fuel tank

Several types of materials are commonly used in the manufacture of the fuel tank. One type is the pressed steel, often coated on the inside with lead/tin to prevent corrosion. They normally have either welded joints and seams or soldered joints (where the tin is cut and bent to shape and the seams are rolled before being soldered). Another type uses expanded synthetic rubber or flame-resistant plastic, moulded to the required shape. This gives a high resistance to damage as it will bend or distort fairly easily. It is also lighter than those made from steel and is rust proof.

The tank is usually fitted with **baffles** (these are partitions inside the tank) to prevent the fuel surging from side to side, especially when the tank is not full. Holes are positioned in the side or top of the tank to allow for the location of the fuel-gauge sensor unit, the supply and return pipes and the filler pipe. A coarse gauze filter is positioned over the fuel feed pipe to prevent large particles of dirt from blocking the pipe. A **vent pipe** is also fitted to relieve the vapour pressure and allow air to enter as fuel is drawn from the tank. The filler tube usually contains the vent pipe and the overflow pipe near to the filler cap end. No petrol vapour is allowed to pass into the atmosphere, so a non-venting filler cap is fitted; it contains a one-way valve

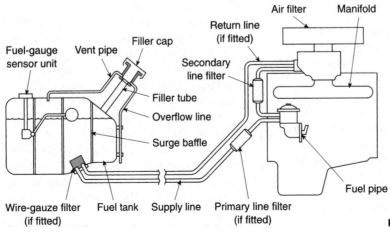

Fig. 3.1 Carburettor-type fuel system

allowing air into the tank and preventing the fuel and its vapour from escaping.

Safety note

If fuel tank repairs are necessary it is recommended that these are only undertaken by the specialist. Any heat could ignite the fuel vapour and cause a serious explosion and possible personal injury or fire, even if the tank has been empty for some months.

Learning task

1. Draw up an operations schedule for removing a fuel tank from a vehicle. Include any safety recommendations.
2. What would be the symptoms of a blocked valve in the filler cap?
3. Remove and refit a fuel tank using the operations schedule you have drawn up, taking special note of any safety precautions that must be observed.

3.4 Fuel pipe lines

Both supply and return lines may be made from plastic or steel piping which is clipped to the body between the tank and the engine. Flexible hoses are used to connect the rigid pipe to the engine, this allows for the movement of the engine on its mountings. Plastic pipes must be protected from heat or there could be a serious risk of fire, especially if welding is to be carried out on the bodywork close to any part of the fuel system.

When the **needle valve** in the carburettor float chamber closes, the fuel passes via the return line back to the tank. A **restrictor** is fitted in the return to prevent too much fuel bypassing the float chamber. In this way a continuous flow of fuel through the system is maintained which reduces the possibility of a **vapour lock** occurring and keeps the fuel at a fairly constant temperature.

3.5 Fuel filters

The idea of fitting a fuel filter is to prevent particles of dirt, water, etc. from entering the main components and creating a blockage, causing excessive wear or premature failure to occur. Although the fuel produced by the petrol companies is clean, it is possible for particles of dust or dirt to enter the tank when filling, and for moisture-laden air to condense into water droplets inside the tank and be drawn into the fuel system. A combination of coarse filters on the inlet and fine filters on the outlet sides of the pump is used, as shown in Fig. 3.1.

Learning tasks

1. At what service intervals should the filter be checked, cleaned or replaced?
2. Inspect a filter and state the material from which it is made. Give reasons for the use of these materials.
3. Fuel filters may be fitted in a variety of different places in the system. Inspect a number of vehicles and list at least two different places giving an advantage for each together with any changes in their servicing requirements.

3.6 Fuel lift pumps

The two main types of fuel lift pumps in common use are:

- The **mechanical** lift pump – fitted to and operated by the engine.
- The **electrical** lift pump – fitted in any convenient position on the vehicle and operated electrically.

Mechanical lift pump

Figure 3.2 shows a **diaphragm-type lever-operated** petrol pump, which draws fuel from the tank and lifts it under pressure to the carburettor float chamber. It is bolted to the engine so that the lever arm rests on an **eccentric** on the **camshaft**. As the lever arm is lifted a diaphragm is operated inside a sealed chamber. This upper chamber contains **inlet** and **outlet valves**. The fuel is drawn from the tank through the inlet valve as the diaphragm is pulled down. On releasing the lever the inlet valve closes and the diaphragm forces the fuel through the outlet valve under pressure from the return spring to the carburettor; as the lever is lifted again the outlet valve closes preventing fuel from passing back into the pump. This sequence is repeated each time the lever is operated until the float chamber is full.

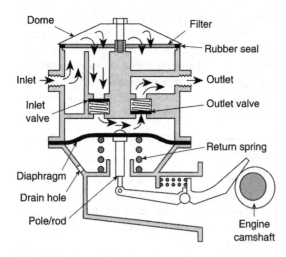

Fig. 3.2 Mechanical lift pump

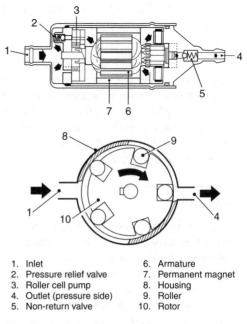

1. Inlet
2. Pressure relief valve
3. Roller cell pump
4. Outlet (pressure side)
5. Non-return valve
6. Armature
7. Permanent magnet
8. Housing
9. Roller
10. Rotor

Fig. 3.3 Electric fuel pump

Electric fuel pumps for petrol injection systems

In petrol injection systems the fuel pump must generate sufficient pressure to pressurise the fuel rail and also be capable of transferring petrol from the fuel tank, to the engine, so that there is an adequate supply of fuel to meet all engine operating conditions. These conditions are met by the use of electric pumps of the type shown in Fig. 3.3.

This petrol pump is driven by a permanent magnet electric motor which may be mounted outside the fuel tank, as shown in Fig. 3.4(a), or inside the tank, as shown in Fig. 3.4(b).

The pump shown in Fig. 3.3 operates as follows: the rotor (10), which is eccentrically placed in the

housing (8) is driven round in a clockwise direction. Rotation of the rotor causes the rollers (9) to move in and out of the slots in the rotor. The increased space at the inlet (1) to the pump sucks fuel into the pump chamber and as the rotor continues to turn, the fuel is forced into the smaller space at the pump outlet (4) via the one way valve (5). The pressure at the pump outlet is approximately 2.5 to 3 bar and it is controlled by the pressure regulating valve. The pressure regulating valve is operated by manifold vacuum and it redirects excess fuel back the fuel tank.

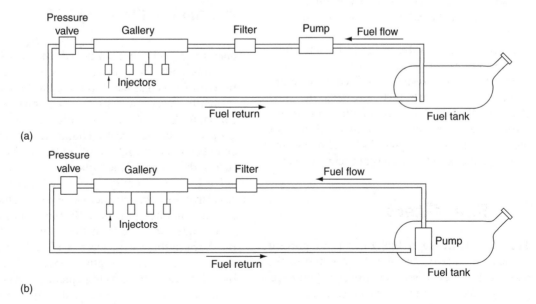

Fig. 3.4 Alternative positions for the petrol pump

3.7 Carburation

Carburation is one of the two methods most commonly used to introduce the fuel into the air that is drawn into the cylinder. The other is fuel injection (see Section 3.9). Before considering any system some explanation of the term carburation, which is the mixing and delivery of, in this case, petrol and air into the cylinders, is necessary.

The mixing of the fuel with the air is done by the carburettor in the **venturi** or **choke tube** (Fig. 3.5). The venturi has a lower pressure at the narrowest section which will draw fuel into the air stream if the outlet from the float chamber is positioned at this point. This drop in pressure is caused by the molecules of air which flow faster at the narrowest section and, at the same time,

move further apart thus lowering the pressure. The fuel in the float chamber has atmospheric pressure acting on its surface and so fuel is forced through the outlet into the air stream in the form of very fine droplets. This mixes with the air. The resulting mixture is drawn into the cylinders through the inlet manifold.

3.8 The carburettor

The function of the carburettor is to control the quantity of fuel and air mixture entering the cylinders, to atomize the fuel into very fine droplets, mix it with the incoming air and vaporize this fine spray into a combustible mixture. The chemically correct ratio of air to fuel is approximately 14:1 by weight, that is 14 parts of air to one part of fuel (this is called the **stoichiometric air–fuel ratio**).

Variable or fixed choke carburettor

Sometimes referred to as **variable vacuum** carburettors, these have a **fixed diameter venturi** and contain jets which enable the engine to operate over a wide range of both engine speeds and loads.

The simple fixed choke carburettor has a float chamber that provides a constant level of fuel, just below the outlet from the jet. This level is maintained by means of the float and needle. The float chamber has a small hole in the top to allow atmospheric pressure to act on the top of the fuel. A choke tube or venturi is fitted in the mixing chamber around the jet outlet to increase the speed of the air at this point and create a depression. This causes the atmospheric pressure acting on the fuel in the float chamber to force fuel via the outlet into the fast moving air stream. The **throttle butterfly valve** regulates the quantity of mixture passing to the cylinders and is operated by the driver pressing the accelerator pedal. However, in some cases it can be operated by hand, as on motor cycles or where a constant engine speed is required, as on some types of heavy vehicles that have a hydraulic pump fitted to operate a crane.

The **simple** carburettor will provide the correct mixture for one engine speed/load only, at higher engine speeds the mixture will progressively become too rich (too much fuel mixed with the air) and at lower engine speeds the

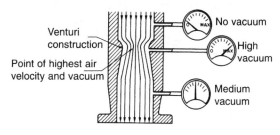

Venturi construction

Point of highest air velocity and vacuum

No vacuum

High vacuum

Medium vacuum

Fig. 3.5 Venturi pipe

mixture becomes too weak (that is not enough fuel is mixed with the air). When starting from cold a smaller proportion of fuel is evaporated and some may condense onto the cold surfaces of the inlet manifold before reaching the combustion chamber. Because of this a mixture richer than the chemically correct ratio is required. As the engine revs are increased the flow rate of the air does not match that of the fuel and consequently the air/fuel mixture becomes progressively too rich. Because of these problems the simple carburettor is not suitable for the modern motor car, although the basic principles of operation still apply. This type of carburettor is shown in Fig. 3.6.

Learning tasks

1. On which types of engines would the simple carburettor be used? Give reasons for your answer.
2. Give two reasons why a simple carburettor is unsuitable for the modern motor vehicle.
3. Dismantle a simple carburettor. State how many jets are fitted and what they are called. Clean and reassemble the carburettor, adjust the slow running jet and maximum and minimum revs to suit the engine.

Mixture correction

The simple carburettor attempts to meter both the fuel and the air. Air, which is a gas, flows very easily; petrol is a liquid which has particles that try to stick together and to the surface of the passages through which it must flow. If a suitable size jet is used for a fixed engine speed/ load then if the speed is increased the larger depression would cause more fuel to flow creating a rich mixture.

To overcome the problem of the mixture getting richer as the engine speed increases an air bleed is fitted in the system. Methods of mixture correction all use the same basic principle, i.e. as the engine speed increases air is introduced into the fuel system to gradually weaken the mixture by reducing the amount of fuel entering the venturi. This gives a fairly constant mixture strength over a wider range of engine speeds and is shown in Fig. 3.7(a).

Air bleed compensation

With this arrangement a well of petrol, fed from the main jet, is subject to the depression in the venturi. As the fuel level is lowered **air bleed holes** are progressively uncovered, thereby admitting more air which reduces the depression acting on the jet which prevents enrichment of the mixture as the depression in the venturi increases. The tube, which contains a number of holes drilled down its length, is sometimes called the **emulsion** or **diffuser tube** as it tends to premix the fuel and air. The level in the well is at the same level as the float chamber when the engine is not running. Some systems have an **air bleed jet** (sometimes called an **air correction jet**) to enable the size of the jet to be altered. This is for tuning purposes and under normal circumstances would not be removed except for cleaning.

Slow-running (idling) systems

When the engine is running slowly, say at tick-over, the depression in the venturi is not high enough to draw fuel out of the main jet and atomize it into very fine droplets. The speed of

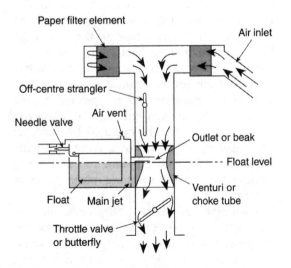

Fig. 3.6 Simple downdraught carburettor

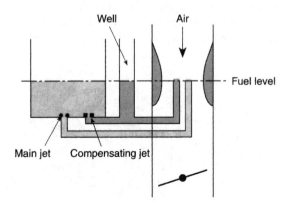

Fig. 3.7(a) Compensating jet action

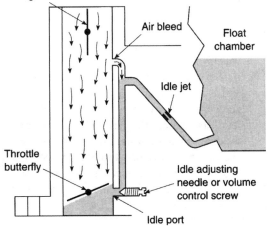

Fig. 3.7(b) Simple idling circuit

the air will also be too slow to keep the fuel in suspension in the air stream (it will tend to fall out of suspension onto the inlet manifold). Because of this the engine will have difficulty in idling and some means of keeping it operating is required. This is the purpose of the slow-running system.

As can be seen from Fig. 3.7(b) the circuit has a jet with a suitable size hole in it and a means of adjusting the volume of air/fuel mixture; and the outlet is positioned on the engine side of the throttle butterfly, where the inlet depression is at its highest. Here it mixes with the air stream as it passes between the throttle butterfly and the wall of the carburettor; in this way the venturi and main jet are bypassed and the engine is

fed through the slow-running system to maintain a fairly constant tick-over.

Air bleed and idling mixture arrangements

When the two systems are incorporated into one carburettor then the system permits the mixture to be controlled over a wide range of engine speeds and loads to give the best possible fuel consumption and the least harmful exhaust gases.

Progression jets and acceleration devices

When the throttle butterfly is opened the depression at the slow running outlet is reduced while at the same time a larger volume of air is passed into the inlet manifold. The depression at the main jet outlet is not enough to draw fuel into the air stream. The result of this is a sudden weakening of the mixture strength, causing the engine revs to drop even further before they begin to pick up again.

To overcome this problem a series of holes (usually two or three) are drilled up the side of the body of the carburettor above the throttle butterfly. As the throttle opens it causes the depression to pass over the progression outlets, bypassing the **volume control screw** and drawing the emulsified fuel directly into the air stream. This enriches the mixture, increasing the engine revs until a point is reached where the depression is sufficient to operate the main jet in the mixing chamber (the venturi). This is shown in Fig. 3.8(a).

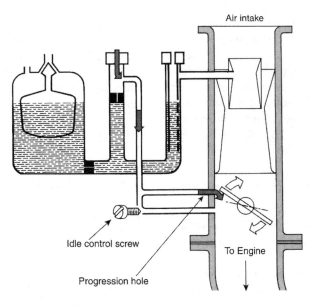

Fig. 3.8(a) The principle of the progression hole

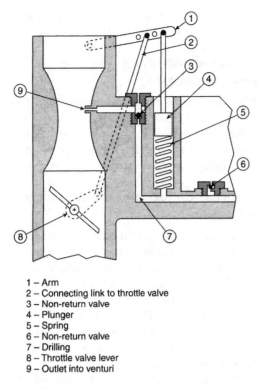

1 – Arm
2 – Connecting link to throttle valve
3 – Non-return valve
4 – Plunger
5 – Spring
6 – Non-return valve
7 – Drilling
8 – Throttle valve lever
9 – Outlet into venturi

Fig. 3.8(b) Mechanically operated acceleration pump

If the throttle is opened suddenly to full throttle neither the slow-running, progression jets or the main jet will operate. This is because there is no depression at the slow-running or progression jets as the throttle is fully open and the depression is not enough to operate the main jet. This causes what is known as a **flat spot** (this is where the driver is thrust forwards and then backwards as the engine almost stalls before the vehicle takes off). To overcome this problem an enrichment device or **acceleration pump** is fitted, linked to the throttle linkage. As the throttle is operated quickly the pump delivers neat fuel directly into the venturi, just at the time when the depression is insufficient to draw fuel from the other systems. If the throttle is opened slowly the accelerator pump will not operate. This is shown in Fig. 3.8(b).

Learning tasks

1. How would the driver notice if there was a hole in item 35 of the Weber carburettor shown in Fig. 3.9.
2. Draw up a service schedule for the carburettor. Identify any safety precautions that should be observed.

---

3. Describe the steps that should be taken when setting up a carburettor that has just been refitted after major overhaul. Itemize any special equipment that you would use.
4. Remove a fixed choke carburettor, remove and clean the following components: slow-running, accelerator, progression and main jets. Reassemble the jets and refit the carburettor to the engine. Run the engine and set tick-over, manual/automatic choke and adjust the mixture according to manufacturer's data. Complete a job card to record all work done.

Cold starting devices

There are two main problems to overcome when starting from cold to ensure the engine will start. First, the speed at which the starter will turn the engine will not be fast enough to create sufficient depression in the venturi to ensure that fuel is drawn from the fuel jet. In cold weather the problem is made worse by the drag of the oil in the bearings and sump, oil being at its thickest when the weather is cold.

Second, there is no heat to assist in vaporizing the fuel which is delivered to the cylinders. The effective mixture of fuel vapour and air will be far too weak to be readily ignited by the spark at the sparking plug.

The simplest way of overcoming these two problems is to temporarily supply an excess of fuel to ensure that a proportion of fuel which will vaporize at this cold temperature will also form an ignitable mixture with the incoming air. To keep the engine running, some excess of fuel will still be required until the engine warms up, and a greater quantity of mixture will be required to off-set the increase in oil drag until the oil warms up. The excess fuel that does not evaporate will be deposited on the cylinder walls. Whilst a small amount of this may be helpful by thinning the oil down, too much will wash the oil film off altogether causing metal-to-metal contact and severe wear to occur. Thus any excess fuel must be no more than is absolutely necessary and must be discontinued as soon as possible.

Cold starting devices may be **automatic** or **manually operated**. The simplest is the manual strangler (Fig. 3.10(a)) where the driver pulls out the choke which is flexibly connected to the carburettor choke control. This has the

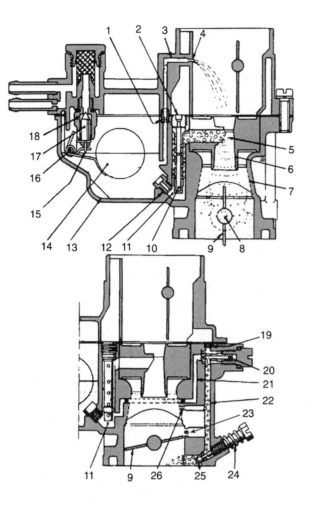

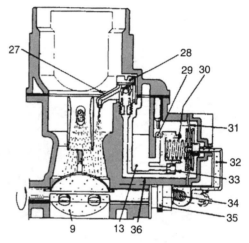

1 – Calibrated bushing	19 – Idle air metering bushing
2 – Air bleed jet	20 – Idle jet
3 – High speed orifice passage	21 – Fuel passage
4 – High speed orifice	22 – Idle passage
5 – Spray nozzle	23 – Transfer orifices
6 – Auxiliary venturi	24 – Idle adjusting screw
7 – Venturi	25 – Idle feed bushing
8 – Throttle shaft	26 – Idle air calibrated orifice
9 – Throttle valve	27 – Pump jet nozzle
10 – Emulsion tube	28 – Delivery valve
11 – Emulsion tube well	29 – Ball valve
12 – Main jet	30 – Intake control spring
13 – Fuel bowl	31 – Diaphragm
14 – Float	32 – Accelerating pump control lever
15 – Hook, needle return to	33 – Calibrated bushing, excess
float tang	fuel discharge
16 – Float hinge pin	34 – Throttle return spring
17 – Valve needle	35 – Throttle control lever
18 – Needle valve	36 – Fuel delivery passage

Fig. 3.9 Weber 32IBA unit

effect of reducing the air supply and enriching the mixture drawn into the engine cylinders.

A simple automatic cold start control (Fig. 3.10(b)) has a **bi-metal spring** wound round the strangler (butterfly) spindle. One end of the spring is attached to the spindle; the other end is attached to the carburettor body in such a way that the butterfly is held in the closed

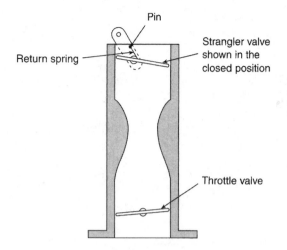

Fig. 3.10(a) Manual choke

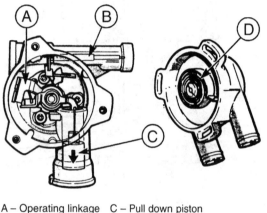

A – Operating linkage   C – Pull down piston
B – Needle valve   D – Bi-metal coil

Fig. 3.10(b) Automatic choke and bi-metal housing assemblies

position when the engine is cold. As the engine warms up the spring gradually unwinds and releases the strangler to the normal run position. The fast idle cam increases the engine speed to enable it to run when the strangler is in operation.

A number of other types of automatic cold start are used by manufacturers. The most common is the **water heated** type where the bi-metal strip is surrounded by a water jacket. As the engine cooling system heats up the heat is transferred to the bi-metal strip. Other methods use **exhaust gas** or **heated air** for this purpose.

Variable-choke carburettor

This is sometimes called the constant-vacuum or constant-depression carburettor. The principle of the variable-choke carburettor is to employ a means whereby the effective choke area moves according to the engine speed as required. This will have the effect of having a constant air velocity (air speed) and therefore a constant depression over the jet. One of the most common carburettors of this type is called the SU carburettor (S. U. stands for Skinners Union as in the original carburettor instead of the dash-pot a diaphragm made from cow hide (leather to you and me) was used).

SU carburettor

This is a **side draught** type and has a tapered needle which is attached to the underside of the piston which moves vertically in a dashpot. The lower part of the piston rests on a flat portion of the venturi (sometimes called the bridge) and forms the moveable part of the variable choke area. The jet is also mounted in the bridge and the tapered needle passes through the middle of the jet. As the engine starts the depression in the inlet manifold is felt above the piston in the suction chamber, through the holes drilled on the engine side of the lower part of the piston. This causes the piston and needle to rise lifting the piston. The resulting air flow over the bridge draws fuel from the jet and mixes it with the incoming air in the correct air/fuel ratio. When the throttle is opened and engine revs increase, the piston lifts higher giving a greater choke area together with a larger jet area. With careful calibration of jet with tapered needle size this type of carburettor can be used on a wide range of engines. It is shown in Fig. 3.11.

Acceleration and mixture enrichment

When the throttle is opened quickly the piston will also rise quickly causing a 'flat spot' to occur just when you wanted the engine to accelerate. To overcome this problem the piston has a **damper** fitted in the central spindle which guides its vertical movement. This is filled with oil. This means that when the throttle is opened quickly the rate at which the piston rises is slowed because the oil has to be displaced through the valve. This will have the effect of increasing the depression over the jet, drawing more fuel for the same amount of air and giving a richer mixture for acceleration purposes.

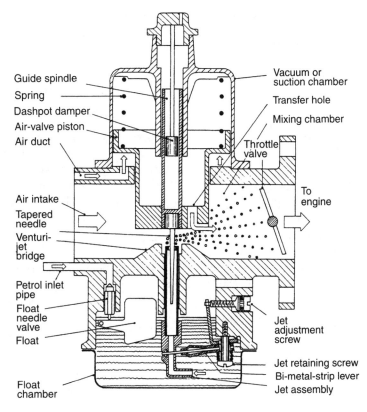

Fig. 3.11 Constant-vacuum, variable-choke carburettor

Cold starting

In cold conditions the driver pulls the choke lever out to operate the cold start. The lever is connected via a flexible cable to a linkage which either lowers the jet in the body to increase the jet area and therefore the amount of fuel, or opens a separate cold start metering valve which bypasses the main jet, introducing fuel into the venturi on the inlet side of the throttle butterfly.

Ford variable-venturi carburettor

This carburettor, shown in Fig. 3.12, uses a **pivoting air valve** to operate a tapered needle, sliding it horizontally in the main jet. It has a separate slow-running system that operates with the main jet at 'tick-over' speeds. It also incorporates an **anti-dieseling valve** that cuts off the slow-running mixture to prevent running-on when the ignition is turned off. A separate system is used for starting purposes when cold.

Learning tasks

1. What would the effects be of a sticking cold starting aid. List the tests that should

be carried out to identify the problem together with the steps taken to put the fault right.
2. List at least two advantages of the variable-choke carburettor over the fixed-choke carburettor. Give one main disadvantage.

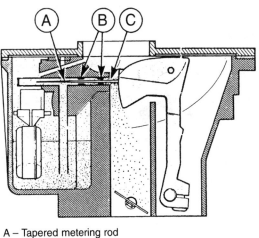

A – Tapered metering rod
B – Main and secondary jets
C – Main fuel outlet

Fig. 3.12 Ford variable venturi carburettor – main jet system

3.9 Computer controlled petrol fuelling systems

Computer controlled petrol injection is now the normal method of supplying fuel – in a combustible mixture form – to the engine's combustion chambers. Although it is possible to inject petrol direct into the engine cylinder, in a similar way to that which is used in diesel engines, the practical problems are difficult to solve and it is still common practice to inject (spray) petrol into the induction manifold. There are, broadly speaking, two ways in which injection into the induction manifold is performed. One way is to use a single injector that sprays fuel into the region of the throttle butterfly and the other way is to use an injector for each cylinder, each injector being placed near to the inlet valve. The two systems are known as single point injection (throttle body injection) and multi-point injection. The principle is illustrated in Fig. 3.13.

Computer controlled multi-point fuel injection system

In petrol injection systems it is often the practice to supply the injectors with petrol under pressure, through a fuel gallery or 'rail'. Each injector is connected to this gallery by a separate pipe, as shown in Fig. 3.14.

The pressure of the fuel in the gallery is controlled by a regulator of the type shown in Fig. 3.15.

This particular pressure regulator is set during manufacture to give a maximum fuel pressure of 2.5 bar. In operation the petrol pump delivers more fuel than is required for injection and the excess pressure lifts the regulator valve

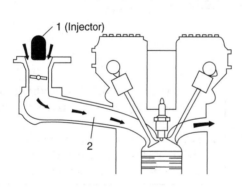

(a)

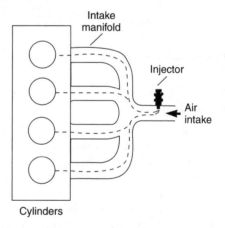

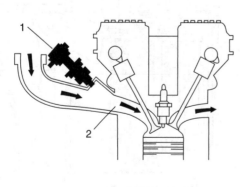

(b)

Fig. 3.13 (a) Single-point injection, (b) Multi-point injection

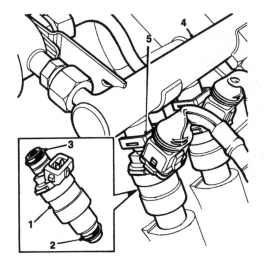

1. Injector
2. 'O' ring seal (manifold)
3. 'O' ring seal (fuel rail)

4. Fuel gallery
5. Retaining clip

Fig. 3.14 The fuel gallery

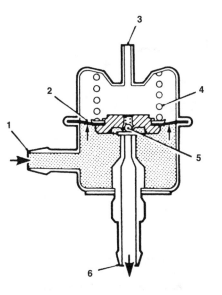

1. Fuel inlet
2. Diaphragm
3. Vacuum pipe connection

4. Pressure control spring
5. Valve
6. Fuel return connection

Fig. 3.15 The fuel pressure regulator

(5) off its seat to allow the excess fuel to return to the fuel tank via the return connection (6).

The internal diaphragm (2) of the regulator is subject to inlet manifold pressure (vacuum) and this permits the diaphragm and spring to regulate fuel pressure to suit a range of operating conditions. Raising the diaphragm against the spring lowers the fuel pressure and this permits a low pressure of approximately 1.8 bar. Lowering the diaphragm in response to higher pressure in the inlet manifold (wider throttle opening) gives a high fuel pressure of approximately 2.5 bar.

With this arrangement, the amount of fuel that each injector sprays into the inlet manifold is determined by the length of time for which the injector valve is opened by its operating solenoid. By varying the length of time for which the injector valve remains open, the amount of fuel injected suits a range of requirements.

Fuelling requirements for a particular engine are known to the designer and are placed in the computer (ECU) memory (ROM). In operation, the ECU receives information (data) from all of the sensors connected with the engine's fuel needs. The ECU compares the input data from the sensors with the data stored in the computer memory.

From this comparison of data, the ECU provides some output data which appears on the injector cables as an electrical pulse that lasts for a set period. This injector electrical pulse time varies from approximately 2 milliseconds (ms), to around 10 ms. The 'duty cycle' concept is based on the percentage of available time for which the device is energised, as shown in Fig. 3.16.

A fuelling map

The performance characteristics of the engine and the driveability of the vehicle are determined by the 'quality' of the input that is put

A typical square waveform is shown in the figure, a single cycle is indicated by 'C', which consists of an ON time 'A', and an OFF time 'B'.

Duty cycle is the length of the ON time 'A' compared to the whole cycle 'C', expressed as a percentage. (Please note, On time can be High and Low on certain systems.)

Using the figure and the time periods, the duty cycle is 25%.

C = 60 ms (100%)
A = 15 ms (1/4 of 60 ms = 25%)
B = 45 ms (3/4 of 60 ms = 75%)

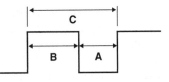

Fig. 3.16 Duty cycle

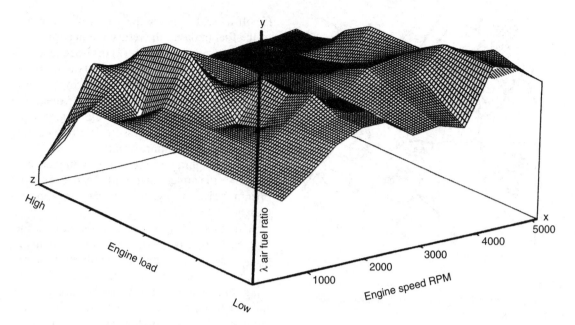

Fig. 3.17 An engine fuelling map

into the design and part of this design is the computer program that is held in the read only memory (ROM) of the ECM. A fuelling map is shown in Fig. 3.17. The variables on the three axis of this graph (map) are engine load, engine speed and air–fuel ratio. For each combination of engine speed and load there is a corresponding air–fuel ratio where the points meet on the surface of the map. Each of these points can be represented by a binary code and it is a range of points from the map that it is stored in the ROM of the ECM. The values stored in the ROM are compared with input signals from sensors in order that the computer can determine the duration of the fuel injection pulse.

Some computerised systems are designed so that the franchised dealership can alter the computer program to match customer requirements. A re-programmable ROM is necessary for this to be done and the work can only be done by qualified personnel acting under the control of the vehicle manufacturer.

Single point injection

In its simplest from, petrol injection consists of a single injector that sprays petrol into the induction manifold, in the region of the throttle butterfly valve, as shown at (4) in Fig. 3.18.

Finely atomised fuel is sprayed into the throttle body, in accordance with controlling actions

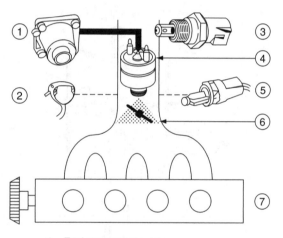

1. Fuel pressure regulator
2. Throttle position sensor
3. Air charge temperature sensor
4. Injector valve
5. Throttle plate control motor (actuator)
6. Throttle plate
7. Engine

Fig. 3.18 Single-point injection details

from the engine computer, thus ensuring that the correct air–fuel ratio is supplied to the combustion chambers to suit all conditions. The particular system shown here uses the speed density method of determining the mass of air that is entering the engine, rather than the air flow meter that is used in some other applications. In order for the computer to work out (compute) the amount of fuel that is needed for a given set of

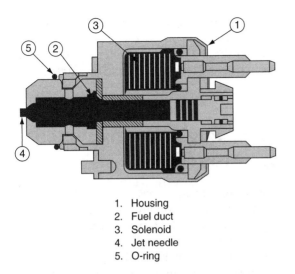

1. Housing
2. Fuel duct
3. Solenoid
4. Jet needle
5. O-ring

Fig. 3.19 The single point CFI (central fuel injection) unit

conditions, it is necessary for it to have an accurate measure of the air entering the engine. The speed density method provides this information from the readings taken from the manifold absolute pressure (MAP) sensor, the air charge temperature sensor and the engine speed sensor.

The actual central injector unit is shown in Fig. 3.19.

The injector valve is operated by the solenoid (3) which receives an electric current in accordance with signals from the engine control computer. When the engine is operating at full or part load, the injector sprays fuel during each induction stroke. When the engine is idling, the injector operates once per revolution of the crankshaft. Because the fuel pressure regulator maintains a constant fuel pressure at the injector valve, the amount of fuel injected is determined by the length of time for which the solenoid holds the valve in the open position.

The throttle plate (butterfly valve) motor is operational during starting, coasting, shutting down the engine and when the engine is idling.

Multi-point injection

Multi-point injection systems commonly use one of two techniques:

1. Injection of half the amount of fuel required, to all inlet ports, each time the piston is near top dead centre;
2. Sequential injection. In this case injection occurs only on the induction stroke.

An example of computer controlled multi-point fuel injection

In the multi-point injection system shown in Fig. 3.20 there is one petrol injector (number 12 in the diagram) for each cylinder of the engine.

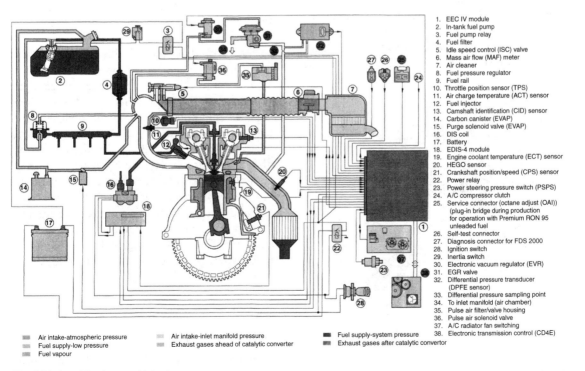

1. EEC IV module
2. In-tank fuel pump
3. Fuel pump relay
4. Fuel filter
5. Idle speed control (ISC) valve
6. Mass air flow (MAF) meter
7. Air cleaner
8. Fuel pressure regulator
9. Fuel rail
10. Throttle position sensor (TPS)
11. Air charge temperature (ACT) sensor
12. Fuel injector
13. Camshaft identification (CID) sensor
14. Carbon canister (EVAP)
15. Purge solenoid valve (EVAP)
16. DIS coil
17. Battery
18. EDIS-4 module
19. Engine coolant temperature (ECT) sensor
20. HEGO sensor
21. Crankshaft position/speed (CPS) sensor
22. Power relay
23. Power steering pressure switch (PSPS)
24. A/C compressor clutch
25. Service connector (octane adjust (OAI))
    (plug-in bridge during production
    for operation with Premium RON 95
    unleaded fuel
26. Self-test connector
27. Diagnosis connector for FDS 2000
28. Ignition switch
29. Inertia switch
30. Electronic vacuum regulator (EVR)
31. EGR valve
32. Differential pressure transducer
    (DPFE sensor)
33. Differential pressure sampling point
34. To inlet manifold (air chamber)
35. Pulse air filter/valve housing
36. Pulse air solenoid valve
37. A/C radiator fan switching
38. Electronic transmission control (CD4E)

▓ Air intake-atmospheric pressure   ▓ Air intake-inlet manifold pressure   ▓ Fuel supply-system pressure
▓ Fuel supply-low pressure   ▓ Exhaust gases ahead of catalytic converter   ▓ Exhaust gases after catalytic convertor
▓ Fuel vapour

Fig. 3.20 A multi-point petrol injection system

Each of the injectors is designed so that it sprays fuel on to the back of the inlet valve. The actual position and angle at which injection takes place varies for different types of engines.

In the system shown in Fig. 3.20, the air flow is measured by the hot-wire type of mass air flow meter. The control computer receives the signal from the air flow meter and uses this signal together with those from other sensors such as engine speed, engine coolant temperature, throttle position, etc., to determine the length of time of the injection pulse.

Sequential multi-point injection

Sequential multi-point injection is a term that is used to describe the type of petrol injection system that provides one injection of fuel for each cylinder during each cycle of operation. To assist in providing the extra controlling input that is required for sequential injection, the engine is fitted with an additional sensor which is driven by the engine camshaft. Hall type sensors and variable reluctance sensors driven by the camshaft are often used for this purpose which is to assist the computer to determine TDC on number 1 cylinder. Figure 3.21 shows one of these sensors which is fitted to an overhead camshaft engine.

Some of the sensors used for fuelling are the same as those used for ignition systems, for example, crank speed and top dead centre sensors, manifold pressure to indicate engine load, etc. Because some of the sensor signals can be

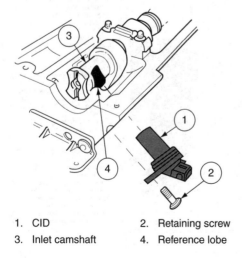

| 1. | CID | 2. | Retaining screw |
| 3. | Inlet camshaft | 4. | Reference lobe |

Fig. 3.21 A cylinder identification sensor

used for both ignition and fuelling, it has become common practice to place them under the control of a single computer and the resulting system is known as an engine management system.

3.10 Engine management systems (EMS)

Engine management systems are designed to make the vehicle comply with emissions regulations as well as to provide improved performance. This means that the number of sensors and actuators is considerably greater than for a simple fuelling or ignition system. The system shown in Fig. 3.20 is fairly typical of modern engine management systems and selected items of technology from the list are picked out to give them closer attention, so that the aim of 'teasing out' features that they have in common with other components is realised. The ignition system also forms part of an engine management system is covered separately in Chapter 10. The intention here is to focus on some of the components in Fig. 3.20. The first component to note is the oxygen sensor at 20. This is a heated sensor (HEGO) and the purpose of the heating element is to bring the sensor to its working temperature as quickly as possible. The function of the HEGO is to provide a feedback signal that enables the ECM to control the fuelling so that the air–fuel ratio is kept close to the chemically correct value where Lambda = 1, this value enabling the catalytic convertor to function at its best. Oxygen sensors are common to virtually all modern petrol engined vehicles and this is obviously an area of technology that technicians need to know about. The zirconia type oxygen sensor is most commonly used and it produces a voltage signal that represents oxygen levels in the exhaust gas and is thus a reliable indicator of the air–fuel ratio that is entering the combustion chamber. The voltage signal from this sensor is fed back to the control computer to enable it to hold Lambda close to 1.

Exhaust catalyst condition monitoring

Vehicles that are equipped to the European On Board Diagnostics (EOBD) standard have

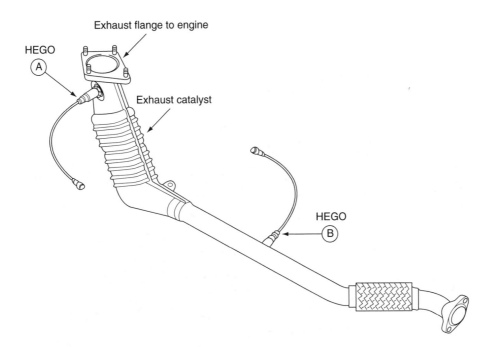

Fig. 3.22 The downstream oxygen sensor that monitors the catalyst

two oxygen sensors, as shown in Fig. 3.22. The signals from these two sensors are compared by the engine management computer and the malfunction indicator lamp (MIL) will be illuminated should the catalyst fail to function correctly.

Exhaust gas recirculation (EGR)

Two items in Fig. 3.20, the electronic vacuum regulator at (30) and the exhaust gas recirculation (EGR) valve at (31), play an important part in this and many other engine management systems and they warrant some attention. In order to reduce emissions of NOx it is helpful if combustion chamber temperatures do not rise above approximately 1800 °C, because this is the temperature at which NOx can be produced. Exhaust gas recirculation helps to keep combustion temperatures below this figure by recirculating a limited amount of exhaust gas from the exhaust system, back to the induction system, on the engine side of the throttle valve and then onwards into the combustion chambers. The principle of operation is described in the details on Fig. 3.23.

In order to provide good performance, EGR does not operate when the engine is cold or when the engine is operating at full load. The inset shows the solenoid valve that controls

the EGR valve and this type of valve is operated on the duty cycle principle. Under reasonable operating conditions, it is estimated that EGR will reduce NOx emissions by approximately 30%.

Computer control of evaporative emissions

Motor fuels give of vapours that contain harmful hydrocarbons, such as benzene. In order to restrict emissions of hydrocarbons from the fuel tank, vehicle systems are equipped with a carbon canister. This canister contains activated charcoal which has the ability to bind toxic substances into hydrocarbon molecules. In the evaporative emission control system, the carbon canister is connected by valve and pipe to the fuel tank, as shown in Fig. 3.24.

The evaporative purge solenoid valve connects the carbon canister to the induction system, under the control of the EMS computer, so that the hydrocarbon vapours can be drawn into the combustion chambers to be burnt with the main fuel air mixture. The control valve is operated by duty cycle electrical signals from the computer which determine the period of time for which the valve is open. When the engine is not running, the vapour from the fuel in the tank passes into

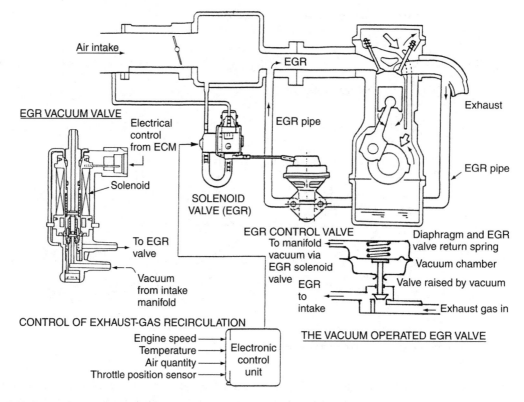

Fig. 3.23 Exhaust gas recirculation system

the carbon canister and when the engine is started up, the engine management computer switches on the solenoid valve so that the vapour can pass into the induction system. After this the frequency of operation of the solenoid valve is dependent on operating conditions.

Evaporative emissions control is part of the emissions control system of the vehicle and it must be maintained in good order.

3.11 Maintenance and repair of petrol fuel systems

Safety

Before commencing any work that entails removal of fuel system components precautions must be taken to prevent fire. In addition to manufacturers recommendations the following should be noted:

- Have a fire extinguisher close to hand;
- Take precautions to avoid sparks;
- De-pressurise the system;
- Avoid petrol spillage;

- Use protective eye wear;
- Familiarise yourself with steps to take in the event of petrol spillage; and
- Ensure that fume extraction equipment is operational.

Carburettor systems

The function of a carburettor system is to supply the engine with a combustible mixture of the correct air–fuel ratio to suit all conditions in which the vehicle is required to operate. Failure to produce the correct type and amount of mixture will lead to a variety of problems, such as poor starting or failure to start, poor idling, hesitation and misfire on acceleration, lack of power, excessive fuel consumption, emissions out of limits, etc.

Fuel system problems are normally associated with the two substances that make up the combustible mixture, namely air and fuel. Too much air and insufficient fuel causes weak mixtures, and too little air and too much petrol leads to rich mixtures. Both of these conditions cause engine operating problems and an understanding of the probable causes of these extreme conditions is a helpful step towards finding a solution to a petrol fuel system problem.

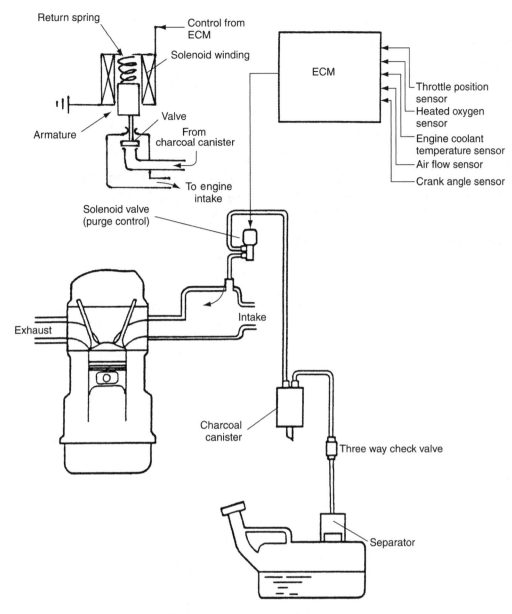

Fig. 3.24 Evaporative emissions control system

Air supply problems – carburettor fuel system

Defect	Possible causes
Insufficient air	Blocked air filter, air filter element fitted badly. Damaged or blocked air intake
Excess air	Air leaks into intake manifold from brake servo pipe or vacuum advance pipe for ignition distributor
	Broken or warped carburettor and or manifold flanges. Loose fixing bolts.
	Broken gaskets.
	Worn throttle butterfly spindle and bushes.
	Defective crankcase ventilation

Fuel supply problems – carburettor fuel system

Defect	Possible causes
Rich mixture – excess fuel	Petrol level high in float chamber.
	Leaking needle valve.
	Excessive petrol pressure.
	Incorrect jets.
	Mixture setting incorrect in constant vacuum carburettor.
	Punctured float
Weak mixture – insufficient fuel	Blocked filter.
	Incorrect mixture setting.
	Pump delivery rate low.
	Fuel pump pressure low

Adjustments – carburettor

Carburettor idle mixture adjustment (constant choke type)

Two types of idle mixture adjustment are used in constant choke carburettors. One type varies the *quality* of the idle mixture by adjusting the amount of air that flows through the idling passage of the carburettor, shown in Fig. 3.25(a). The other type, shown in Fig. 3.25(b), fixes the air–fuel ratio of the idle mixture by means of petrol and air jets and the idle adjustment varies the *quantity* of mixture entering the engine.

Outline procedure for adjustment of idling speed

To adjust engine idle settings:

- The engine must be at normal operating temperature;
- Workshop exhaust extraction operating;
- Tachometer and gas analyser connected;

- Cold start choke in off position;
- Throttle stop screw should be adjusted to give the correct speed;
- Idling mixture screws should be adjusted to achieve the correct CO setting; and smooth operation of the engine.

When the CO setting is correct it may be necessary to reset the idling speed.

Adjusting CO level and idle speed – petrol injection [Early models of vehicles]

When engines have been in use for some time, and when new parts have been fitted, it may be necessary to adjust the engine idling settings to ensure that CO levels in the exhaust gas are within the manufacturer's specification. The procedure for adjusting CO settings and idle speed on petrol injection systems varies considerably across the range of modern vehicles and it is

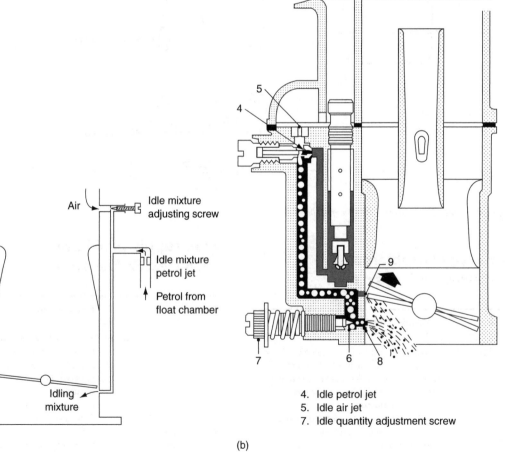

4. Idle petrol jet
5. Idle air jet
7. Idle quantity adjustment screw

(a)   (b)

Fig. 3.25 Quality control of idle mixture

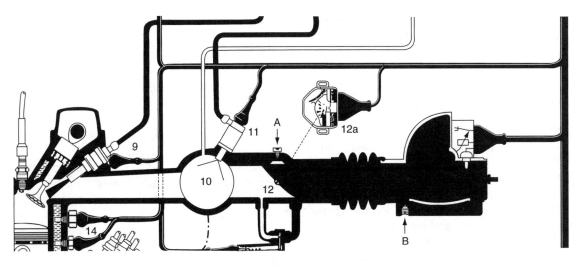

Fig. 3.26 Idling adjustments

essential that details of the procedure for any given vehicle are known, or are readily available. The following example outlines the procedure for one particular system and it is intended to provide an insight into the procedure.

The two screws marked A and B in Fig. 3.26 are provided so that idle speed and mixture strength (CO level) may be adjusted. The screw A varies the size of the throttle by-pass passage through which the air for idling passes and thus controls engine idle speed. The screw at B varies the size of the air flow meter by-pass and effectively controls the mixture strength for idling purposes.

Equipment required

- Exhaust extraction system
- Manufacturer's special tools and data should be to hand
- Tachometer
- Exhaust gas analyser
- Small tools

Procedure

- Observe safety precautions;
- Have tools and equipment to hand;
- With the engine at normal operating temperature and electrical loads switched off, or set as per instructions, and the tachometer and gas analyser connected, the engine is first checked for engine speed and CO reading. If adjustment is required, the screws at A and B are adjusted to give correct engine speed and CO level, with checks being made to ensure that the engine is running smoothly.

> *Learning task*
>
> Study the section of a workshop manual that describes the procedure for adjusting the CO level on one type of vehicle that is commonly worked on at your place of work.

Idling settings on engine management systems (EMS)

In the type of engine management system shown in Fig. 3.20, engine idle conditions are controlled by the computer (EEC IV module). In response to sensor signals, the computer adjusts the idle control valve and alters the ignition setting to maintain stable engine operation. Over a period of time, component wear may cause the EMS computer to adopt new settings for idling speed and ignition timing. Should this occur and new components, such as an idle control valve, be fitted, it will be necessary to use a manufacturer's dedicated equipment to re-set the computer memory so that correct engine idling is restored.

Some other petrol fuel system maintenance procedures

Carburettor system

Routine maintenance

Fuel system maintenance is part of the routine maintenance schedule of a vehicle. Factors such as cleaning fuel filters and replacing air filters, checking and servicing carbon canisters, together with visual checks of electrical connections, normally ensure that the fuel system is kept in good working condition.

Petrol injection system

Routine maintenance

Routine maintenance should include visual checks of electrical circuits and connections and also similar checks of fuel lines. Mileage based servicing will include renewal of air filters, servicing the carbon canister, inspecting and cleaning or renewing and fuel filters.

Since the introduction of European On Board Diagnostics (EOBD), all emissions related functions, such as the fuel system, are constantly monitored by the engine computer (ECM). In the event of a defect in part of the fuel system, the malfunction indicator lamp (MIL) will alert the driver and fault codes will be stored in the computer. If the fuel system is properly maintained in accordance with the manufacturer's recommendation, problems with the fuel system should be a rare occurrence.

Other maintenance and repair

Whilst fault codes may not provide a complete answer to a fuel system problem, they are extremely useful in guiding one to the general area of the fault. In the event of a petrol injection fuel system failure, it may be necessary to conduct further tests. The tests shown here deal with mechanical aspects namely, fuel pressure and injector condition. Electrical tests are covered in the electrical sections of this book.

Safety

Please refer to Section 3.9

Fuel gallery pressure

The quantity of fuel injected is dependent on the fuel pressure being accurately maintained at the level specified by the manufacturer. The pressure may be checked by placing an accurate pressure gauge in the main fuel line.

Before attempting to connect the pressure gauge, the system fuel system must be depressurised and all precautions taken to prevent spillage. This procedure varies from vehicle to vehicle and the manufacturer's advice must be observed. Once the gauge is connected and all traces of petrol have been removed, the test procedure may then be followed and the pressure should fall within the prescribed limits, of the order of 2.5 bar to 3 bar. This figure is only a guide; the manufacturer's figures are the ones that must be used in any particular case. Figure 3.27 shows a pressure gauge positioned at the outlet from the fuel filter.

Petrol injector testing

Figures 3.28(a),(b) show examples of two tests that are possible on a vehicle. Again, these should only be undertaken by competent personnel who are in possession of the necessary instructions and who are careful to take all necessary safety precautions.

Figure 3.28(a) illustrates the type of test that can be performed to check for injector leakage. The injectors are removed from the manifold and securely placed over drip trays. Note that they remain connected to the fuel and electrical system. The approved procedure is then applied

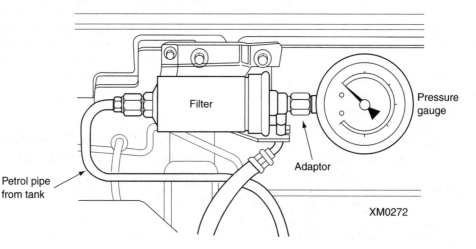

Fig. 3.27 Fuel line pressure test

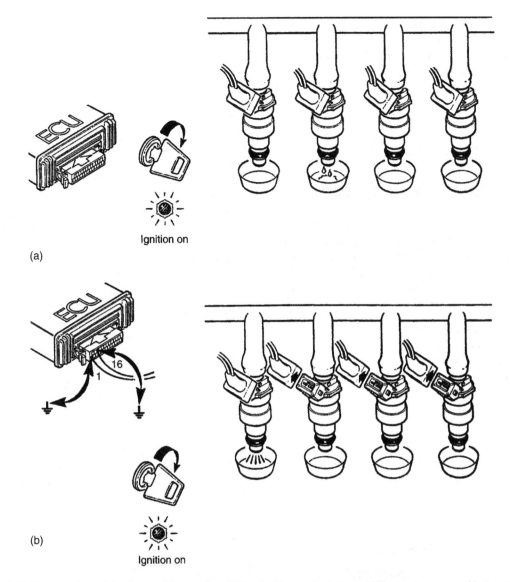

Ignition on

(a)

Ignition on

(b)

Fig. 3.28 The general principle of a petrol injector test. (a) Testing injectors for leakage, (b) Testing the amount of fuel per injector

to energise the injectors and a careful visual inspection will reveal if there are any leaks. In the case shown, it is recommended that any injector that leaks more than two drops of fuel per minute should be replaced.

Figure 3.28(b) shows the same injectors. In this case, three of them are disconnected and the amount of fuel delivered by the injector, which is still connected, is carefully measured. It is possible to obtain special injector test benches on which these tests can be performed, in ideal conditions.

In many applications, the petrol injectors can be activated, by the diagnostic tester, through the serial port. This makes it possible to hear the injectors 'click' as they are pulsed. If the fuel supply is active during this process, a certain amount of fuel will enter the intake manifold.

In order to prevent damage to the exhaust catalyst, steps must be taken to prevent the fuel passing through the engine and one procedure advocates cranking the engine over with the spark plugs removed, to expel the unburnt fuel, on completion of the injector test.

Learning tasks

1. Draw up a service schedule for the complete fuel system of a petrol injection vehicle. Itemize any specialized equipment that should be used to complete the task.
2. Complete an exhaust gas test on a petrol engine vehicle fitted with fuel injection. Adjust the CO to give the correct reading according to the manufacturer's data.

3. Carry out the service schedule shown at (I) above and itemize any faults found or recommendations to be made on the job sheet.
4. Remove a petrol injector unit and test for correct operation, spray pattern and dribble from nozzle. Pressure-test the distribution circuit and check results against manufacturer's data.

3.12 Diesel fuel systems

The purpose of the injection equipment in a diesel fuel system is to supply quantities of fuel oil into the combustion chamber in the form of a very fine spray at precisely timed intervals. To achieve this the following components are usually employed:

- a fuel tank or tanks
- a fuel feed pump
- a fuel filter or filters
- high- and low-pressure fuel supply lines
- an injector pump
- injectors
- a timing device
- a governor

Systems and components vary in design and performance; however, layouts of the components which might be found in a typical system are shown in Figs 3.29 and 3.30. The purposes of the fuel tank and low-pressure pipes are similar to those of the petrol engine.

Lift pump

One of two types is used. The first is the diaphragm-type, similar in operation to the petrol fuel system except that it commonly has a **double diaphragm** fitted that is resistant to fuel oil. These are used in the low-pressure systems and deliver fuel at a pressure of approximately $34.5 \, kN/m^2$. The other type used is the **plunger operated** pump where higher delivery pressures are required. The plunger is backed up by a diaphragm to prevent fuel leakage. These deliver at approximately $104 \, kN/m^2$.

Fuel filters

Working clearances in the injector pump are very small, approximately $0.0001 \, mm$ ($0.000 \, 04$ inch); therefore the efficiency and life of the equipment depends almost entirely on the cleanliness of the fuel. The fuel filter therefore performs a very important function, that of removing particles of dirt and water from the fuel before they get to the injector pump. After much research, it was found that specially impregnated paper was the best filtering material, removing particles down to a few microns in size. The element consists of

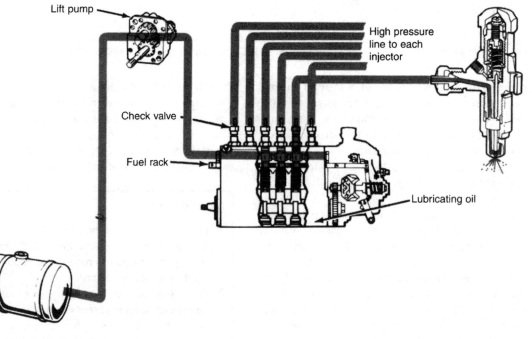

Fig. 3.29 In-line fuel-injection-pump system

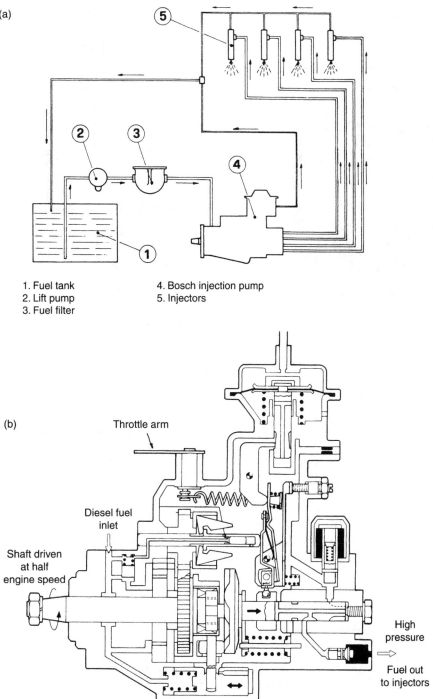

(a)

1. Fuel tank
2. Lift pump
3. Fuel filter
4. Bosch injection pump
5. Injectors

(b)

Throttle arm

Diesel fuel inlet

Shaft driven at half engine speed

High pressure

Fuel out to injectors

Fig. 3.30 (a) The basic components of a diesel fuel system are a fuel tank, lift pump with priming lever, fuel filter, injection pump, injector pipes and injectors. Additonal pipes connected to the injection pump and injectors return excess fuel to the fuel tank, (b) Distributor-type fuel injection pump

the specially treated paper wound around a central core in a spiral form, enclosed in a thin metal canister giving maximum filtration within minimum overall dimension.

Most filters are of the **agglomerator** type (Fig. 3.31), i.e. as the fuel passes through the element, water, which is always present, is squeezed out of the fuel and agglomerates (joins together) into larger droplets which then settle to the base of the filter by sedimentation. Choking of the filters is caused not only by the solid matter held back by the element but also by the sludge

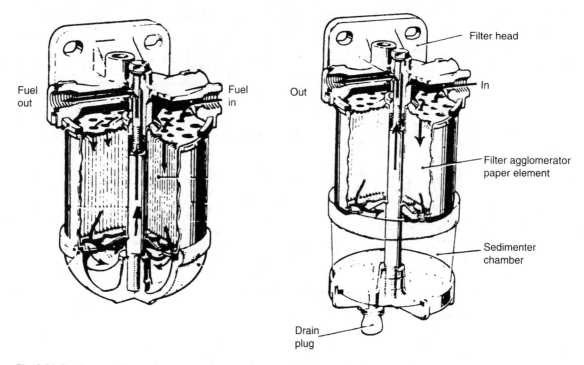

Fig. 3.31 Bowless-type filter-agglomerator-sedimenter showing agglomerator flow through element

and wax in the fuel which, under very cold conditions, form a coating on the surfaces of the fuel filter element thus reducing the rate of fuel flow. Where this becomes too much of a problem a simple sedimenter-type filter, i.e. one which does not incorporate an element, is fitted between the fuel tank and the lift pump. The object of a simple sedimenter is to separate the larger particles of dirt, wax and water from the fuel; they are then periodically drained off. Provision is made for the venting of the filter of air (this is commonly called **bleeding** the system, where all the air in the fuel system is removed). If air enters the high-pressure fuel lines then the engine will not run.

Learning tasks

1. Remove and refit the fuel filter on a diesel engine vehicle, bleed the system and run the engine.
2. Remove and refit the lift pump of a diesel system, test the operation of the pressure/vacuum valves, record results and check with the manufacturer's data. Give reasons for any differences and make recommendations to the customer on serviceability.
3. Draw up a simple checklist for tracing leaks on a diesel fuel system.
4. Make a list of personal safety/hygiene precautions that a mechanic should consider when working on a diesel fuel system.

---

5. What are the symptoms of a blocked fuel filter?
6. What are the service intervals for the diesel fuel system and what would it involve?

Injector pump

The function of the injector pump is to:

- deliver the correct amount of fuel;
- at the correct time;
- at sufficiently high a pressure to enable the injector to break up the fuel into very fine droplets to ensure complete combustion (that is complete burning of all the fuel injected).

The **multi-element** or the **DPA** (distributor pump application) injector pumps are the ones most commonly used.

Multi-element injector pump

This consists of a casing containing the same number of pumping elements as there are cylinders in the engine. Each element consists of a **plunger and barrel** machined to very fine tolerances, and specially lapped together to form a mated pair. A **helix** (similar to a spiral) is formed on the outside of the plunger which communicates with the plunger crown, either by a drilling in the centre of the plunger or a slot machined in the side. There are two ports in the barrel, both of which connect to a common fuel gallery feeding all the elements. The plunger is operated by a cam and follower tappet and returned by a spring. This system is shown in Fig. 3.32.

When the plunger is at BDC, fuel enters through the barrel ports filling the chamber above, and also the machined portion forming the helix of the plunger. As the plunger rises, it will reach a point when both ports are effectively cut off (this is known as **spill cut-off** point and is the theoretical start of injection).

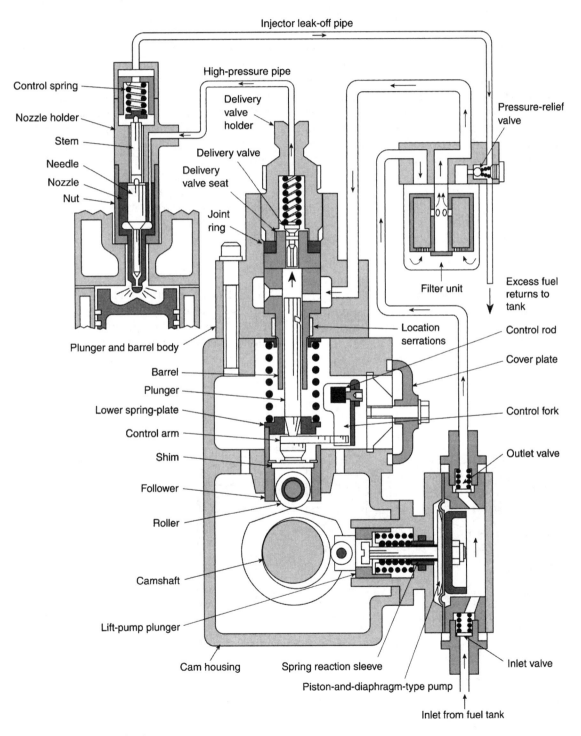

Fig. 3.32 CAV ('Minimec') in-line injection pump

Further upward movement of the plunger forces the fuel through the **delivery valve**, injector pipes and injectors into the combustion chamber. With the plunger stroke being constant, any variation in the amount of fuel being delivered is adjusted by rotating the plunger causing the helix to uncover the spill port sooner or later depending on rotation. Immediately the helix uncovers the spill port, the fuel at high pressure above the plunger spills back to the common gallery, the delivery valve resumes its seat and injection stops without any fuel dribbling from the injector. Each injector is rotated simultaneously by a rack or **control rod**. No fuel or engine stop position is obtained by rotating the plunger so that the helix is always in alignment with the ports in the barrel, in this position, pressure cannot build up and hence no fuel will be delivered. Figure 3.33 illustrates this.

Delivery valve

The purpose of a delivery valve (Fig. 3.34) being fitted above each element is to:

- prevent fuel being drawn out of the injector pipe on the downward stroke of the plunger;
- ensure a rapid collapse of the pressure when injection ceases, thus preventing fuel from dribbling from the injector;
- maintain a residual pressure in the injector pipes.

The valve and guide of the delivery valve are machined to similar tolerances as the pumping elements. Approximately two thirds of the valve is machined to form longitudinal grooves. Above the grooves is the **unloading collar**; immediately above the collar is the valve seat. When the pump is on the delivery stroke, fuel pressure rises and the delivery valve moves up until the fuel can escape through the longitudinal grooves. Immediately the plunger releases the fuel pressure in the barrel, the delivery valve starts to resume its seat under the influence of the spring and the difference in the pressure above and below the valve. As the unloading collar enters the guides dividing the element from the delivery pipe, further downwards movement increases the volume above the valve (by an amount equal to the volume between the unloading collar and the valve seat). The effect of this increase in volume is to suddenly reduce the pressure in the injection pipe so that the nozzle valve 'snaps' closed onto its seat thus instantaneously terminating injection without dribble.

Excess fuel device

Multi-element injection pumps are normally fitted with an excess fuel device which, when operated, allows the pumping elements to deliver fuel in excess of normal maximum. This ensures that the delivery pipes from the injection pump to the injectors are quickly

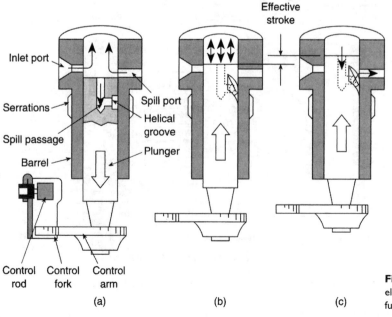

Fig. 3.33 Plunger and barrel. (a) Pumping element filling, (b) Injection, (c) Spill: no fuel out

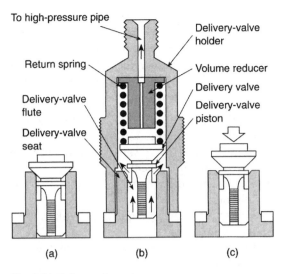

Fig. 3.34 Delivery-valve action

primed if the engine has not been run for some time or if assistance is needed for easy starting in cold conditions. When the device is operated (with the engine stationary), the rack or control rod of the fuel injection pump moves to the excess position under the pressure of the **governor spring**. On operating

the starter, excess fuel is delivered to the engine. As soon as the engine starts, governor action moves the control rod towards the minimum fuel position making the device inoperative.

Cam shapes

To prevent the possibility of reverse running, it is normal practice to fit a camshaft with profiles designed so that the plunger is at TDC for approximately two-thirds of a revolution. In the event of a back-fire the engine will not run. Certain CAV pumps are fitted with reversible type camshafts. In order to prevent reverse running, a spring-loaded coupling, similar to the **pawl-type free-wheel** is fitted between the pump and the engine. Figure 3.35 shows a cam profile.

Phasing and calibration

Phasing is a term used when adjustment is made to ensure injection occurs at the correct time, i.e. on four-cylinder engines each element injects at 90° intervals while on a six-cylinder

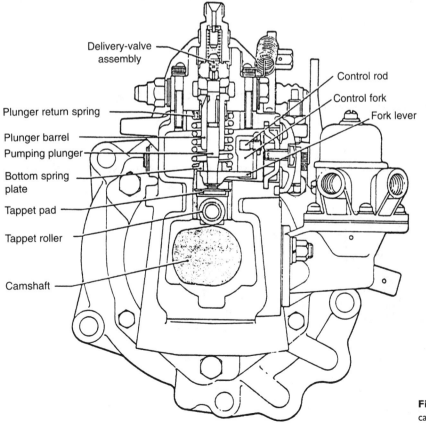

Fig. 3.35 Sectioned view showing cam profile

engine each element injects at 60° intervals. This adjustment is carried out by raising or lowering the plunger so the spill cut-off point is reached at the correct time. Simms pumps have spacers in the tappet blocks (Fig. 3.34). CAV pumps have normal tappet adjustment. Phasing should not be confused with spill timing (this is when the injection pump is timed to the engine).

Calibration refers to the amount of fuel that is injected. Correct calibration ensures that the same amount of fuel is injected by each element at a given control rod setting. It is effected by rotating the plunger independently of the control rod. Both phasing and calibration can only be carried out on proper equipment and using data sheets to obtain speed and fuel delivery settings for any given injection pump. When settings are adjusted correctly the maximum fuel stop screw is sealed and must not be adjusted under any circumstances.

Lubrication

The delivery valves and pumping elements are lubricated by the fuel oil, a small quantity of which leaks past the plunger and barrel into the cambox. The camshaft, bearings, tappets, etc. are lubricated by engine oil contained in the cambox. To prevent a build-up of oil due to the fuel leaking past the elements, the level plug incorporates a leak-off pipe.

Spill timing the multi-element pump to the engine

After mounting the injection pump to the engine and checking that the alignment and drive coupling clearance are correct, it is necessary to adjust the pump to ensure that number one cylinder is on compression. Refer to the workshop manual for the correct static timing (e.g. it may read 28° BTDC). The correct method would be as follows.

1. Set the engine to 28° BTDC with number one cylinder on compression stroke.
2. Remove the delivery valve from number one cylinder pump element.
3. Replace the delivery valve body and fit the spill pipe.
4. Loosen the pump coupling and fully retard the pump.
5. Ensure that the stop control is in the run position.
6. Operate the lift pump; fuel will now flow from the spill pipe.
7. Whilst maintaining pressure on the lift pump, slowly advance the injection pump when a reduction in the flow of fuel from the spill pipe will be noticed as the plunger approaches the spill cut-off point. Continue advancement until approximately one drop every ten to fifteen seconds issues from the spill pipe.
8. Tighten the coupling bolts, remove the spill pipe and refit the delivery valve. The pump is now correctly timed in relation to the engine.

Spill timing is shown in Fig. 3.36.

Bleeding the fuel system

Bleeding the fuel system refers to removing all the air from the pipes, lift pump, filters, injection pump and injectors. This operation must be carried out if the fuel system is allowed to

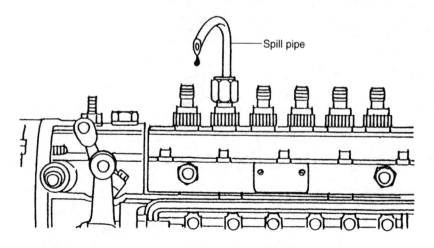

Fig. 3.36 Spill timing no 1 pumping element to no 1 cylinder

run out of fuel, any part of the system is disconnected or the filter elements are changed. Assuming the system has been disconnected for some reason the correct method is as follows.

1. Disconnect the pressure side of the lift pump, operate the lift pump until fuel, free from air bubbles, flows from the outlet. Reconnect the fuel line.
2. Slacken off the bleed screw of the fuel filter and operate the lift pump until all the air is expelled from the filter Re-tighten the bleed screw.
3. Open the bleed screw (sometimes called the vent screw) on the injection pump and operate the lift pump again. When fuel free from air bubbles comes out re-tighten the screw.
4. It may also be necessary to bleed the high-pressure pipes to the injectors by slackening the union at the injector and operating the starter until small amounts of fuel can be seen to be coming from the union. Re-tighten the unions and operate the starter to run the engine. Small amounts of air in the fuel system, though not necessarily enough to prevent the engine starting, may cause loss of power and erratic running. It is therefore necessary to carry out this operation methodically and with care.

Learning tasks

1. Explain in your own words the difference between phasing, calibration and spill timing. When would each term be used in the course of servicing the diesel fuel system?
2. Remove a multi-element injection pump from a diesel engine, turn both the pump and engine over, re-time the pump to the engine, bleed the fuel system and run the engine.

The DPA or rotary fuel injection pump

The DPA fuel injection pump (Fig. 3.37) serves the same purpose as the multi-element type and offers the following advantages:

- it is smaller, more compact and can be fitted in any position not just horizontal;
- it is an oil-tight unit, lubricated throughout by fuel oil;

- only one pumping element is used, regardless of the number of cylinders to be supplied;
- no ball or roller bearings are required and no highly stressed springs are used;
- no phasing is required; calibration once set is equal for all cylinders;
- an automatic advance device can be fitted.

In this pump the fuel at lift pump pressure passes through a nylon filter, situated below the inlet union, to the **transfer pump**. Fuel pressure is increased by the transfer pump, depending on the speed of rotation of the pump and controlled by the **regulating valve**. The regulating valve maintains a relationship between pump speed and transfer pressure, which at low revolutions is between 0.8 and $1.4 \, \text{kg/cm}^2$ ($11–20 \, \text{lbs/inch}^2$) increasing to between 4.2 and $7.0 \, \text{kg/cm}^2$ ($60–100 \, \text{lbs/inch}^2$) at high revolutions. From the transfer pump, fuel flows through a gallery to the **metering valve**. The metering valve, which is controlled by the **governor**, meters the fuel passing to the rotor depending on engine requirements. The fuel is now at metering pressure, this being lower than transfer pressure. As the rotor rotates, the inlet ports come into alignment and fuel enters the rotor displacing the plungers of the pumping elements outwards until the ports move out of alignment. Further rotation brings the outlet ports of the rotor into alignment with one of the outlet ports which are spaced equally around the hydraulic head. At the same time, contact between the plunger rollers and the cam ring lobes forces the pumping elements inwards. Fuel pressure between the plungers increases to injection level and fuel is forced along the control gallery, through the outlet port to the injector pipe and injector. As the next charge port in the rotor aligns with the metering valve port, the cycle begins again. Figure 3.38 illustrates this.

The inside of the cam ring has as many equally spaced lobes as there are cylinders in the engine. Each lobe consists of **two peaks**, the recess between being known as the retraction curve. As the pumping element rollers strike the first peak, injection takes place. On reaching the retraction curve, a sudden drop in pressure occurs and injection stops without fuel dribbling from the injector. Further movement of the rotor brings the rollers into contact with the second peak which maintains residual line pressure until the outlet port moves out of

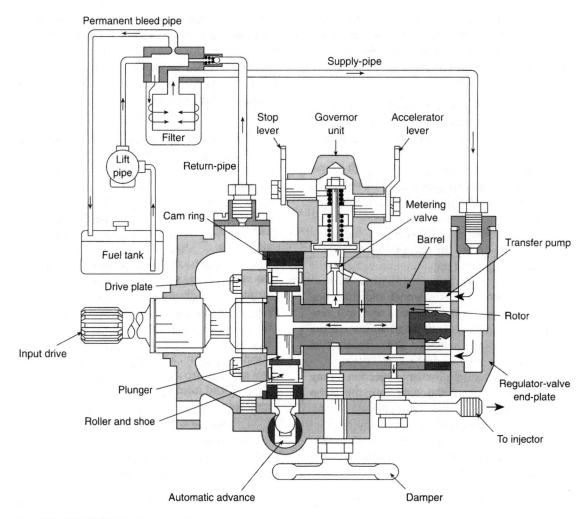

Fig. 3.37 CAV ('DPA') distributor-type injection pump

alignment. The cam ring rotates within the pump housing varying the commencement of injection. Movement is controlled by the advance/retard device.

It should be remembered that this type of pump is lubricated by the fuel oil flowing through the pump and if it runs out of fuel at any time, due to a new pump being fitted, any parts of the fuel system being disconnected or the filter elements having been changed, then to prevent damage to the pump occurring the following method should be used to bleed the system of air.

1. Slacken the vent screw on the fuel filter. This may be a banjo union of the 'leak-off' pipe. Operate the lift pump until fuel free from air flows from the filter. Tighten the vent screw.
2. Slacken the main feed pipe union nut at the pump end and operate the lift pump until

fuel free from air flows from the union. Tighten the union.
3. Slacken the vent screw of the governor control housing and the vent screw in the hydraulic head. Operate the lift pump until fuel free from air flows from the vent screws. Tighten the hydraulic head vent screw and governor control housing vent screw.
4. Crank the engine by hand one revolution and repeat the operations listed at 3.
5. Slacken all high-pressure injection pipes at the injector end and turn the engine over with the starter until fuel free from air flows from unions. Tighten unions and start engine.

Note: Always ensure that the stop control is in the start position and the throttle is wide open when bleeding the pump.

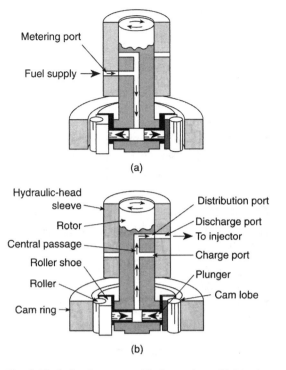

Fig. 3.38 Cycle of operation (a) charge phase, (b) injection phase

Figure 3.39 shows one type of injector pump. This is fitted with a mechanical means of governing the maximum and minimum revs of the engine. The automatic advance operates when the engine is stationary, enabling injection to take place earlier than normal when starting the engine. This gives easier starting and reduces the amount of smoke being passed through to the atmosphere.

Injectors

The injector can be considered as an **automatic valve** which performs a number of tasks. It may vary in design but all conform to the following requirements:

- it ensures that injection occurs at the correct pressure;
- it breaks up the fuel into very fine droplets in the form of a spray which is of the correct pattern to give thorough mixing of the fuel with the air;
- it stops injecting immediately the injection pump pressure drops.

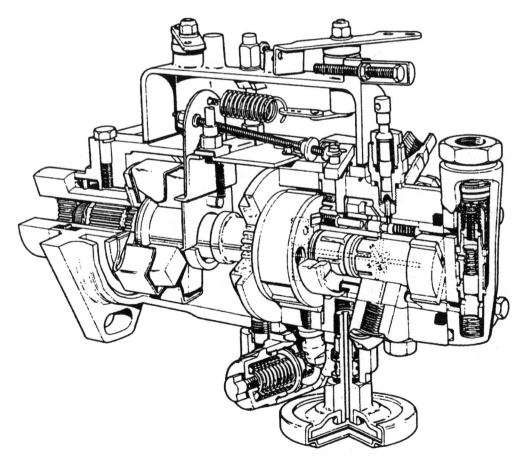

Fig. 3.39 DPA pump with mechanical governor and automatic advance

The **nozzle** is the main functional part of the injector. It consists of a **needle valve** and **nozzle body** machined to fine tolerances and lapped together to form a mated pair. Extreme care must be exercised when handling this component.

Lubrication is achieved by allowing a controlled amount of fuel to leak past the needle valve. Provision is made for this **back-leakage** to return either to the filter or fuel tank, it also circulates heated fuel to help overcome the problem of the fuel freezing in very cold weather.

Adequate cooling of the injector is most important and is catered for by careful design and positioning in the cylinder head. To prevent over-heating, it is essential that the correct injector is fitted and that any joint washers are replaced each time the injector is removed (do make sure the old joint washer is removed before fitting a new one). No joint washer is required on those engines fitted with copper sleeves which form part of the injector housing. In addition to the normal copper sealing washer, a corrugated type steel washer is sometimes located between the nozzle and the heat shield. It is essential that the outside edge faces *away* from the nozzle, otherwise serious overheating of the injector will occur. To prevent the washer turning over whilst fitting it should be fed down a long bladed instrument such as a screwdriver or a suitable length of welding rod.

Method of operation

The needle is held on its seat by spring pressure acting through the spindle. Fuel, when delivered at high pressure from the injector pump, acts on the shoulder at the lower end of the needle. When the fuel pressure exceeds the spring tension the needle lifts off its seat and fuel is forced through the hole(s) in a finely atomized spray. The spring returns the needle valve back onto its seat at the end of each injection. The spring tension is adjustable by releasing the lock nut and screwing the spring cap nut in or out as required. This determines at what pressure injection commences. The pressures at which the injector operates are very high, typically between 125 and 175 atmospheres. It should be noted that the spring tension only determines at what pressure injection starts. Pressure may momentarily increase up to a maximum of approximately 420 atmospheres (to start the

needle moving), depending on engine speed and load. At these high pressures it is very important that the testing is done with the proper equipment and in the correct way. If the hand was placed in front of the injector whilst it was being tested then it is possible for the fuel to be injected directly into the blood stream with serious consequences and possible death. Use of safety equipment such as gloves and goggles is essential to reduce any possible risks of injury to a minimum.

Types of injector

These are classified by the type of nozzle fitted and fall into four main groups

- single-hole
- multi-hole
- pintle
- pintaux

Single- and multi-hole injectors

The **single-hole** nozzle has one hole drilled centrally in its body which is closed by the needle valve. The hole can be of any diameter from 0.2 mm (0.008 in) upwards. This type is now rarely used in motor vehicle engines. **Multi-hole** injectors (Fig. 3.40) have a varying number of holes drilled in the bulbous end of the nozzle beneath the needle valve seating. The actual number, size and position depends on the requirements of the engine concerned. There are usually three or four. This is the type that is fitted to the **direct injection** engine which, due to the larger combustion chamber, requires the fuel to be injected in a number of sprays at high pressure to ensure even distribution and good penetration of fuel into the rapidly moving air stream. They are often of the long stem type to give good cooling of the injector.

Pintle

This nozzle is designed for use with **indirect injection** combustion chambers. The needle valve stem is extended to form a pintle which protrudes beyond the mouth of the nozzle body. By modifying the size and shape of this pintle the spray angle can be altered from parallel to a 60° angle or more. A modified pintle nozzle, known as the **delay** type, gives a reduced rate of injection at the beginning of delivery. This gives quieter running at idling speed on certain engines.

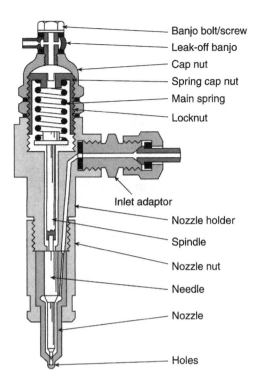

Fig. 3.40 Multi-hole injector

Pintaux nozzles

A development of the pintle-type nozzle, these have an auxiliary hole to assist starting in cold conditions. At cranking speeds, the pressure rise is slow and the needle valve is not lifted high enough for the pintle to clear the main discharge port. The fuel passing the seat is sprayed from the auxiliary hole towards the hottest part of the combustion chamber (that is within the area of the heater plug). At normal running speeds, the rapid pressure rise lifts the pintle clear of the main discharge port allowing the fuel to form the appropriate spray pattern. Approximately 10% of the fuel continues to pass through the auxiliary hole at normal running speeds to keep it free from carbon.

The efficiency of the injector deteriorates with prolonged use making it necessary to service the nozzles at periodic intervals. The frequency of maintenance depends on factors such as operating conditions, engine condition, cleanliness of fuel, etc.

Examples of nozzles are shown in Figs 3.41(a) and 3.41(b).

Cold starting aids

Some form of cold starting aid is usually fitted to CI engines to assist starting in cold

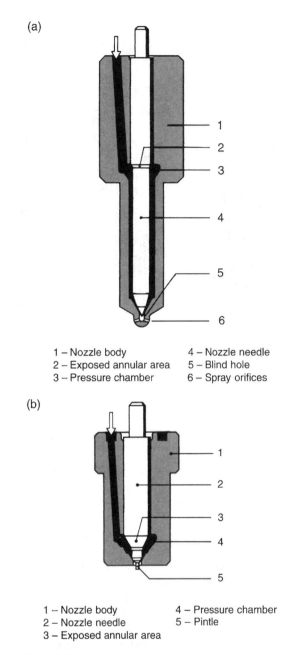

(a)

1 – Nozzle body   4 – Nozzle needle
2 – Exposed annular area   5 – Blind hole
3 – Pressure chamber   6 – Spray orifices

(b)

1 – Nozzle body   4 – Pressure chamber
2 – Nozzle needle   5 – Pintle
3 – Exposed annular area

Fig. 3.41 (a) Hole-type nozzles, (b) pintle-type nozzles

conditions. There are numerous types the two most common being:

- heater plugs
- thermo-start unit

Heater plugs

These are located in the cylinder head, the heating element of the plug being located just inside the combustion chamber. When an electrical current is supplied to the plug, the element heats up thus heating the air trapped in

(a)   (b)

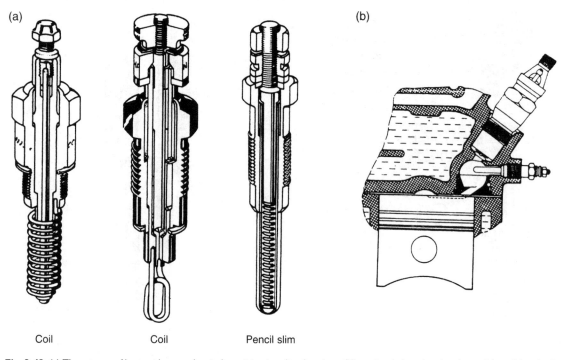

Coil   Coil   Pencil slim

Fig. 3.42 (a) Three types of heater plugs used on indirect injection diesel engines, (b) sectional view showing the position of the plug in the cylinder head

the chamber. Typical heater plugs are shown in Fig. 3.42.

The **'pencil'**-type heater plug is similar in design to the **'coil'**-type except that the heating coil is contained in a tube which when heated glows red. The plugs are wired in **parallel** and operate at 12 V with a loading of approximately 60 W. This gives a maximum element temperature of 950–1050 °C. When the heater plug has been in service some time the small air gap between the element sheath and the cylinder head becomes filled with carbon. This reduces the efficiency making starting more difficult in very cold weather. There is also a risk of the element burning out through over-heating. This can be avoided by removing the heater plugs and cleaning out the plug hole from time to time. Before refitting the plug remove any particles of carbon which may have lodged in the conical seating in the cylinder head. A faulty plug may be located by removing each feed wire in turn and fitting a test lamp or ammeter in the circuit. If the plug is in working order, the lamp will light or the ammeter will show a reading of approximately five amps when the heater circuit is in operation. A quick test can be carried out when the engine is cold. Operate the heater circuit and

after approximately 30 seconds each plug should feel warm to the touch.

In the double-coil type the plugs are connected in **series**, each one operating at 2 V. A resistance unit is wired into the circuit to reduce battery voltage to plug requirements. Apart from keeping the exterior of the plug and electrical connections tight, no other servicing or maintenance is required.

Thermo-start unit

The unit, shown in Fig. 3.43, is screwed into the inlet manifold below the butterfly valve. Fuel is supplied to the unit from a small reservoir fed from the injector leak-off pipe. The thermostart comprises a valve surrounded by a heater coil, an extension of which forms the igniter. The valve body houses a spindle which holds a ball valve in position against a seat, preventing fuel entering the device. When an electric current is supplied to the unit, the valve body is heated by the coil and expands. This releases the ball valve from its seat, allowing fuel to enter the manifold where it is vaporized by the heat. When the engine is cranked, air is drawn into the manifold and the vapour is ignited by the coil extension thus heating the air being drawn into the engine.

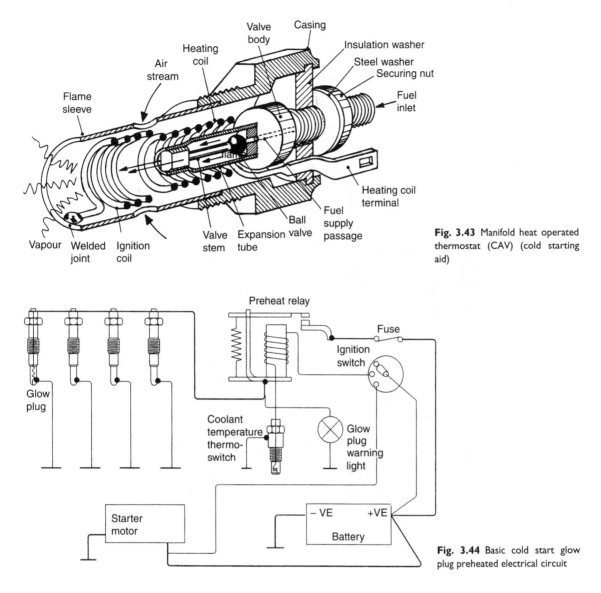

Fig. 3.43 Manifold heat operated thermostat (CAV) (cold starting aid)

Fig. 3.44 Basic cold start glow plug preheated electrical circuit

On switching off the current to the unit, the valve body contracts and the spindle returns the valve to its seat cutting off the fuel supply. The reservoir must be positioned 10–25 cm above the thermo-start unit to provide a positive fuel supply. This unit gives very little trouble in service provided the preheat time before operating the starter does not exceed that recommended (approximately 15 seconds).

Electrical connections

The heater plugs are connected so that when the ignition is switched on the warning lamp on the dashboard is illuminated; this warns the driver that the heater plugs are operating (warming up). The starter should not be operated until the warning lamp goes out which happens automatically after approximately 10–20 seconds. The heater plugs will now be at their correct temperature for cold starting. This is shown in Fig. 3.44.

3.13 Common rail diesel fuel system

Another recent development in computer controlled diesel systems is the common rail system shown in Fig. 3.45.

In this common rail system, the fuel in the common rail (gallery) is maintained at high pressure (up to 1 500 bar). A solenoid operated control valve that is incorporated into the head of each injector is operated by the ECM. The point of opening and closing of the injector

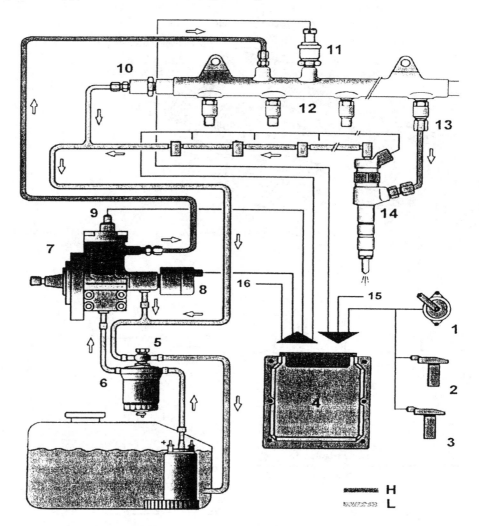

1. Accelerator
2. Engine speed (crank)
3. Engine speed (cam)
4. Engine control module
5. Overflow valve
6. Fuel filter
7. High pressure pump
8. Pressure regulating valve
9. Plunger shut-off
10. Pressure limiting valve
11. Rail pressure sensor
12. Common rail
13. Flow limiter
14. Injector
15. Sensor inputs
16. Actuator outputs

Fig. 3.45 The Rover 75 common rail diesel fuel system

control valve is determined by the ROM program and the sensor inputs. The injection timing is thus controlled by injector control valve and the ECM and the quantity of fuel injected is determined by the length of time for which the injector remains open and this also determined by the ECM.

The high injection pressures used and electronic control of the amount of fuel that is injected means that combustion efficiency is good, thus leading to improved fuel economy and reduced emissions.

Common rail diesel system injectors

Figure 3.46 shows two views of an electronically controlled injector that is used in typical Common Rail diesel fuel system. In diagram A, the control valve is closed and no fuel is being injected while, in diagram B the control valve is open and fuel is being injected into the engine cylinder.

Operation of injector in the closed position (diagram A)

The solenoid (1) is not energised and the control valve (2) is held on its seat. Fuel from the common rail fills the upper chamber (4) and the lower chamber (7). The pressures in the upper and lower chambers are equal. The injector needle valve (8) is held firmly on its seat by control spring (6) and no injection occurs.

Operation of the injector in the open position (diagram B)

An electrical pulse from the ECM energises the solenoid (1), lifting the control valve (2) and releasing pressure in the upper chamber (4). The imbalance of pressure between the upper and lower chambers causes the injector needle valve (8) to be lifted against the spring (6) and atomised fuel is injected into the engine cylinder.

Learning tasks

1. Remove a DPA fuel injection pump from a diesel engine. Turn both the pump and engine, time the pump to the engine, bleed the fuel system and run the engine.
2. Draw up a service schedule for a diesel system fitted with a DPA fuel pump. Base the schedule on either mileage or time.
3. State how the timing of a DPA fuel injection pump is adjusted. When would it be necessary to carry out this adjustment?
4. Remove the heater plugs, clean and check their resistance. Record your results and compare them with the manufacturer's recommendations. How would the driver notice if the heater plugs were not operating correctly?
5. Write out a logical test procedure for identifying the fault of a diesel engine producing black smoke when operating under load.
6. Complete the above test and produce a report to show the minimum and maximum engine revs, static timing, amount of advance on acceleration and amount of smoke allowable from the exhaust for a diesel engine. Compare your results with the manufacturer's data and produce a set of recommendations for the customer's report sheet. Name any specialized equipment used together with printouts that were produced in the test.
7. Remove a set of injectors from both direct injection and indirect injection engines. Using the correct equipment, test for correct opening pressure, back leakage and spray pattern. Refit the injectors and run the engine.
8. Dismantle both types of injectors and identify the main differences. After cleaning all the components reassemble the injectors, set the pressure and check back leakage and spray pattern.

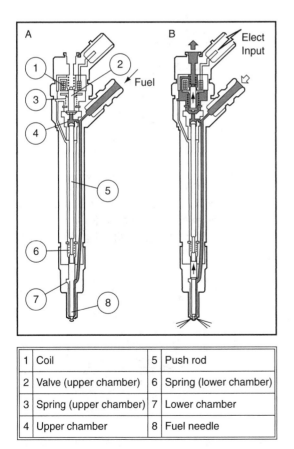

1	Coil	5	Push rod
2	Valve (upper chamber)	6	Spring (lower chamber)
3	Spring (upper chamber)	7	Lower chamber
4	Upper chamber	8	Fuel needle

Fig. 3.46 A common rail fuel injector

9. On undertaking a smoke meter test on a diesel engine, it was found that the smoke produced was excessive. Draw up a simple logical test procedure for tracing and rectifying the fault.

Practical assignment – SU carburettor

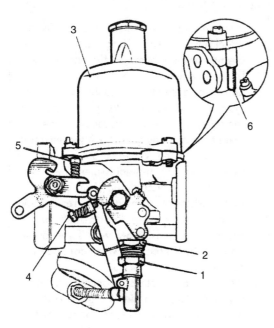

The type HS carburettor

1 – Jet adjusting nut	4 – Fast-idle adjusting screw
2 – Jet locking nut	5 – Throttle adjusting screw
3 – Piston/suction chamber	6 – Piston lifting pin

Activity – tuning single carburettors

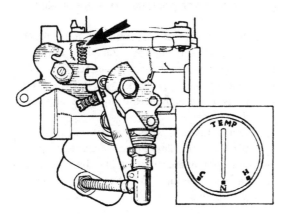

1. (a) Warm engine up to normal temperature.
   (b) Switch off engine.
   (c) Unscrew the throttle adjusting screw until it is just clear of its stop and the throttle is closed.
   (d) Set throttle adjusting screw $1\frac{1}{2}$ turns open.

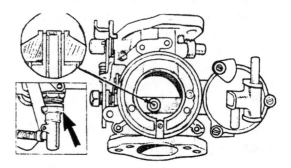

2. (a) Mark for reassembly and remove piston/suction chamber unit.
   (b) Disconnect mixture control wire.
   (c) Screw the jet adjusting nut until the jet is flush with the bridge of the carburettor or fully up if this position cannot be obtained.

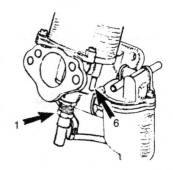

3. (a) Replace the piston/suction chamber unit as marked.
   (b) Check that the piston falls freely onto the bridge when the lifting pin (6) is released.
   (c) Turn down the jet adjusting nut (1) two complete turns.

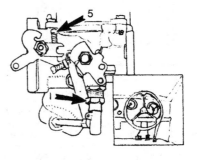

4. (a) Restart the engine and adjust the throttle adjusting screw (5) to give desired idling as

indicated by the glow of the ignition warning light.

(b) Turn the jet adjusting nut (I) up to weaken or down to richen until the fastest idling speed consistent with even running is obtained.

(c) Readjust the throttle adjusting screw (5) to give correct idling if necessary.

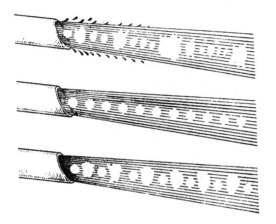

5. The effect of mixture strength on exhaust smoke
   (a) Too weak: Irregular note, splashy misfire, and colourless.
   (b) Correct: Regular and even note.
   (c) Too rich: Regular or rhythmical misfire, blackish.

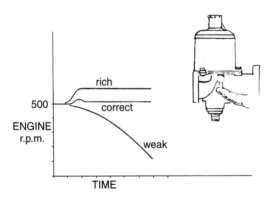

6. (a) Check for correct mixture by gently pushing the lifting pin up about $\frac{1}{32}$ in (0.8 mm) after free movement has been taken up.

   (b) The graph illustrates the effect on engine r.p.m. when the lifting pin raises the piston, indicating the mixture strength:
   rich mixture   RPM increase considerably
   correct mixture RPM increase very slightly
   weak mixture   RPM immediately decrease

   (c) Readjust the mixture strength if necessary.

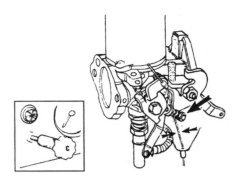

7. (a) Reconnect the mixture control wire with about $\frac{1}{16}$ in (16 mm) free movement before it starts to pull on the jet lever.

   (b) Pull the mixture control knob until the linkage is about to move the carburettor jet and adjust the fast-idle screw to give an engine speed of about 1000 RPM when hot.

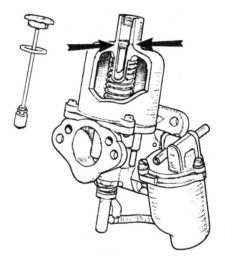

8. Finally top up the piston damper with the recommended engine oil until the level is $\frac{1}{2}$ in (13 mm) *above* the top of the hollow piston rod.

   Note: On dust-proofed carburetters, identified by a transverse hole drilled in the neck of the suction chambers and no vent in the damper cap, the oil level should be $\frac{1}{2}$ in (13 mm) below the top of the hollow piston rod.

Self assessment questions (diesel)

I. The electronically controlled injectors on a certain common rail diesel injection system operate at 80 volts with a current

of 20 amperes. The power of these injectors is:
(a) 16 kW
(b) 1 600 Watts
(c) 4 Watts
(d) 0.25 kW

2. In indirect injection diesel engines:
(a) the injectors spray fuel into the induction manifold
(b) the injectors spray fuel into a pre-combustion chamber
(c) the fuel is injected into a cavity in the piston crown
(d) the injection pressure is 10 bar

3. Diesel injectors are lubricated:
(a) by a separate supply of engine oil
(b) by means of a grease nipple mounted on the injector nozzle
(c) by leak back of diesel fuel
(d) by coating the needle valve with PTFE

4. In a diesel fuel system equipped with an in-line injection pump, dribble from the injectors is prevented by:
(a) a strong injector spring
(b) reducing the pressure at the lift pump
(c) a cylindrical collar on the delivery valve
(d) a sharp cut-off profile on the injection pump camshaft

5. Checking the phasing of an in-line fuel injection pump is carried out to:
(a) ensure that fuel injection starts at the correct number of degrees after top dead centre
(b) ensure that each element commences delivery at the correct angular interval
(c) ensure that the engine cannot run backwards
(d) ensure that the lift pump is working

6. The low pressure, diaphragm type lift pump operates at:
(a) 345 bar
(b) 34.5 kN/m^2
(c) 104 kN/m^2
(d) 0.104 bar

7. Spill timing is a process that:
(a) permits the operator to determine the point at which an element of the injector pump starts to inject so that the pump can be timed to the engine
(b) determines the amount of time that fuel flows through the spill port
(c) sets the amount of fuel that each pump element delivers
(d) sets the time constant of an electronic control unit

8. After fitting a new fuel filter element, the fuel system should be:
(a) completely drained down
(b) bled to remove any air that may be trapped
(c) bled by slackening the connections at the injectors
(d) put back into commission by starting the engine

9. The excess fuel device on an injection pump:
(a) is used to increase pulling power on steep hills
(b) comes into operation when the turbocharger is operating
(c) is a cold start aid
(d) increases the engines thermal efficiency

10. Diesel fuel injectors operate at:
(a) approximately 180 bar
(b) 7.5 lbf/in^2
(c) 0.05 Mpa
(d) 18 kN/m^2

Self assessment questions (petrol fuel systems)

1. In the mixture control idling system of a carburettor, the CO content of the exhaust gas at idling speed can be adjusted. Which one of the following is accurate?
(a) turning the idle mixture adjustment screw clockwise weakens the mixture
(b) turning the idle mixture adjustment anti-clockwise weakens the mixture
(c) to make this adjustment the engine must be cold
(d) the choke must be operating

2. When performing a test to check fuel gallery pressure on a petrol injection system a pressure test gauge may be temporarily connected into the petrol supply line. In order to perform this test:
(a) the fuel system must be de-pressurised before connecting the pressure test gauge
(b) the gallery pressure should not exceed 1 bar
(c) the gallery pressure should be several hundred bar
(d) the manifold absolute pressure sensor should be disconnected

3. In an EOBD system, the exhaust oxygen sensor that is placed downstream from the catalytic convertor is used to:
(a) act as a back in case the upstream oxygen sensor fails

(b) monitor the catalyst performance

(c) control emissions while the engine is cold

(d) operate the exhaust gas re-circulation valve

4. In an EOBD type engine management system, the engine idle speed is controlled by:

(a) the electronic idle valve, or stepper motor that is under the control of the engine control module

(b) the exhaust gas re-circulation valve

(c) the electronic idle valve and ignition timing in response to outputs from the engine control module

(d) the engine temperature sensor

5. Exhaust gas re-circulation helps to control emissions by:

(a) heating up the catalyst when the engine is cold

(b) reducing fuel consumption

(c) reducing temperatures in the combustion chamber so that the production of NOx is reduced

(d) increasing the volumetric efficiency

6. The test data for the supply voltage at a manifold absolute pressure sensor, states that the voltage should be 5 volts, plus or minus 4%. During a test of such a sensor, the voltage reading should be:

(a) 5.4 volts

(b) 4.6 volts

(c) between 4.8 volts and 5.2 volts

(d) $5 \times 100/4$ volts

7. The carbon canister in the fuel system reduces emissions because:

(a) the evaporative purge control valve permits fuel vapour from the petrol tank to be passed from the carbon canister to the engine for consumption

(b) the carbon enriches the fuel

(c) the carbon absorbs petrol vapour so that the carbon canister can be emptied periodically

(d) the carbon canister lowers fuel pressure in the gallery

8. In a throttle body petrol injection system:

(a) there is a separate injector for each cylinder

(b) there is a single injector at the throttle body

(c) air flow is not metered

(d) a catalytic convertor is not required for control of emissions

9. In a petrol engine, the air–fuel ratio for chemically correct combustion is:

(a) 28:1

(b) 14.7:1

(c) 21:1

(d) 19.6:1

4
Engine air supply and exhaust systems

Topics covered in this chapter

Engine air supply – air filters – manifolds
Exhausts systems – silencers
Oxygen sensor
Air flow meters
Turbo-charger

Liquid fuels such as petrol and diesel oil contain the energy that powers the engine. **Hydrogen** and **carbon** are the main constituents of these fuels and they need **oxygen** to make them burn and release the energy that they contain. The oxygen that is used for combustion comes from the atmosphere. Atmospheric air contains approximately 77% nitrogen and 23% oxygen by weight. For proper combustion, vehicle fuels require approximately 15 kg of air for every 1 kg of fuel. When the fuel and air are mixed and burned in the engine, chemical changes take place and a mixture of gases is produced. These gases are the exhaust. The principal exhaust gases are **carbon dioxide, nitrogen** and **water** (steam).

> ### Safety note
>
> You should be aware that combustion is often not complete and carbon monoxide is produced in the exhaust. Carbon monoxide is deadly if inhaled. For this reason, engines should never be run in confined spaces. Workshops where engines are operated inside must be equipped with adequate exhaust extraction equipment.

If all conditions were perfect the fuel and air would burn to produce carbon dioxide, super-heated steam, and nitrogen (Fig. 4.1). Unfortunately these 'perfect' conditions are rarely achieved and significant amounts of other harmful gases, namely **carbon monoxide CO, hydrocarbons HC, oxides of nitrogen NOx** and **solids** (particulates like soot and tiny metallic particles) are produced (Fig. 4.2).

Most countries have laws that control the amounts of the harmful gases (emissions) that are permitted and engine air and exhaust systems are designed to comply with these laws.

4.1 Engine air supply systems (carburettor)

Figure 4.3 shows the main features of a petrol (spark ignition) engine **air supply system**. Our main concern, at the moment, is the **air cleaner** and the **intake pipe** (induction tract) leading to the combustion chamber. The **blow-by filter** and the **positive crankcase ventilation** (PCV) system are part of the **emission control system**.

Air filter

The purpose of the air filter (Fig. 4.4) is to remove dust and other particles so that the air reaching the combustion chamber is clean. Pulsations in the air intake generate quite a lot of noise and the air cleaner is designed so that it also acts as an **intake silencer**.

The porous paper element is commonly used as the main filtering medium. Intake air is drawn through the paper and this traps dirt on the surface. In time this trapped dirt can block the filter and the air filter element is replaced at regular service intervals. Failure to keep the air filter clean can lead to problems of restricted air supply which, in turn, affects combustion and exhaust emissions.

Temperature control of the engine air supply

To improve engine 'warm up' and to provide as near constant air temperature as possible, the air cleaner may be fitted with a device that heats up the incoming air as and when required.

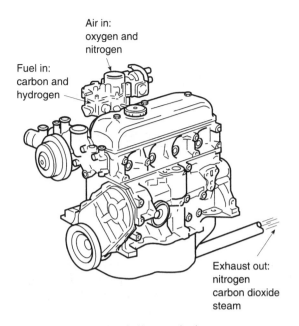

Air in:
oxygen and
nitrogen

Fuel in:
carbon and
hydrogen

Exhaust out:
nitrogen
carbon dioxide
steam

Fig. 4.1 Engine showing fuel/air in and exhaust out

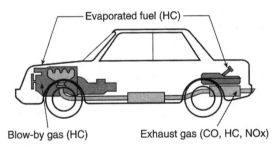

Evaporated fuel (HC)

Blow-by gas (HC)

Exhaust gas (CO, HC, NOx)

Fig. 4.2 Various bad emissions from engine and fuel system

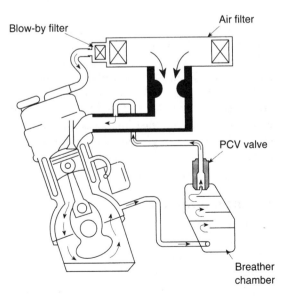

Blow-by filter

Air filter

PCV valve

Breather
chamber

Fig. 4.3 An engine air supply system

Air filter

Fig. 4.4 An air cleaner

Figure 4.5 shows the layout of a temperature controlled air intake.

It shows that the **manifold vacuum** has lifted the **intake control door** off its seat and air is drawn into the air cleaner over the surface of the **exhaust manifold**. The **bi-metallic spring** has pushed the **air bleed valve** on to its seat. When the air in the **air cleaner** reaches a temperature of 25–30 °C, the bi-metal spring lifts the air bleed valve from its seat. This removes the vacuum from above the **air control door** diaphragm, thus closing the intake control door. The intake air now enters direct from the atmosphere. During cold weather the temperature-control system will be in constant operation.

Checking the operation of the air control valve

Figure 4.6 shows the type of practical test that can be applied to check the operation of the air control diaphragm and the air door. The flap is raised against the spring and a finger is placed over the vacuum pipe connection. The air door valve is then released and the valve should remain up off its seat.

Fuel injected engine air intake system

Figure 4.7 shows an air intake system for a modern engine fitted with electronic controls.

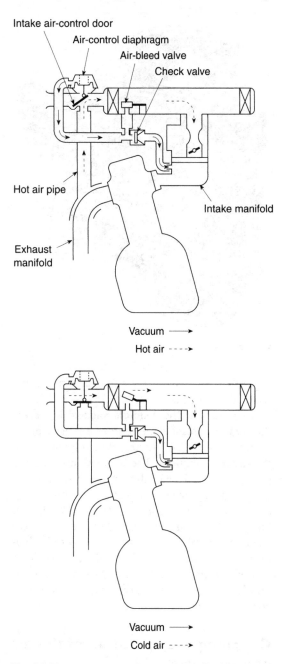

Fig. 4.5 Temperature control of intake air

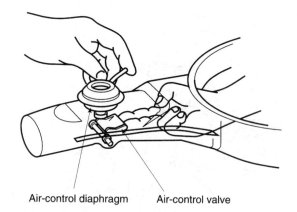

Fig. 4.6 Checking the air control valve

movement of the flap produces a voltage at the sensor that is an accurate representation of the amount of air entering the engine. This voltage signal is conducted to the electronic control unit (**ECU**), where it is used in 'working out' the amount of fuel to be injected.

Throttle position sensor (potentiometer)

This is the sensor (shown as 6 on Fig. 4.7) that 'tells' the ECU the angular position of the throttle butterfly valve. The principle of a throttle position sensor is shown in Fig. 4.9. The voltage from the throttle position sensor is also conducted to the ECU for use in the process of determining the amount of fuel to be injected.

Idle air control

When the driver's foot is removed from the accelerator pedal, the throttle butterfly valve is virtually closed. In order that the engine will continue to run properly (idle) the intake system is provided with a separate system that provides an air supply for idling purposes. Figure 4.10 shows an electronically controlled air valve.

Air valves of the type shown in Fig. 4.10 are controlled by the ECU. The ECU will admit extra air to provide fast-idle speed so that the engine keeps running even when extra load is imposed with the throttle closed, such as by switching on the headlights.

Air intake manifold

On many engines the intake air is supplied to a central point on the intake manifold as shown in Fig. 4.11. The manifold is provided with ports that connect to the individual cylinders of the engine. The seal, often a gasket, between the intake manifold and the cylinder head must be secure and it is

This system has several components that we need to consider briefly.

Air flow meter

The air intake for a fuel injected petrol engine includes an **air flow meter** (Fig. 4.8). The purpose of the air flow meter is to generate an electrical signal that 'tells' the electronic controller how much air is flowing into the combustion chambers. It is also known as an 'air flow sensor'.

The intake air stream causes a rotating action on the sensor flap, about its pivot. The angular

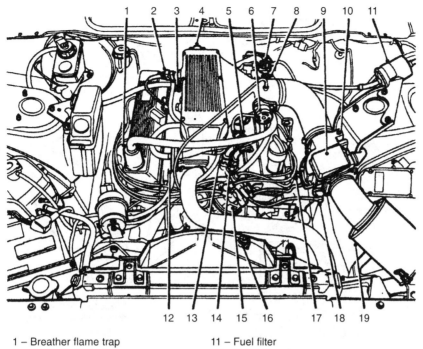

1 – Breather flame trap
2 – Cold start injector
3 – Fuel pressure regulator
4 – Overrun valve
5 – Throttle potentiometer connection
6 – Throttle potentiometer
7 – Engine air breather
8 – Idle speed adjustment screw
9 – Air flow meter
10 – Idle mixture adjustment screw

11 – Fuel filter
12 – Extra air valve
13 – Coolant temperature switch
14 – Thermo-time switch
15 – Distributor
16 – Diagnostic plug
17 – Spark plug – No.1 cylinder
18 – Ignition coil
19 – Air cleaner

Fig. 4.7 Air supply system – petrol injection engine

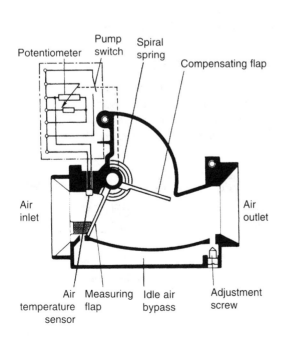

Fig. 4.8 An air flow meter

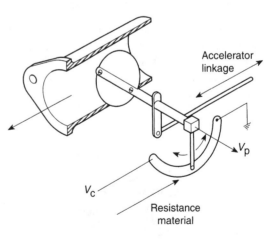

V_c = Constant voltage supply from computer
V_p = Voltage giving position of throttle

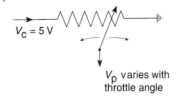

V_c = 5 V

V_p varies with throttle angle

Fig. 4.9 Throttle position sensor

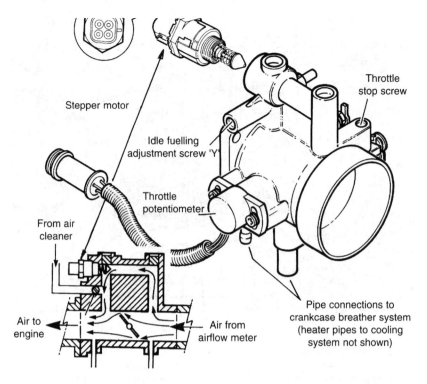

Fig. 4.10 An air control valve

advisable to check the tightness of the securing nuts and bolts periodically. These nuts and bolts must be tightened in the proper sequence, shown for a particular engine in Fig. 4.12.

Variable length induction tract

The length of the induction tract (tube) through which the engine's air supply is drawn has an effect on the operating efficiency of the engine. For low to medium engine speeds a long induction tract is beneficial whereas a shorter induction tract is beneficial at high engine speed. Figure 4.13(a)

shows a simplified variable length induction system. An electronically controlled valve between the throttle valve and the engine switches air flow between the long and short tracts as required by engine speed. At low to medium engine speed this air valve is closed as shown in Fig. 4.13(b). As the engine speed rises the electronic control unit causes the air valve to open, as shown in Fig. 4.13(c) and the main mass of air enters the engine through by the short route.

Turbo charging

Engines that rely on atmospheric air pressure for their operation are known as naturally aspirated engines. The power output of naturally aspirated engines is limited by the amount of

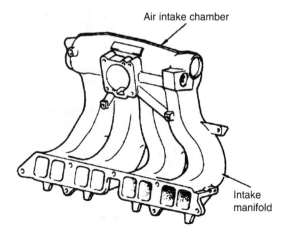

Fig. 4.11 An air intake manifold

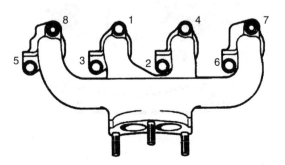

Fig. 4.12 Tightening sequence for exhaust manifold retaining hardware (nuts)

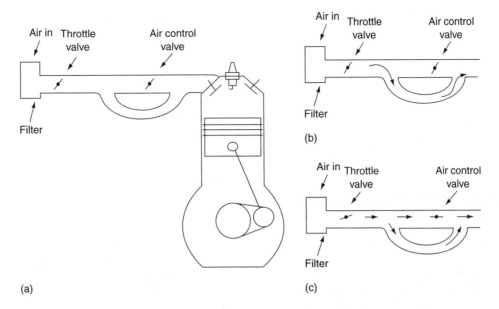

Fig. 4.13 (a) Variable length induction tract, (b) idling to medium speed, (c) high speed

air that can be induced into the cylinder on each induction stroke. Engine power can be increased by forcing air into the cylinders under pressure. The turbo charger is commonly used on modern engines to boost the pressure of the incoming air. The turbo charger consists of two main units, the turbine and the compressor mounted on a common shaft. The turbine is driven by exhaust gas and utilises energy that would otherwise be wasted by being expelled through the exhaust system into the atmosphere.

The general principle of turbo charging is shown in Fig. 4.14. Exhaust gas energy is directed to the small turbine which is

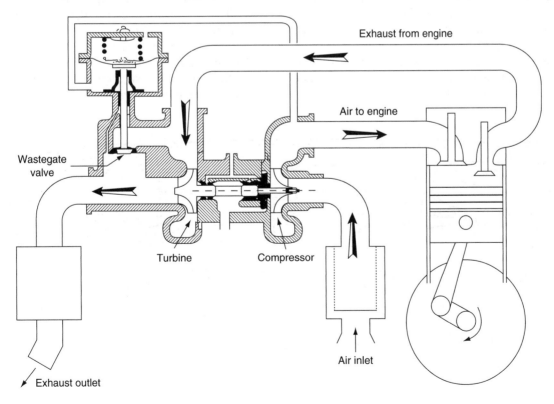

Fig. 4.14 A turbo charger system

connected to the shaft that also drives the compressor. After driving the turbine, the exhaust gas travels through the exhaust system and out through the tail pipe. Intake air is drawn into the compressor through the air filter. The compressor raises the pressure to approximately 0.5 bar above atmospheric pressure and the pressure is controlled by the waste gate valve that allows the exhaust gas to by-pass the turbine.

Service attention to the air supply system

A normal service requirement is to renew the air filter element at regular intervals. Should the **filter element** become 'clogged' it will restrict the flow of air to the engine and cause the engine to 'run rich'. This will obviously affect the engine performance and it will also cause the emission control system to malfunction. To guard against problems in this area, all vehicles should be properly serviced before they are tested for compliance with the emissions regulations, for example, a pre-MOT service and check-over.

Other items that should receive regular attention include: sensor connections, condition of flexible hoses and tightness of joints and securing brackets. Air flow meters, throttle position sensors, etc. are normally quite reliable. In the event of problems with these items, it will normally be necessary to 'read out' fault codes, either through the 'on-board' diagnostic system or by the use of diagnostic code readers and other tools.

4.2 Exhaust system

A fundamental purpose of the **exhaust system** is to convey exhaust gas away from the engine and to expel it into the surrounding atmosphere at some point that is convenient to vehicle occupants. On motor cars this usually means through the tail pipe at the rear of the vehicle. On trucks it is often at some other point, often just behind the driver's compartment. Another function of the exhaust system is to deaden the exhaust sound to an acceptable level. This sound level is regulated by law.

The exhaust system shown in Fig. 4.15 has a number of sections. Starting at the front of the vehicle the sections are: the **down pipe**, or **front pipe**; the unit marked 1, this is the **catalytic convertor**, an interconnecting pipe to the **first silencer** (2); a further length of pipe which is curved to take it round other units on the vehicle; this pipe connects to the **rear silencer** (3); and the gas finally leaves through the **tail pipe**.

The exhaust system is mounted firmly to the exhaust manifold at the front end, but it is flexibly mounted to the underside of the vehicle for the remainder of its length. The flexible mountings are required to allow the system to move independently of the main vehicle structure. This allows for the fact that the engine itself is normally flexibly mounted, and the type of exhaust mountings used prevent noise and vibration being transmitted to the vehicle structure. In order to provide for ease of maintenance and economy when replacing parts, the

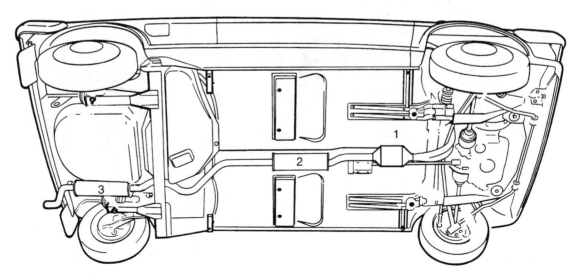

Fig. 4.15 Underside view of an exhaust system

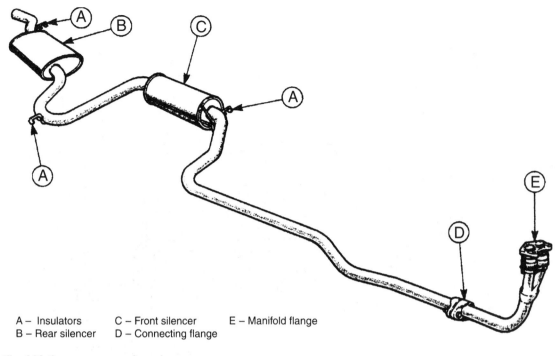

A – Insulators   C – Front silencer   E – Manifold flange
B – Rear silencer   D – Connecting flange

Fig. 4.16 Component parts of an exhaust system

exhaust system is made up from a number of separate components. The number of such parts varies from vehicle to vehicle, but Fig. 4.16 shows a typical exhaust system, together with its fixings.

The exhaust catalyst is dealt with at the end of this section. The two silencers may be of different types, **expansion**-type and **absorption**-type. The expansion-type is sometimes called a resonator because of the way that it works. Figure 4.17 shows the basic principles of the two types of silencer commonly used.

Broken or damaged exhaust systems

In addition to the noise factor, which is often the most evident sign of a broken exhaust system, there is also the danger of leaking exhaust gas finding its way into the passenger compartment. Broken exhaust systems should be repaired as a matter of urgency and, when an exhaust has been repaired, it should be thoroughly checked to ensure that there are no leaks. In order to prevent leaks, all pipe joints, flanges and their gaskets and other fixings must be made secure, using sealant where it is recommended.

Catalytic convertor

Function

The purpose of the catalytic convertor is to enable the vehicle to comply with emissions regulations. The principal emissions of concern are oxides of nitrogen (NOx), hydrocarbons (HC), carbon monoxide (CO) plus particulates

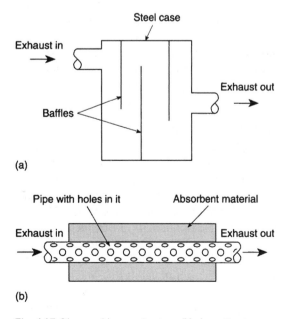

Fig. 4.17 Silencers (a) expansion-type, (b) absorption-type

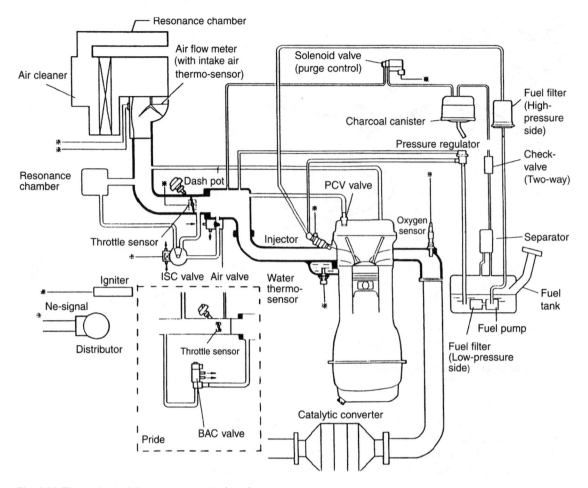

Fig. 4.18 The catalyst and the oxygen sensor in the exhaust system

(pieces of soot and metallic particles). Figure 4.18 shows a catalytic convertor in the exhaust system. It is mounted near to the engine. This helps to maintain it above its minimum working temperature of approximately 300 °C. Further up the front pipe, near the engine, is an **oxygen sensor**. The oxygen sensor plays an important part is the successful operation of the catalyst.

Method of operation

A catalyst is a material that produces chemical changes in substances that come into contact with it, but is not itself affected by the chemical changes. In the exhaust catalyst the three harmful pollutants, NOx, HC and CO, are changed (by **oxidation** or **reduction**) into nitrogen, carbon dioxide and water. A catalytic convertor that deals with these three pollutants is called a three-way catalyst (TWC). The catalytic convertor will only function correctly when the engine is supplied with a chemically correct (stoichiometric) air/fuel ratio.

For petrol engines this air/fuel ratio is approximately 15:1 by weight. This ratio is given the name **LAMBDA** which has a value of 1. If the air/fuel ratio is 12:1 (a rich mixture) LAMBDA is about 0.9. If the air/fuel ratio is 17:1 LAMBDA is about 1.1. The **exhaust oxygen sensor** (EGO) 'measures' the oxygen content of the exhaust gas and this gives an accurate measure of the air/fuel ratio of the mixture entering the engine, and that is why it is sometimes known as a LAMBDA sensor.

The commonly used catalyzing agents are the rare metals **platinum**, **rhodium** and **palladium**. These metals are spread thinly over a large surface area so that the exhaust gas can be made to come into contact with them. This large surface area is achieved by the use of **pellets** covered with the metals, or a **ceramic honeycomb** covered with catalysts. The two different methods are shown in Fig. 4.19. In order to function correctly, the catalyst must reach approximately 300 °C and it is becoming common practice to fit catalysts with a **pre-heating element** which makes them more

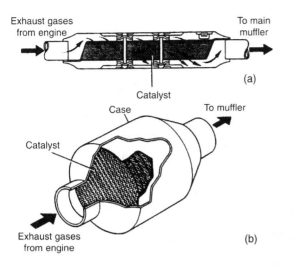

Fig. 4.19 Exhaust catalysts. (a) Pellet-type catalytic converter. Catalysts supported on aluminas, carbons and other speciality materials are produced for use in many chemical processes. The cost-effectiveness and catalytic selectivity of finely dispersed platinum group metal catalysts greatly exceeds those of alternative processes: indeed, many modern processes depend exclusively on these catalysts, (b) Honeycomb- or monolithic-type catalytic converter. Autocatalysts utilize honeycomb supports to assure the maximum reactive surface within the smallest possible volume, with consequent high catalytic activity and minimum engine power loss. Whilst the majority of catalysts use a ceramic support, some specialized applications demand the added durability of steel supports

effective during cold starts. A note of warning too. Leaded fuel will '**poison**' a catalyst and prevent it from working. Leaded fuel must never be used in an engine fitted with a catalytic convertor.

Oxygen sensor

The exhaust oxygen sensor is a critical part of the engine emissions control system. The oxygen sensor transmits an electrical signal to the ECU and the ECU then makes the necessary adjustments to the amount of fuel injected so that LAMBDA is maintained within the design limits. The active element (a ceramic) of the oxygen sensor is sensitive to temperature and it must be hot for it to work properly. For this reason, oxygen sensors are often fitted with an electric heating element. This type of sensor is then known as a **heated exhaust gas oxygen sensor** (HEGO).

At the time of writing, there are two different types of exhaust oxygen sensors in common use. One type uses **zirconium oxide**; this produces a small voltage at its terminals. The other type uses **titanium oxide**; this relies on changes of conductivity for its operation and is supplied with voltage. It is important to recognize this

because the test procedures for the sensors involve different procedures. Exhaust oxygen sensors are shown in Fig. 4.20.

Exhaust emissions test

Exhaust emissions are tested by means of an **exhaust gas analyser**. Failure to 'pass' the test does not necessarily mean that the catalyst has failed. As mentioned in the section on air supply, a choked air filter can upset the whole system. Exhaust catalysts are expensive items and it is important to make sure that all other parts of the engine emissions system are in good working order before condemning the catalyst.

Maintenance

Exhaust systems are subject to corrosive attack from the outside and the inside environment. The outside corrosion arises from water, salt and other corrosive substances that may be present on the running surface. The internal corrosion arises, in part, from the water that is produced by combustion of the fuel. Each litre of petrol burned produces almost a litre of water. When the exhaust system is hot, the water leaves the tail pipe as super-heated steam but, when the exhaust system is cold and the atmospheric conditions are right, some of the steam condenses to water inside the exhaust. To extend the life of exhaust silencers they are sometimes coated, internally and externally, with a thin layer of aluminium. The exterior of the whole exhaust system should be checked for signs of holes and all pipe clamps and joints should checked for leaks. The flexible mountings should also be checked from time to time and any that are found to be in danger of breaking should be replaced, in order to prevent more expensive work at some later date.

Learning tasks

1. Draw up a list of the equipment that you have used to perform an emissions test for an MOT. Describe the procedure and give details of the acceptable MOT emissions levels for any modern vehicle.
2. Draw up a test procedure for checking the oxygen sensor. Itemize any precautions that should be observed, e.g. corrosion of terminals and care when removing and refitting.
3. Remove a complete exhaust system from a vehicle, including the exhaust manifold. Replace the manifold gasket and any

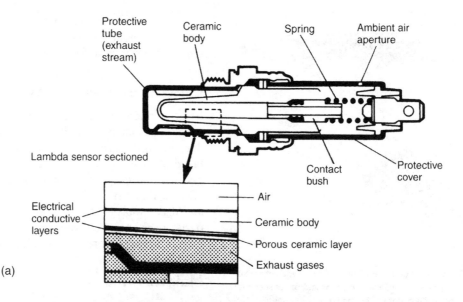

Fig. 4.20 Exhaust oxygen sensor (a) construction of zirconia-type oxygen sensor, (b) the oxygen sensor in the exhaust pipe

other components that are corroded and mountings that are separating. Replace/refit the exhaust system. How should the system be checked for leaks after refitting?

4. Remove the inlet manifold complete with carburettor. Check the operation of the one-way valve to the brake servo cylinder. Refit the manifold using a new gasket. State the purpose of using water to heat/cool the inlet manifold.

5. Remove the air cleaner and check the operation of the air control valve. Replace the filter element and air cleaner.

6. What are the symptoms of a leaking inlet manifold gasket? What tests would you use to identify this fault?

7. Describe the procedure that you have used to fit a replacement air filter to an engine. Explain how a blocked air filter may affect the exhaust emissions from an engine.

8. Examine an engine inlet manifold and note down the type of material that it is made from. State why it is necessary to follow a sequence when tightening an inlet manifold when refitting it to an engine.

9. Examine an exhaust oxygen sensor installation. Note down the number of wires that lead into the sensor and state whether or not it is a type that carries its own heating element.

10. After fitting a new exhaust system it is necessary to check that there are no leaks in the system. Describe the practical methods that you use, in your workplace, to carry out this check.

11. Exhaust fumes are very dangerous. Write down the precautions that must be observed when working on running engines, in a confined space. Make a list of the exhaust extraction equipment fitted in your workshop.

Practical assignment – air supply and exhaust systems

Objective

The purpose of this assignment is to allow you to develop your knowledge and skill in this important area of motor vehicle work and to help you to build up your file of evidence for NVQ.

Activity

The nature of this assignment is such that you will need to work under supervision, unless you have previously been trained and are proficient. The assignment is intended to be undertaken during a routine service and MOT test on a petrol engined vehicle that is equipped with electronically controlled petrol injection.

The assignment is in two parts.

Part I

1. Locate the idle control valve. Describe the purpose of this valve and describe any adjustments that can be made to it.
2. Specify the effects on engine performance and exhaust emissions that may arise if this valve is not working correctly.

Part 2

1. Perform the emissions test part of the MOT test.
2. Record the results and compare them with the legal limits.
3. Write short notes on the test procedure and describe any work that should be carried out before starting the emissions test.
4. Give details of remedial work that can be performed if a vehicle should fail the emissions test.
5. Make a list of the tools and equipment used.

Self assessment questions

1. A blocked air filter may cause:
   (a) low fuel consumption
   (b) the engine to run rich
   (c) increased cooling system pressure
   (d) excess oxygen in the exhaust
2. An air leak through the intake manifold flange gasket may cause:
   (a) the engine to run rich
   (b) the mal-function indicator lamp to illuminate
   (c) the positive crankcase ventilation system to fail
   (d) excess Nox in the exhaust
3. Turbocharged engines:
   (a) are less efficient than naturally aspirated ones because power is absorbed in driving the turbo charger
   (b) are more efficient than naturally aspirated engine because the make use of energy in the exhaust gas that would otherwise be wasted
   (c) are normally diesel engines
   (d) have high fuel consumption
4. An air flow sensor:
   (a) provides an electrical signal that tells the ECU how much air is flowing to the engine
   (b) controls the air flow to the air conditioning system
   (c) allows the exhaust gases to by-pass the turbo charger
   (d) restricts air flow to prevent engine overload
5. The potentiometer type throttle position sensor:
   (a) generates electricity that is used as a signal to inform the ECU of the amount of throttle opening
   (b) receives a supply of electricity and sends a signal to the ECU that is an accurate representation of throttle position
   (c) is always situated on the accelerator pedal
   (d) always operates on the variable capacitance principle
6. Exhaust gas oxygen sensors:
   (a) are used to detect oxygen levels in the exhaust gas
   (b) operate affectively at all temperatures
   (c) always generate electricity
   (d) are only needed on direct injection engines
7. During combustion, hydrogen in the fuel combines with oxygen from the air:
   (a) and produces water in the exhaust system
   (b) and produces Nox in the exhaust gas

(c) and produces carbon monoxide in the exhaust gas

(d) and causes the ignition to be retarded

8. When an engine fails the emissions test it means that:

(a) the catalytic convertor is not working

(b) the emissions control system is not working and this may be caused by a range of engine problems

(c) the oxygen sensor has failed

(d) the air filter is blocked

9. Exhaust systems may be protected against corrosion by:

(a) coating them inside and outside with a layer of bitumen

(b) a thin layer of aluminium applied to the inner and outer surfaces

(c) reducing the temperature of the exhaust gases

(d) coating them with underseal

10. Atmospheric air contains approximately 77% nitrogen and 23% oxygen by mass:

(a) the nitrogen is a hindrance because it is not used in combustion

(b) the nitrogen helps to form water

(c) the nitrogen forms part of the gas that expands in the cylinder to make the engine work

(d) the oxygen and nitrogen makes Nox at very low temperatures

5
Transmission systems

Topics covered in this chapter

Clutch
Gearbox
Fluid flywheel and torque convertor
Transmission systems – drive shafts
Hub types – semi-floating, fully floating, three-quarter floating
Gear ratios – torque multiplication

5.1 The clutch

Most clutches used in the modern motor car are called **friction clutches**. This means that they rely on the friction created between two surfaces to transmit the drive from the engine to the gearbox. The clutch fulfils a number of different tasks. The three main ones are:

- it connects/disconnects the drive between the engine and the gearbox;
- it enables the drive to be taken up gradually and smoothly;
- it provides the vehicle with a temporary neutral.

The three main component parts of a clutch are:

- the **driven plate**, sometimes referred to as the **clutch**, **centre** or **friction plate**;
- the **pressure plate** which comes complete with the clutch cover, springs or diaphragm to provide the force to press the surfaces together;
- the **release bearing** which provides the bearing surface which, when the driver operates the clutch pedal, disconnects the drive between the engine and the gearbox.

Through the centre of the pressure plate the **input shaft** of the gearbox (sometimes called the **spigot**, **first motion** or **clutch shaft**) is splined to the middle of the driven plate. On the conventional vehicle layout it is located on a bearing (called the **spigot bearing**) in the flywheel.

Multi-spring clutch

The driving members of the multi-spring clutch consist of a **flywheel** and pressure plate (both made of cast iron) with the driven plate trapped between. The pressure plate rotates with the flywheel, by means of projections on it locating with slots in the clutch cover which is bolted to the flywheel. A series of springs located between the cover and the pressure plate force the plate towards the flywheel trapping the clutch plate between the two driving surfaces. The **primary** (or input) shaft of the gearbox is splined to the hub of the clutch disc and transmits the drive (called **torque** which means turning force) to the gearbox. The drive is disconnected by withdrawing the pressure plate and this is achieved by the operation of the release levers. Figure 5.1 shows a multi-spring clutch.

Diaphragm-spring clutch

The diaphragm-spring clutch (shown in Fig. 5.2) is similar in many ways to the coil spring type. The **spring pressure** is provided by the **diaphragm** which also acts as the lever to move the pressure plate. On depressing the clutch pedal the release lever is forced towards the flywheel thus pulling the pressure plate away from the flywheel. This type has a number of advantages over the coil spring type.

- It is much simpler and lighter in construction with fewer parts.
- It has a lighter hold-down pressure and is therefore easier to operate thus reducing driver fatigue.
- The clutch assembly is smaller and more compact.
- It provides almost constant pressure throughout the life of the driven plate.

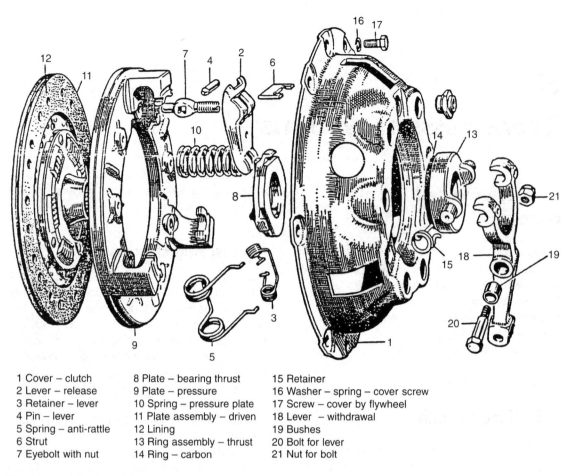

1 Cover – clutch	8 Plate – bearing thrust	15 Retainer
2 Lever – release	9 Plate – pressure	16 Washer – spring – cover screw
3 Retainer – lever	10 Spring – pressure plate	17 Screw – cover by flywheel
4 Pin – lever	11 Plate assembly – driven	18 Lever – withdrawal
5 Spring – anti-rattle	12 Lining	19 Bushes
6 Strut	13 Ring assembly – thrust	20 Bolt for lever
7 Eyebolt with nut	14 Ring – carbon	21 Nut for bolt

Fig. 5.1 The coil spring clutch

- Unlike the coil spring the diaphragm spring is not affected by centrifugal force at high engine speeds. It is also easier to balance. One type of diaphragm spring, the strap drive, gives almost frictionless movement of the pressure plate inside the clutch cover.

Friction plate construction

The important features incorporated in the design of a driven plate can best be seen by considering the disadvantages of using a plane steel plate with a lining riveted to each side. A number of serious problems would very soon be encountered.

- Buckling of the plate can occur due to the heat that is generated when the drive is taken up.
- The drive may not be disconnected completely (called **clutch drag**), caused by the plate rubbing against the flywheel when the clutch should be disconnected.
- Very small movement of the clutch pedal. The clutch is of the 'in or out' type, with very little control over these two points. This may cause a sudden take up of the drive (called **clutch judder**).

To overcome these problems the plate is normally **slotted** or 'set' in such a way as to produce a flexing action. Whilst disengaging, the driven plate will tend to jump away from the flywheel and pressure plate to give a clean break. Whilst in this position the linings will be held apart and air will flow between them to take away the heat. During engagement, spring pressure is spread over a greater range of pedal movement as the linings are squeezed together. This gives easier control and smoother engagement. In most cases the hub is mounted independently in the centre of the clutch plate, allowing it to twist independently from the plate. To transmit the drive to the plate either springs or rubber, bonded to the hub and plate, are used. This flexible drive absorbs the torsional (twisting) shocks due to the engine vibrations on clutch take-up, which could otherwise cause transmission noise or rattle. These features can be seen in the clutch drive (H) in Fig. 5.2.

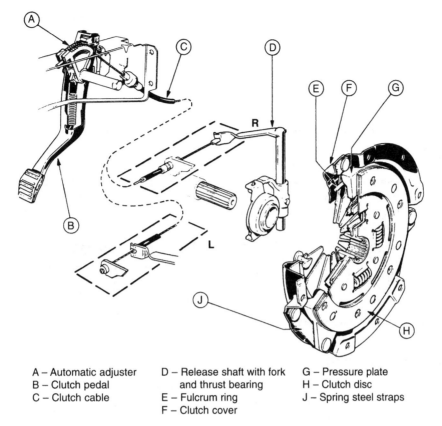

A – Automatic adjuster
B – Clutch pedal
C – Clutch cable

D – Release shaft with fork
    and thrust bearing
E – Fulcrum ring
F – Clutch cover

G – Pressure plate
H – Clutch disc
J – Spring steel straps

Fig. 5.2 Diaphragm spring clutch components (L = LHD, R = RHD)

Friction plate materials

Coefficient of friction

This is the relationship between two surfaces rubbing together. On a clutch both static and sliding friction are necessary. Sliding friction occurs when the drive is being taken up and static friction when the clutch is fully engaged.

Wear properties

This relates to the ability of a material to withstand wear. The surfaces of the flywheel and pressure plate should not become scored or damaged due to friction as the surfaces slide over each other. If they do become damaged then any number of clutch faults could become apparent; under these circumstances they would be replaced. A typical coefficient of friction for a motor car would be approximately 0.3.

Linings

These should have a high and stable coefficient of friction against the flywheel and pressure plate surfaces over a wide a range of temperatures and speeds. They should also have good wear resistance and not score or cause thermal damage to any of the surfaces with which they are in contact (e.g. flywheel or pressure plate). The material used for the linings is normally an **asbestos resin** mix which gives the best compromise to meet the above demands. A thermosetting resin reinforced by **asbestos fibres** and small amounts of **copper**, **brass** or **aluminium** is added to improve the wear properties. Two types of linings are in use, woven linings and moulded linings.

Woven linings

These are formed from a roll of loosely woven asbestos containing brass or zinc wires which is soaked in resin and other ingredients. The resin is then polymerized (which hardens the resin mixture). This has the following advantages:

- it is very flexible compared to other types;
- it is easier to drill;
- it can be cut and bent to shape.

They also have a high coefficient of friction when used in conjunction with cast iron surfaces. They are not suitable for heavy duty use where the temperature is much above 200 °C (this gives a loss of friction and a high wear rate).

Moulded linings

These are more popular now. In this type the various components are mixed together and cured in dies at high temperatures and pressures to form a very hard material. These can withstand high temperatures and pressures when in use. Different types of linings can be made having widely different coefficients of friction and wear properties because of the much wider range of **polymers** that can be used in their manufacture. They require more careful handling during drilling and riveting as the material is more brittle than the woven type. They must also be formed to the required radius as they cannot be cut or bent. Heavy duty clutches often use **ceramic materials** as these can withstand higher operating temperatures and pressures. Since the end of November 1999, asbestos has been banned by law in the manufacture of new clutch linings. (**Refer to section on Health and Safety**.)

Operating mechanisms

The clutch release bearing is fitted to the clutch release fork. When the clutch pedal is pressed down by the driver, pressure is exerted against the diaphragm spring fingers by the release

bearing forcing them towards the flywheel, this action disengages the drive from the engine.

The pedal linkage may be either **mechanical** or **hydraulic**. It must be flexibly mounted as the engine (which is mounted on rubbers) may move independently of the body/chassis.

Hydraulic linkage

In this arrangement, shown in Fig. 5.3, oil is displaced by the movement of the piston in the **master cylinder** to the **slave cylinder** to operate the release fork. The reservoir on the master cylinder tops up the system as wear in the clutch takes place.

Mechanical linkage

This operates on the principle of levers and rods or cable. The movement ratio and force ratio can be arranged to give a large force acting on the clutch with a small force acting on the pedal. Figures 5.2 and 5.4 shows this system.

Adjustment

Adjustment of the hydraulic operating mechanism is automatic. As wear of the linings takes place the piston in the slave cylinder takes up that amount of wear.

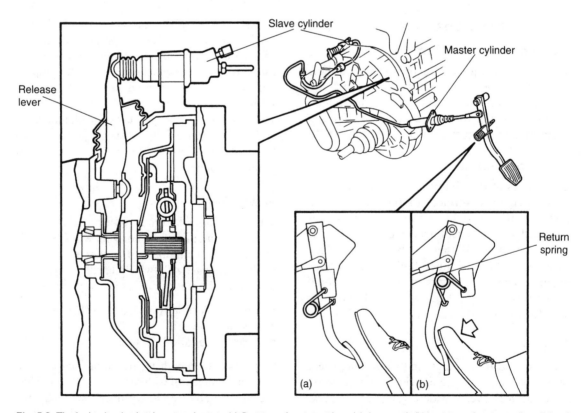

Fig. 5.3 The hydraulic clutch release mechanism. (a) Position of spring with pedal depressed, (b) position of spring in the off (rest) position

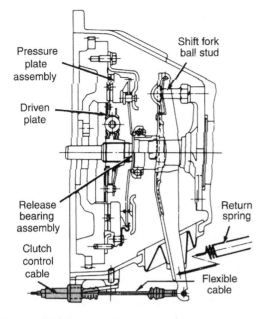

Fig. 5.4 Cable-operated clutch arrangement

The mechanical system is usually a flexible cable and may have an automatic self-adjusting mechanism as in Fig. 5.2 or else it may need to be checked and adjusted at regular service intervals.

Clutch faults

These are the most common faults found on the clutch.

Clutch slip

This occurs when the clutch fails to transmit the power delivered by the engine. It may be caused by one of the following faults:

- insufficient 'free' movement on the clutch pedal or at the withdrawal (release) lever;
- oil or grease on the friction linings;
- friction linings worn out;
- scored faces on the flywheel or pressure plate.

Clutch drag

This produces some difficulty in engaging and changing gear. It may be cause by one of the following faults:

- insufficiently effective pedal travel (excessive 'free' movement on the pedal or slave cylinder push rod);
- hydraulic system defective (due usually to the loss of most if not all of the hydraulic oil);

- release lever broken, incorrectly adjusted or seized;
- release bearing defective;
- clutch plate seized on the gearbox input shaft splines;
- clutch shaft spigot bearing (used where the gearbox input shaft is located in the flywheel) is seized in the flywheel;
- clutch linings are cracked or buckled;
- oil or grease on the linings.

Vibration or judder

This occurs when the drive is gradually taken up. It can be caused by one of the following faults:

- cracked or distorted pressure plate;
- loose or badly worn linings;
- loose or protruding rivets;
- misalignment of the gearbox with the engine, caused by distortion of the bell housing or the dowels not being located correctly;
- engine or gearbox mountings defective;
- excessive backlash (play) in the universal joints (UJs) of the drive/prop shafts, (bolts may be loose on the UJ flange);
- flywheel bolts loose on the crankshaft flange;
- defective release mechanism.

Other types of clutches
Multi-plate clutches

Used in certain specialized applications, e.g. agricultural tractors, where a continuous drive is required at the power take-off (PTO); heavy-duty vehicles such as those used by the armed forces; and earth moving and large plant vehicles. The most common application that the motor mechanic is likely to see is that of the motor bike where a small multi-plate clutch can transmit a large amount of power due to the large surface area of all the clutch plates added together.

Fluid clutches
The fluid flywheel

The fluid flywheel is a form of automatic clutch. It consists of two main parts, the pumping element that is attached to the engine crankshaft and the turbine element that is attached to the gearbox input shaft. Both parts are encased in a container that carries the oil that allows the unit to function. Both the pumping element and the turbine element contain a number of segments as shown in Fig. 5.5(a). As the engine crankshaft rotates, oil is directed towards the outer edges of the pumping elements and from there into the outer edges of the turbine element, as shown in Fig. 5.5(b). As the engine speed increases, the force exerted by the oil issuing from the pumping element causes the turbine element to rotate thus transmitting the drive from the engine to the gearbox, as shown in Fig. 5.5(c).

The fluid flywheel permits the drive to be taken up smoothly as engine speed is gradually increased. As the speed of the turbine element begins to reach the speed of the pumping element, the oil returning from the turbine begins to match the speed of the oil leaving the pumping element. This causes a small amount of slip to occur, which is a feature of fluid couplings and vehicle design allows for this. Fluid flywheels do not multiply torque, merely acting as automatic clutches.

The torque convertor

The torque convertor is also a form of automatic clutch with a third element, known as the stator, being added. The stator is mounted on an extension of the gearbox casing and acts to re-direct oil flow from the pumping element to the turbine and, in doing so, increases the engine torque being transmitted from the crankshaft to the gearbox. The stator is also mounted on a one way clutch, similar in operation to a bicycle free-wheel, allowing the stator to rotate with the other two elements of the torque convertor, as the speed of the turbine element begins to reach the same speed as the pumping element. When all three elements of the convertor are rotating together, the unit acts as a fluid flywheel and there is no torque multiplication. When the torque conversion effect is fully

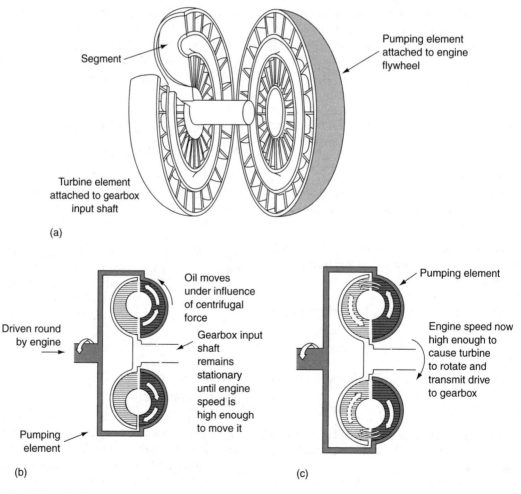

Fig. 5.5 A fluid flywheel

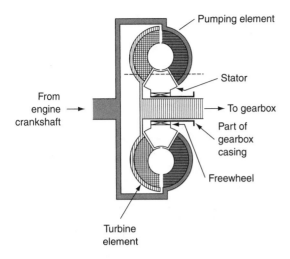

Fig. 5.6 Torque convertor

operational, the torque multiplication is approximately 2:1.

Stalling speed

The stalling speed of a torque convertor is the maximum speed that the engine can reach when the transmission is engaged and the vehicle is prevented from moving. At stalling speed, the energy from the engine is absorbed in a shearing action of the oil which, in the process, becomes heated. A stalling speed test is one of the tests that can be performed to check the working condition of an automatic transmission system.

Learning tasks

1. By reference to manuals and technology books, draw up a typical load/deflection graph for coil and diaphragm spring clutches. From the graphs state the load for both new and worn linings in the engaged and disengaged positions. From your results can you identify a reason why the diaphragm spring is now the most common type of clutch used in light vehicles?
2. Remove a clutch from an engine and inspect the main components. Note the amount of wear on the following units:
   (a) the flywheel
   (b) the pressure plate
   (c) the release bearing
   (d) the driven plate, including the linings
   (e) the operating mechanism.
   Draw up your recommendations giving reasons for your findings.
3. What are the service recommendations for the clutch? Produce a simple service schedule that would enable your recommendations to be recorded.
4. Describe the *symptom* (how would the driver notice) and the *cause* (what caused the problem to occur) of the following:
   (a) clutch slip
   (b) clutch drag
   (c) clutch judder

5.2 The gearbox

The purpose of the gearbox is:

- to multiply the torque being transmitted by the engine;
- to provide a means of reversing the vehicle;
- to provide a permanent position for neutral.

The gearbox can be described as a collection of levers (the gears) that are arranged so as to multiply the turning effort (the torque) of the engine to suit the needs of the vehicle.

Transmission layouts

The position of the engine in relation to the rest of the transmission is shown in Fig. 5.7. The advantages of each layout are taken into consideration by the manufacturer when designing the vehicle.

Types of gears used in the gearbox

Several types of gear teeth shape are used in the modern gearbox depending on the application. All are called **spur** teeth.

Straight cut teeth

Shown in Fig. 5.8 these have a rolling/sliding action as they mesh together. They are rather noisy when running together and generate high forces and temperatures between the gear teeth which tends to force them apart when high loads are transmitted. They are almost always used for reverse gear. They can also be found in the gearboxes of some racing cars.

Helical cut teeth

These have the teeth cut at an angle across the gear producing a sideways force which is normally taken by thrust washers on the shaft. The

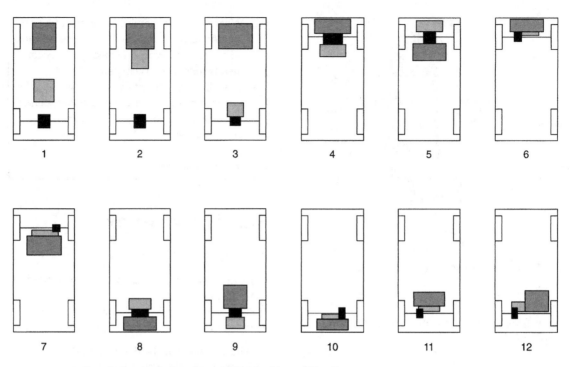

1. A centrally mounted gearbox, used on the Morgan Plus Four
2. The most common layout for RWD vehicles with the gearbox mounted directly on the engine
3. For better weight distribution, the gearbox is mounted with the differential; this arrangement is used on the Alfretta and Ferrari Daytona
4. This layout has been used by Citroen with the engine at the extreme front thus allowing ample passenger space
5. This rather unusual layout has been used by Renault
6. This is the layout of Issigonis in the Mini. This also gives lots of passenger space
7. This variation of the FWD is used by Peugeot
8. This rear engine layout is used on the famous VW air-cooled engine cars
9. This mid-engined layout is used on all Formula One cars and also the Tomaso Pantera
10. No cars now use this rear transverse mounted engine
11. This is the arrangement used by the Ferrari Dino and Boxer
12. Another unusual layout this time employed on the Lamborghini Urraco

Fig. 5.7 Transmission arrangements showing the position of the engine, gearbox and final drive

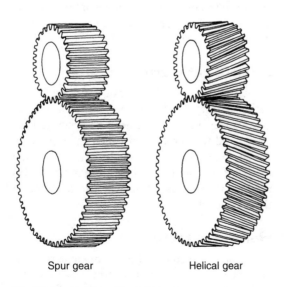

Spur gear   Helical gear

Fig. 5.8 Gear teeth profile straight cut

helical shape is used on the forward gears as they are quieter running, have more than one tooth in mesh at any one time and can therefore carry a greater load. **Double helical** gears are sometimes used to overcome the side force and can transmit even greater loads but they are more expensive to produce.

Sliding mesh gearbox

In this type of gearbox, shown in Fig. 5.9, the **mainshaft gearwheels** slide on splines in the direction of the arrows to mesh with the appropriate layshaft gear for first, second and third gears. Top gear is a **dog clutch** connection joining input and output shafts to give a 1:1 ratio. For reverse the 'compound' idler gear slides along the shaft to mesh with mainshaft and layshaft first

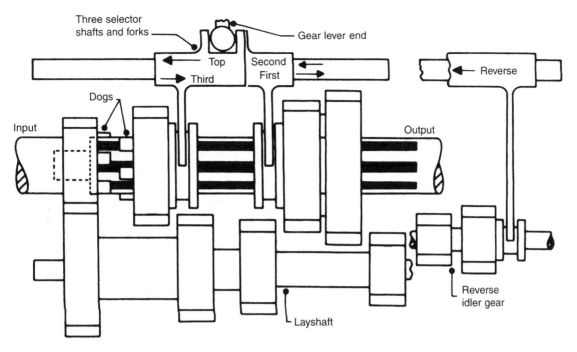

Fig. 5.9 Three speed sliding mesh gearbox

gearwheels. To enable all this to happen all the gear teeth are cut straight. This type is no longer used in the modern motor vehicle, as the gears are difficult to change without some noise occuring.

Constant mesh gearbox

In a constant mesh gearbox (Fig. 5.10) the main-shaft gearwheels rotate on bushes and are in constant mesh with the **layshaft gears**. The appropriate gearwheel may then be locked to the output shaft and made to revolve with it by a dog clutch splined to

the shaft and slid along it by the same sort of selector fork and collar as was used in the sliding mesh gearbox. This has the following advantages:

- it allows the use of helical gears;
- it is quieter in operation than spur-type teeth;
- it is stronger than the spur type as there is more than one tooth in engagement at any one time;
- it makes gear changing easier as the gearwheels have to be rotating at the same speed before engagement can take place (this is achieved through the use of a **synchromesh** device).

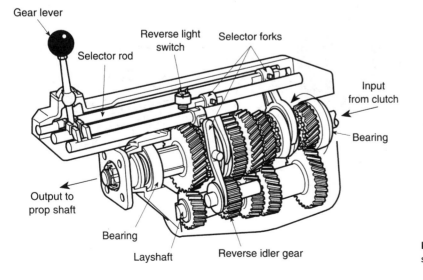

Fig. 5.10 The constant mesh four speed gearbox

Synchromesh gearboxes

If the dog clutches could be replaced by some kind of friction clutch, perfect synchronization of the output shaft and the selected gearwheel could be achieved rapidly and smoothly. A friction clutch strong enough to transmit full torque would be far too big and heavy, but a small clutch that had to do no more than overcome the inertia of a freely rotating gearwheel and

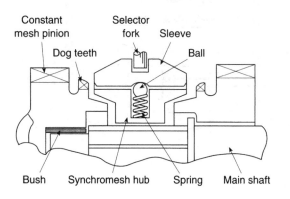

Fig. 5.11 Constant load synchromesh gearbox

layshaft assembly could be quite small. The two synchromesh devices used in the gearbox are the **constant load** (Fig. 5.11) and the **baulk ring** (Fig. 5.12).

Constant load synchromesh

The **female cone** of the clutch is formed in the hub, which has internal and external splines. A series of spring loaded balls are carried in radial holes in the hub, and these push outwards into a groove machined in the sleeve. Movement of the selector fork carries the sleeve and hub on splines along the mainshaft towards the gear selected and allows the cones to contact. At this point, the friction between the cones adjusts the speed of the gearwheel to suit the hub and mainshaft. Extra pressure on the lever will allow the sleeve to over-ride the spring loaded balls and positively engage with the dogs on the gear. If the gear change is rushed there will not be enough time for the gearwheels to synchronize their speeds and the change will be noisy. The time taken for the speeds to equalize is governed

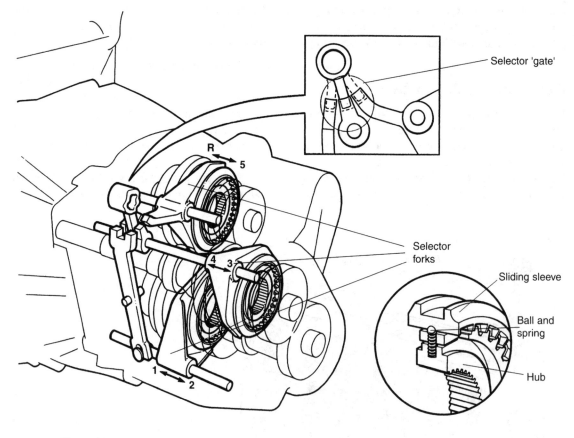

1–5 – First to fifth gears
R – Reverse gear

Fig. 5.12 Baulk ring synchromesh

by the frictional force at the cone faces, which in turn is governed by:

- the total spring strength;
- the depth of the groove in the sleeve;
- the angle of the cones;
- the coefficient of friction between the cones.

> *Note*: If any of these factors are reduced, due to mechanical defects, synchronization will take longer and noise will probably be heard. Because of this the constant load will only be in older types of gearboxes. The baulk ring synchromesh has almost universally taken its place and is the most common to be found in the modern gearbox.

Baulk ring synchromesh

This is a development of the constant load type. The main advantages of the baulk ring synchromesh over the constant load synchromesh are:

- the synchronization of the dog clutches is quicker, thus allowing a quicker gear change;
- the dog clutches cannot engage until their speeds are equal and therefore noise will be eliminated.

When the gear selector moves the outer sleeve the baulk ring cone contacts the gearwheel cone. Rotation of the baulk ring is limited to the clearance between the shifting plate and the slot in the baulk ring; this clearance is exactly half the pitch of the dog teeth. This limited rotation of the baulk ring therefore prevents the outer hub sleeve from meshing with the gearwheel dog clutch. When the hub and gearwheels are synchronized (running at the same speed) the baulk ring centralizes in the slot to allow dog clutch engagement between the sleeve and the gear dogs. If the driver applies a greater force on the gear lever the outer sleeve, which is unable to slide over the baulk ring, will apply a greater force to the baulk ring thus increasing the frictional force which, in turn, speeds up the synchronization of the dog clutch members.

Path of the drive through the gearbox

Figure 5.13 shows the path taken by the drive through the four forward gears and reverse of a trans-axle (front engine FWD) gearbox.

Gear selection mechanism

This allows the driver to select and engage a gear. It is achieved by the use of a gear lever which engages in the **selector gate**, and a number of **selector rods** with forks. The rods slide in the gearbox housing and the forks attached to the rods locate in grooves in the outer part of the sliding gear hub (the **sleeve**). A means of locking

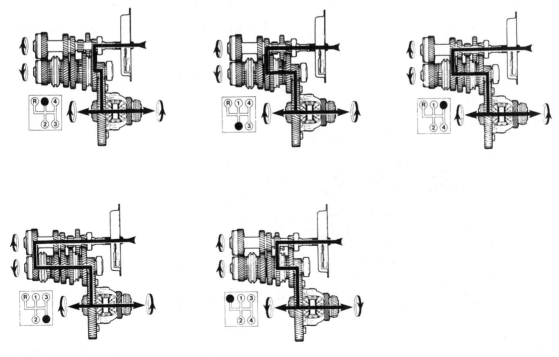

Fig. 5.13 Path of the drive through a five speed gearbox

the gear in position and also preventing the selection of more than one gear at once is provided in the selector arrangement.

Interlocking mechanism

One type of locking mechanism is shown in Fig. 5.14. The shift locking plate rotates with the guide shaft to prevent the movement of the other gear shift rods, ensuring that only one gear is engaged at any one time.

Another arrangement uses plungers between the selector rods (Fig. 5.15). The caliper plate interlock operates in a similar way to the shift locking plate, preventing movement of the selector rods when a gear is engaged.

Direct acting gear lever

In the direct acting arrangement, the gear lever is mounted directly in the top of the gearbox. As

Fig. 5.14 Locking mechanism, A guide spring, B shift locking plate

the lever is operated by the driver, it pivots on a ball and socket near the base of the lever and acts directly on the selector gate.

Remote acting gear lever

In the remote control type, the gear lever operates through a number of rods and levers before it moves the selector in the gearbox. This is shown in Fig. 5.10. The gear lever may be mounted on the steering column (**column gear change**) or positioned between the driver's seat and passenger's seat, for convenience. This type is the most commonly used as most vehicles are of the transverse engine, trans-axle layout. Its main advantage is that it allows the gearbox to be positioned in what is the most advantageous position for weight distribution and drive line layout, as well as being the most comfortable position for the driver.

Trans-axle

This is an American term used where the transmission and axle are mounted together in one unit, as can be seen in most front engine FWD vehicles. The trans-axle may also be used where the vehicle is a front engine with RWD and the gearbox is located with the rear axle to give better weight distribution together with more room in the passenger compartment.

Learning tasks

1. Draw up a service schedule for the gearbox. In it identify the following:
   (a) type of oil used and the quantity required
   (b) position of drain and level plugs

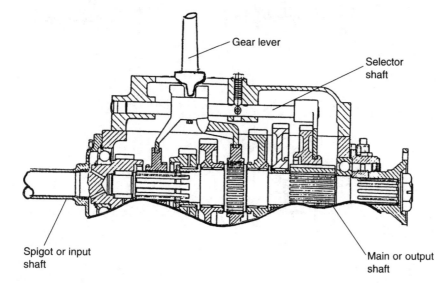

Gear lever

Selector shaft

Spigot or input shaft

Main or output shaft

Fig. 5.15 Direct acting gear lever

(c) security and soundness of gearbox mountings

(d) oil leaks and remedial work required

(e) noise coming from gearbox when in operation

(f) ease of gear engagement and play in gear linkage.

2. Draw up a simple checklist/worksheet for the inspection of a dismantled gearbox. Include the inspection of gear teeth, shafts, bearings, selector mechanism, operation of interlocking mechanism, oil seals and impurities in the gear oil. Ensure that you have space to write your recommendations.

3. Remove and dismantle either a FWD or RWD gearbox and complete your worksheet as you inspect the components. Produce your recommendations for your supervisor or the customer.

4. (a) Give two symptoms of a worn output shaft bearing.

   (b) Name one simple test procedure for identifying the fault.

   (c) How should the fault be put right?

5. Name one simple test to identify the difference between a clutch with drag (difficulty in engaging gear) and a worn synchromesh device.

6. On checking the gear oil it was found to be contaminated with silvery looking bits. What would you suspect the fault to be and how should it be rectified?

5.3 Epicyclic gear train

Purpose

This is used to produce different gear ratios within the **overdrive** and **automatic** transmissions. The output from the engine or gearbox is now the input into the **planetary set** of gears. The four main components in the simple epicyclic gear train are:

- the **sun** gear
- the **annulus**
- the **planet** gears
- the **planet gear carrier**

These are shown in Fig. 5.16.

All the gears are helical in design and are always in **constant mesh**. To select different gear ratios the sun, annulus or planet carrier is held. This will give a number of forward as well as reverse gears.

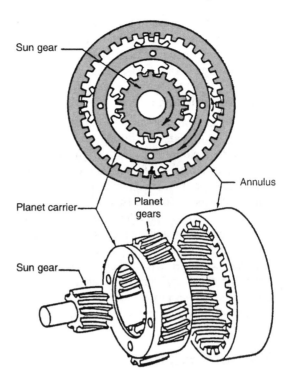

Fig. 5.16 Simple epicyclic gear train

Method of operation

A simple epicyclic gear train can perform the following functions in a motor vehicle gearbox.

Neutral condition

If any one member is rotated and the remaining two members are allowed to run free, no drive will be transmitted through the gear train and the whole unit will merely idle in neutral.

Direct drive

If any two of the three members are locked together, the third member will be carried round by the teeth of the two locked members at the same speed and in the same direction. The gear train has in effect become solid, and therefore acts in the same manner as a direct mechanical coupling with no increase in torque.

Forwards reduction

If the annulus gear is held stationary and the sun gear is rotated, the planet pinions will be compelled to 'walk' in the annulus. The planet carrier will therefore rotate in the same direction as the sun gear, but at a much reduced speed and with a corresponding increase in torque. Conversely, if the sun gear is held stationary and the annulus is rotated, the planet pinions will be compelled to walk round the sun gear. The planet carrier will therefore rotate in the same direction as the annulus gear, but at a less reduced speed and with not so great an increase in torque.

Reverse reduction

If the planet carrier is held stationary and the sun gear is rotated, the 'idling' planet pinions will rotate the annulus gear in the opposite direction. The speed of the annulus gear will be reduced and its torque increased.

It is not possible to have two forward with one reverse in a single epicyclic gear train as one member must always be connected to the output and one to the input. This gives one forward gear and one reverse gear. But when two simple epicyclic gear trains are connected together in series it is possible to gain more gear ratios and this is the arrangement used in the automatic gearbox.

5.4 Overdrives

An overdrive (means to drive faster) exists when the input shaft or gear rotates more slowly than the output shaft or gear. A typical overdrive arrangement is shown in Fig. 5.17.

Purpose

It allows the engine speed to be reduced for a given road speed. This has the effect of reducing fuel consumption, wear and engine noise.

The main units in an overdrive are a simple epicyclic gear train, a hydraulically operated friction clutch and a uni-directional (one way) clutch.

Method of operation

When **direct drive** is selected (overdrive not operating) the springs move the cone clutch along the splined sun gear into contact with the annulus thus locking the epicyclic gear train. The drive is transmitted through the one-way clutch and the cone friction clutch transmits the drive on overrun and in reverse.

When overdrive is engaged (Fig. 5.18), oil under pressure from a pump driven by the input shaft acts on pistons which force the **cone clutch** against spring pressure into contact with the casing. This prevents the sun from rotating. The planet carrier, which is the drive input, causes the planet wheels to rotate around the locked sun gear. This action will drive the annulus faster than the input shaft (planet carrier) to give overdrive. During this condition the one-way clutch is free-wheeling. The overdrive may be engaged without the use of the main clutch, thus making the operation quick and easy.

The uni-directional clutch will transmit a drive when rotated in one direction and will 'freewheel' when turned in the opposite direction. The one-way clutch shown in Fig. 5.19(a) has rollers that move up the ramps and the wedging action provides a drive from the inner to the outer member. If the outer member rotates faster, then the rollers will move down

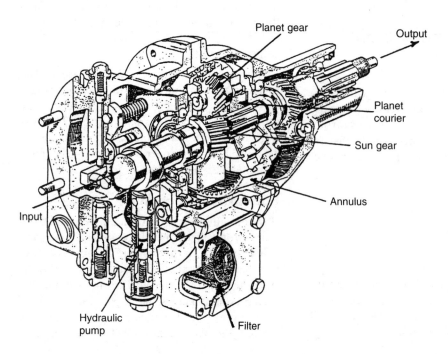

Fig. 5.17 The Laycock-de-Normanville overdrive unit as fitted to the main output shaft of the gearbox

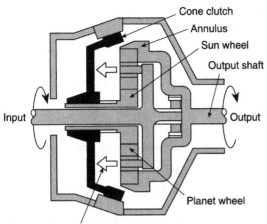

Fig. 5.18 Overdrive engaged

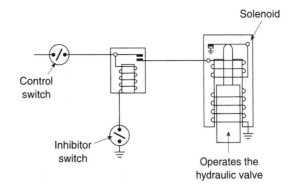

Fig. 5.20 Wiring diagram for an overdrive unit

operation when top or third gear is engaged; at all other times, even if the switch is on, the electrical system will not operate the solenoid to engage the overdrive. A wiring diagram is given in Fig. 5.20.

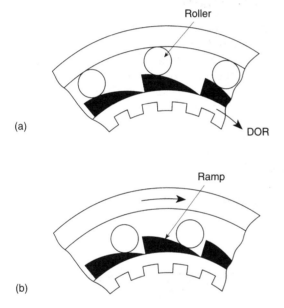

Fig. 5.19 The one-way clutch (a) drive – inner rotates faster than outer, (b) free-wheel – outer rotates faster than inner

---

> ### Learning tasks
>
> 1. What type of oil should be used to lubricate the overdrive unit and why?
> 2. Draw up a simple service schedule for the overdrive unit. Include checks that should be made on the electrical circuit.
> 3. How should the overdrive be checked for correct operation? Name any test equipment that you would use.
> 4. How would the driver notice if the one-way clutch was slipping? What test would you undertake to identify the fault from a slipping engine clutch?

the ramps to the wider section and no drive will be passed to the inner member (Fig. 5.19(b)).

Another type of one-way clutch is the **sprag clutch**. The same sort of wedging action to give a drive takes place.

The main disadvantages of an overdrive are:

- increase in cost;
- additional weight carried;
- extra space required to accommodate the unit;
- added complication.

Electrical circuit

The overdrive is operated by a switch usually situated in the top of the gear lever. The circuit has an inhibitor switch so that the overdrive is only in

5.5 Propeller shafts and drive shafts

Prop shafts

On a vehicle of conventional layout (front engine RWD), the purpose of the **prop shaft** is to allow a drive to be transmitted through a varying angle from the gearbox to the final drive at the rear axle. The usual type of layout used is the **Hotchkiss open shaft drive** arrangement as in Fig. 5.21.

Universal joints (UJs) are used to allow for the variation in the drive angle as the road wheels rise and fall. As the rear axle swings up and down the distance between the axle and the gearbox changes. It is therefore necessary to allow for this change in length by the use of a **sliding joint** in the prop shaft assembly.

Figure 5.22 shows how the prop shaft varies in length and that either the sliding joint can be

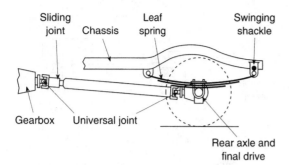

Fig. 5.21 The Hotchkiss open shaft drive

positioned on the shaft (Fig. 5.21) or the shaft may slide in and out of the gearbox on the main output shaft. The shaft will also vary in length as the drive is taken up and the suspension twists clockwise, or as the brakes are applied and the suspension twists anti-clockwise as can be seen in Fig. 5.22.

The prop shaft is made **tubular** to:

- reduce weight;
- reduce out-of-balance which would cause **whirling** as the shaft rotates;

- make it similar in strength to a solid shaft and easier to balance.

Because of the problems of weight and 'whirling' the shaft is often divided into two and a centre bearing is fitted. This arrangement is called a **two-piece prop shaft** (Fig. 5.23). The centre bearing is chassis mounted and provides support for the shaft thus effectively reducing its length. The advantages of this are:

- vibration is reduced;
- smaller diameter shafts can be used;
- it is more suitable for high prop shaft speeds;
- the rear axle can be mounted further back from the engine.

Drive shafts

In the front engine FWD, drive shafts are used instead of a prop shaft. These are normally fitted with constant-velocity (CV) joints to enable the drive to pass through a greater angle without the problems associated with universal joints. Figure 5.24 illustrates this.

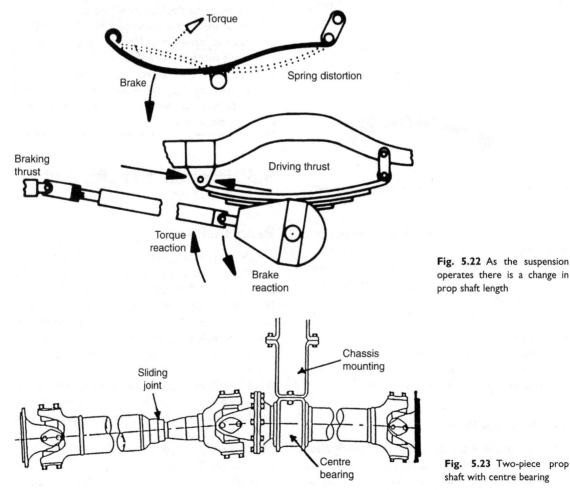

Fig. 5.22 As the suspension operates there is a change in prop shaft length

Fig. 5.23 Two-piece prop shaft with centre bearing

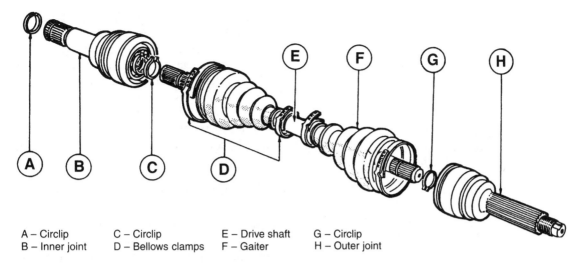

A – Circlip   C – Circlip   E – Drive shaft   G – Circlip
B – Inner joint   D – Bellows clamps   F – Gaiter   H – Outer joint

Fig. 5.24 Drive shaft assembly and CV joints as used on front engine FWD

5.6 Constant-velocity universal joints

Variation in speed

When a single Hooke-type joint is used to transmit a drive through an angle, you will find that the output shaft does not rotate at a constant speed. During the first 90° of its motion, the shaft will travel faster, and on the second 90° slower. Correction of this speed variation is normally done by a second coupling, which must be set so that when the front coupling increases the speed the rear coupling decreases the speed (Fig. 5.25). When the two Hooke-type couplings are aligned correctly, the yoke at each end of the propeller shaft is fitted in the same plane. When fitted to a vehicle, the two or more joints cannot always be operated such that the two angles remain equal. To allow for angle variation and shaft wind-up, especially on larger vehicles such as LGVs, the joints are sometimes set slightly out of phase.

Apart from the Hooke joint a number of other types are used. The doughnut and the Layrub coupling are similar in that they are made from

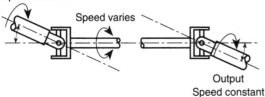

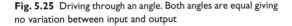

Fig. 5.25 Driving through an angle. Both angles are equal giving no variation between input and output

steel and rubber bonded together. So long as the angle of drive is not too large these will give satisfactory service with the added advantage of not requiring any lubrication.

Constant-velocity (CV) joints

A single Hooke-type joint driving through a comparatively large angle would give severe vibration due to the inertia effect produced by the speed variation. This variation increases as the drive angle enlarges, for example, the percentage variation in speed is about nine times greater at 30 mph than at 10 mph. Any drive shaft subjected to speed changes during one revolution will cause vibration, but provided the drive angle is small and the shaft is light in weight, then drive-line elasticity will normally overcome the problem. Large drive angles, as used in FWD vehicles, need special compact joints which maintain a constant speed when driving through an angle – these joints are called CV joints.

Tracta CV joints

The need for CV joints was discovered in 1926 when the first FWD vehicle was made in France by Fenallie and Gregoire – the Tracta (traction-avant) car. When the second car was made, the drive-line incorporated CV joints, and the type used on that car is now known as a Tracta joint. This type is now made by Girling and details are shown in Fig. 5.26.

Reference to the figure shows that the operating principle is similar to two Hooke-type joints: the angles are always kept constant and the yokes are set in the same place.

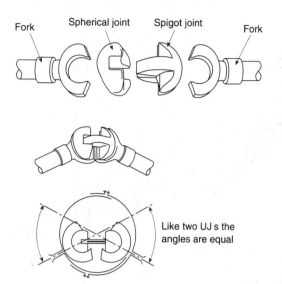

Fig. 5.26 The Tracta joint

The joint is capable of transmitting a drive through a maximum angle of about 40° and its strong construction makes it suitable for agricultural and military vehicles, but the friction of the sliding surfaces makes it rather inefficient. The joint is also too large to be accommodated in a small FWD vehicle with wheel diameters of less than approximately 15 inches (38 cm).

Rzeppa-type CV joint

This joint was patented by A H Rzeppa (pronounced Zeppa) in America in 1935. The development of this type is in common use today; it is called a Birfield joint and is made by Hardy Spicer. Figure 5.27 shows the construction of a Birfield joint.

Constant velocity is achieved if the device (**steel balls** in this case) connecting the driving shaft to the driven shaft rotates in a plane that bisects the angle of drive: the Birfield joint achieves this condition. Drive from the inner to outer race is by means of longitudinal, elliptical grooves which hold a series of steel balls (normally six) that are held in the bisecting plane by a cage. The balls are made to take up their correct positions by offsetting the centres of the radii for inner and outer grooves.

A Birfield joint has a maximum angle of about 45°, but this angle is far too large for continuous operation because of the heat generated by the balls. Lubrication is by grease – the appropriate quantity is packed in the joint 'for life' and a synthetic rubber boot seals the unit.

Plunge joints

These universal joints allow a shaft to alter its length due to the up and down movement of the suspension. A Birfield joint has been developed to give a plunge action (in and out) as required. It will be seen that the grooves holding the balls in place are straight instead of curved; this allows the shaft length to vary. Positioning of the balls in the bisecting plane is performed by the cage. Since about 20° is about the maximum drive angle, this type of joint is positioned at the engine end of the drive shaft used for FWD vehicles.

Weiss CV joint

This type was patented in America in 1923 by Weiss and later developed by Bendix. it is now

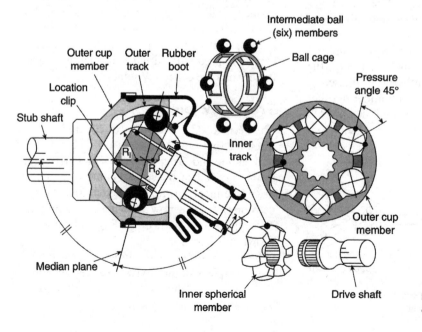

Fig. 5.27 Birfield Rzeppa-type constant-velocity joint

known as the Bendix–Weiss and produced in this country by Dunlop. The two forks have grooves cut in their sides to form tracks for the steel balls; there are four tracks so four balls are used to transmit rotary motion. A fifth ball is placed in the centre of and locates the forks, resisting any inward force placed on the drive shaft. The driving balls work in compression, so two balls take the forward drive and the other two operate when reverse drive is engaged. The complete joint is contained in a housing filled with grease. The maximum angle of drive is approximately 35°. Constant velocity is achieved in a manner similar to the Rzeppa joint – the balls always take up a position in a plane that bisects the angle of drive.

Learning tasks

1. How would the driver notice if there was severe wear in the UJ of the prop shaft?
2. Draw up a schedule for removing and refitting the CV joint on a front engine FWD vehicle. Make a special note of any safety precautions that should be observed.
3. What checks would be made on the drive-line of a front engine RWD or FWD vehicle when undertaking an MOT?

---

4. What should you look out for when reassembling a divided-type prop shaft. What would be the result if these were ignored?
5. If the vehicle on being driven round a corner made a regular clicking noise, what component in the drive-line would you suspect as being worn and why? How would you put the fault right?

5.7 Axles and axle casings

Axles

Axles may be divided into two types, the **live axle** and the **dead axle**. The difference is that a dead axle only supports the vehicle and its load, whereas a live axle not only supports the vehicle and its load but also contains the drive.

Axle casings

Three main types of casings are in use.

- **Banjo** Normally built up from steel pressings and welded together. The crown wheel assembly is mounted in a malleable housing which is bolted to the axle. This is the most common type fitted to light vehicles and is shown in Fig. 5.28.

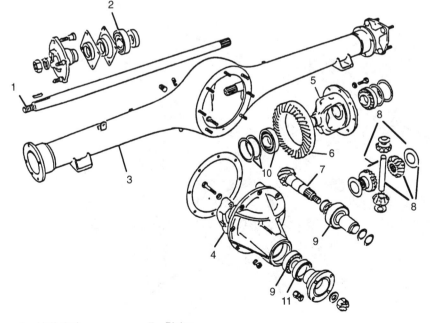

1 – Half-shaft
2 – Half-shaft bearing
3 – Axle casing
4 – Final drive casing
5 – Differential housing
6 – Crown wheel
7 – Pinion
8 – Final drive assembly
9 – Pinion bearings
10 – Differential housing bearings
11 – Pinion oil seal

Fig. 5.28 Banjo axle casing

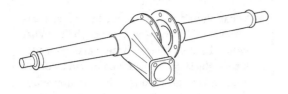

Fig. 5.29 Split axle casing

- **Split casing**   These are formed in two halves and bolted together to contain the final drive and differential (see Fig. 5.29). They are used more in heavy vehicle applications as they are more rigid in construction and can withstand heavy loads.
- **Carrier**   This is more rigid than the banjo casing. The final drive assembly is mounted directly into the axle and the axle tubes are pressed into the central housing and welded into place. A cover is fitted to the rear of the housing to allow for access and repair. It is used in off-road 4 × 4 and LGVs and is shown in Fig. 5.30.

5.8 Final drive

The purpose of the final drive is two-fold: first to transmit the drive through an angle of 90°, and second to provide a permanent gear reduction and therefore a torque increase, usually about 4:1 in most light vehicles. Several types of bevel gearing have been used in the drive between the pinion and the crown wheel. These are the straight bevel, the spiral bevel, the hypoid and the worm and wheel.

Straight bevel

The straight bevel (Fig. 5.31(a)) is not designed for continuous heavy-duty high-speed use as in the normal crown wheel and pinion. This is because only one tooth is in mesh at any one time. The gearing is noisy and suffers from a high rate of wear, but the design is the basis from which the final drives are formed.

Spiral bevel

Shown in Fig. 5.31(b) this has more than one tooth in mesh at a time. It is quieter, smoother, can operate at much higher loadings and is used where the shaft operates centre-line as the crown wheel.

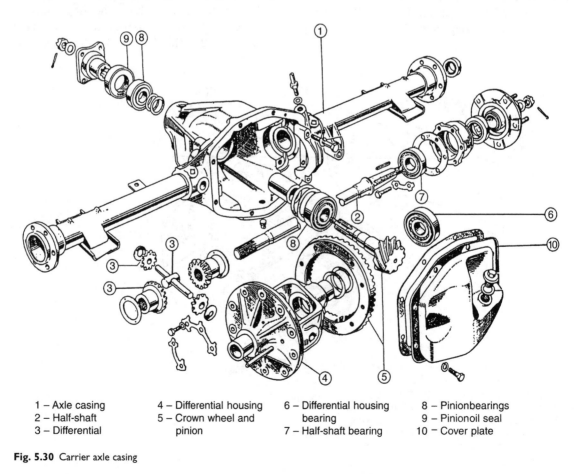

1 – Axle casing	4 – Differential housing	6 – Differential housing	8 – Pinionbearings
2 – Half-shaft	5 – Crown wheel and	bearing	9 – Pinionoil seal
3 – Differential	pinion	7 – Half-shaft bearing	10 – Cover plate

Fig. 5.30 Carrier axle casing

(a)

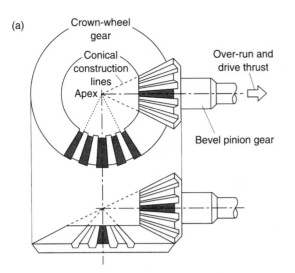

Fig. 5.31(a) Geometry of straight-tooth bevel-gear crown wheel and pinion

(b)

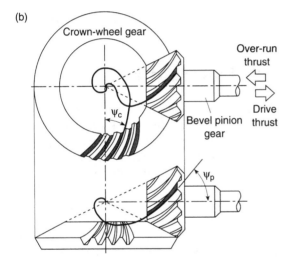

Fig. 5.31(b) Geometry of spiral-tooth bevel-gear crown wheel and pinion

(c)

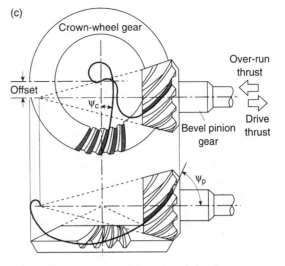

Fig. 5.31(c) Geometry of hypoid-tooth bevel-gear crown wheel and pinion

(d)

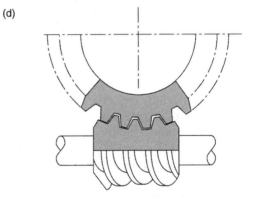

Fig. 5.31(d) Hourglass worm and worm-wheel final drive

Hypoid bevel

This is similar to the spiral bevel in action but the shaft operates at a lower level than the crown wheel. This has the advantage of the prop shaft being lower and not intruding into the floor space of the passenger compartment. It is stronger and is the quietest in operation. Extreme pressure (EP) oil must be used as a lubricant because of the frictional forces and high loadings generated between the gear teeth (a disadvantage of this arrangement). The materials used in their manufacture are a nickel–chrome alloy which is carburized after machining and case-hardened to give long life. Figure 5.31(c) shows a hypoid bevel.

Worm and wheel

The worm in this arrangement, shown in Fig. 5.31(d), is driven by the prop shaft and looks like a very coarse screw thread. The wheel has a toothed outer edge with which the screw thread of the worm meshes. There are a number of advantages with the worm and wheel the two main ones being: first it provides a very large gear reduction and therefore a large torque increase in one gear set; and second it gives a high prop shaft, this is especially useful for off-road vehicles and LGVs.

Servicing and adjustments

Excessive wear will result on the gear teeth if the backlash between the crown wheel and pinion is set incorrectly.

To adjust the backlash, shims must be added or removed to obtain the correct readings. The crown wheel teeth shown in Fig. 5.32 give the different

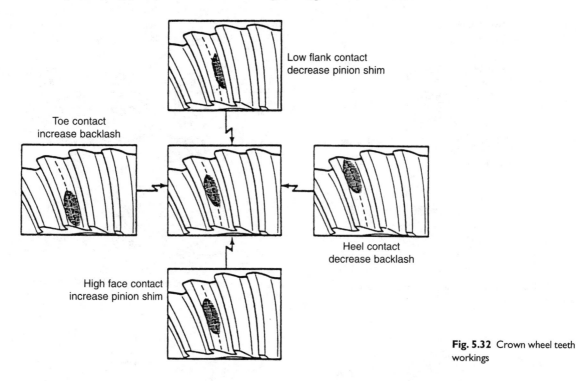

Low flank contact
decrease pinion shim

Toe contact
increase backlash

Heel contact
decrease backlash

High face contact
increase pinion shim

Fig. 5.32 Crown wheel teeth workings

markings, the centre diagram showing where the marks should occur. To obtain the markings on the gear teeth a little engineers' blue is placed on the pinion teeth and the pinion rotated until the crown wheel has completed one revolution.

A dial test indicator (DTI) should be used to check the crown wheel and pinion backlash (this is the amount of play between the gear teeth). It should be positioned as shown in Fig. 5.33.

The task of setting up the crown wheel and pinion is quite skilled. This is why the exchange unit comes complete with the differential and all the settings are done by the manufacturer. All the mechanic has to do when changing the final drive is fit the unit to the axle.

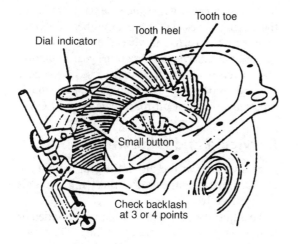

Dial indicator

Tooth heel

Tooth toe

Small button

Check backlash
at 3 or 4 points

Fig. 5.33 Position of the DTI

5.9 Differential

The purpose of the differential is to allow the wheels to rotate at different speeds, whilst still transmitting an equal turning force (torque) to both wheels. The half-shafts are splined to the sun gear, the planets transmitting the motion from one sun gear to the other when the vehicle is turning a corner. This is shown in Fig. 5.34.

When both drive shafts are travelling at the same speed the planet gears orbit (rotate) with the sun gears but do not rotate on their shafts. The whole unit acts as a solid drive. See Fig. 5.35.

If one shaft is stopped the planet gears turn on their shafts, orbiting round the stationary sun gear and driving the other sun gear but twice as fast.

In the final drive the differential is in a housing (sometimes called a cage) to which the crown wheel is bolted. When the car is travelling in a straight line the planet gears orbit, but do not rotate on their shafts, and the unit drives both half-shafts at the same speed as the crown wheel and with the same turning force.

When turning a corner, the sun gear on the inner half-shaft turns more slowly than the crown wheel, the outer half-shaft, driven by the other sun gear, turns correspondingly faster. The crown wheel turns at the average of the half-shaft speeds.

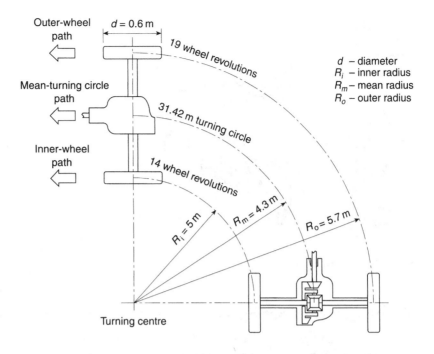

Fig. 5.34 Illustration to show the distance travelled by each wheel. The differential allows for this difference

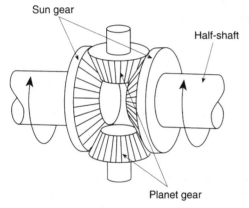

Fig. 5.35 Operation of the differential when the vehicle is travelling in a straight line

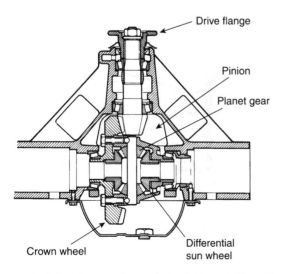

Fig. 5.36 Final drive gear. Crown wheel and pinion and differential

A sectioned view of the complete final drive and differential arrangement is shown in Fig. 5.36. Both the crown wheel and pinion are held in the housing by taper roller bearings. The cross pin acts as the drive for and the shaft on which the planet gears rotate.

5.10 Rear hub bearing and axle shaft arrangements

Three layouts are now commonly used and are subjected to various forces acting on the axle shaft, for example, **twisting** when accelerating or braking, **bending** due to cornering and some of the load and **shear** due to the vertical load imposed on the axle shaft.

Semi-floating axle

This hub arrangement, shown in Fig. 5.37(a), is used on many small light vehicles. The road wheel is attached to the **half-shaft** rather than the hub and the bearing is fitted between the half-shaft and the **axle casing**. Therefore, if a break should occur in the half-shaft inside the axle casing the wheel will tilt and very often become detached due to the lack of support at the inner end of the half-shaft. As can be seen the shearing point is positioned between the shaft and the axle casing, and the shaft is subject to shearing, bending and twisting forces.

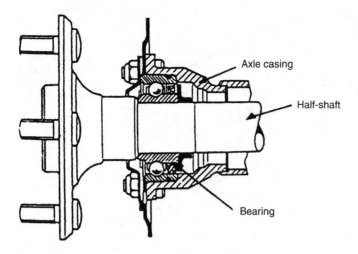

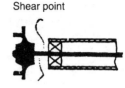

Fig. 5.37(a) Semi-floating axle

Three-quarter floating axle

Used on cars and light vans, the main difference is the position of the bearing, now shown on the outside of the axle between the hub and the outside of the axle casing. The casing therefore takes most of the weight of the vehicle and its load. The shaft is still subject to twisting and bending forces. This type is shown in Fig. 5.37(b).

Fully floating axle

In this arrangement (Fig. 5.37(c)) the bearings (normally taper roller) are fitted between the hub and the outside of the axle casing. In this way the only force to which the axle shaft is subject is a twisting action.

Lubrication

To prevent oil from leaking between the shaft and housing, lip-type oil seals are pressed into the housing. In some cases an oil slinger washer is located just inside and next to the bearing to prevent flooding of the seal. Pressure build-up due to temperature changes during operation is prevented by a breather in the top of the axle casing.

The action of the gear teeth meshing together tends to break down the oil film on the gear teeth (due to the very high pressures and forces created between the gear teeth). The oil used in the axle therefore must prevent a metal-to-metal contact from taking place. An EP additive in the oil reacts with the metal surfaces at high

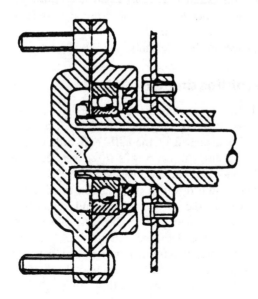

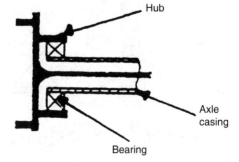

Fig. 5.37(b) Three-quarter floating axle

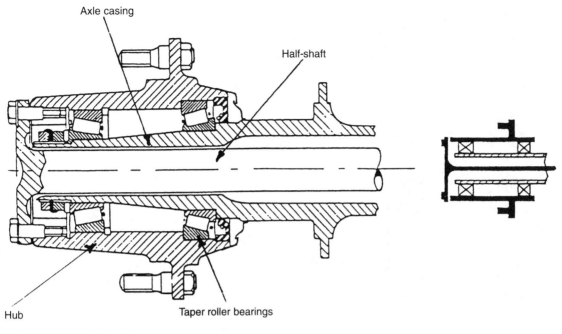

Axle casing

Half-shaft

Hub

Taper roller bearings

Fig. 5.37(c) Fully floating axle

temperatures to produce a low-friction film or coating on the teeth. This prevents scuffing and rapid wear from taking place; in effect the additive becomes the oil. It is essential that no other type of oil is used as this could cause rapid wear and early failure of the final drive to take place.

5.11 Types of bearings used in the transmission system

Plain bearings

A plain bearing can be described as a round hole lined with bearing metal or fitted with a bearing bush. These metals reduce friction in the bearing and also enable easy replacement as wear takes place. Typical examples of where these are used are in the engine (big-end and main bearings), in the gearbox (main shaft and lay shaft bearings) and in the flywheel (spigot shaft bearing).

Ball and roller bearings

Ball and roller bearings reduce friction by replacing sliding friction with rolling friction. There are several types of ball and roller bearings in use.

Single-row deep groove radial or journal bearing

In this bearing (Fig. 5.38(a)) the outer 'race' is a ring of hardened steel with a groove formed on its inner face. The inner race is similar to the outer race except it has a groove on its outer face. Hardened steel balls fit between the two rings. The balls are prevented from touching each other by the cage. Although designed for radial loads, journal bearings will take a limited amount of end-thrust, and they are sometimes used as a combined journal and thrust bearing.

Double-row journal bearing

This bearing is similar to the single-row journal bearing and is shown in Fig. 5.38(b). There are two grooves formed in each race together with a set of balls and cages. These are able to take larger loads and are often used where space is limited. A typical example would be in the wheel bearing or hub bearing. Some axle shafts are fitted with this type.

Double-row self-aligning journal bearing

This bearing (Fig. 5.38(c)) has a double row of grooves and balls. The groove of the outer race is ground to form part of a sphere whose centre is on an axis with the shaft. This allows the shaft

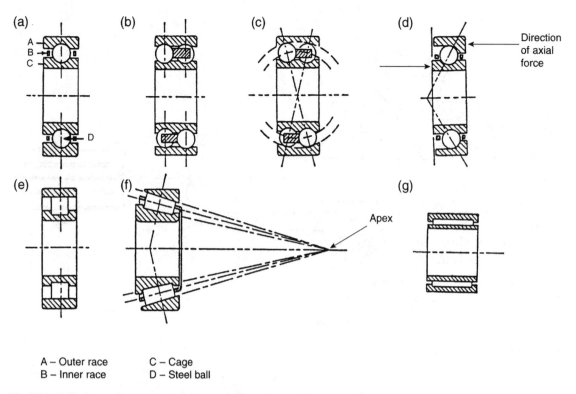

A – Outer race   C – Cage
B – Inner race   D – Steel ball

Fig. 5.38 (a) Single-row deep groove radial or journal bearing, (b) double-row radial or journal bearing, (c) double-row self-aligning bearing, (d) radial and one-way thrust bearing, (e) parallel-roller journal bearing, (f) taper roller bearing, (g) needle roller bearing

to run slightly out of alignment which could be caused by shaft deflection. This is commonly used as the centre bearing in the two/three piece prop shaft.

Combined radial and one-way thrust bearing

This type (Fig. 5.38(d)) is designed to take end-thrust in one direction only, as well as the radial load. They are normally mounted in pairs and are commonly used in the rear axles and clutch withdrawl thrust bearings. It is very important that these are fitted correctly otherwise the bearing will fall to pieces. They are also used as wheel bearings fitted back to back to enable them to take the side-thrust imposed on the wheel.

Parallel-roller journal bearing

This bearing (Fig. 5.38(e)) consists of cylindrical rollers having a length equal to the diameter of the roller. These can withstand a heavier radial load than the corresponding ball bearing, but cannot withstand any side-thrust. They are used in large gearboxes and in some cases have replaced the plain bearing.

Taper roller bearing

The rollers and race of this bearing (Fig. 5.38(f)) are conical (tapered) in shape with a common apex (centre) crossing at a single point on the centre line of the shaft. This ensures that the rollers do not slip. The rollers are kept in position by flanges on the inner race. This type of bearing can withstand larger side-thrusts than radial loads and they are normally used in pairs facing each other. They are most often used in the final drive, rear and front wheel hub assemblies. Some form of adjustment is normally used to reduce movement to a minimum or to provide some pre-load to the bearing.

Needle roller bearing

In this bearing (Fig. 5.38(g)) the rollers have a length of at least three roller diameters, packed in cages between inner and outer sleeves. The bearings can be found in universal joints and gearboxes especially where space is limited.

Learning tasks

1. Draw up a simple service schedule for checking the axle and final drive for correct

operation. Include such things as type, grade and amount of oil used, and time/mileage base for changing/checking oil level.

2. How should the differential be checked for abnormal wear? What would be the main signs of early failure?
3. If a steady whining noise was heard as the vehicle was driven at a constant speed along a level road what would you suspect as being the fault and what would your recommendations be to put it right?
4. What main safety precautions would you observe when removing a live axle from a vehicle?
5. Why is it that most vehicles now use the hypoid type of gearing on the final drive?
6. Draw up the procedure for replacing the half-shaft on a live axle, assume the vehicle has been brought into the workshop, and the half-shaft has sheared between the final drive and the hub bearing.
7. Investigate the differences between a front engine FWD and a front engine RWD final drive and differential. Itemize the differences in servicing and adjustments that may be necessary when removing, stripping, rebuilding and refitting.
8. An axle requires a new crown and pinion fitting. Remove and dismantle an assembly using the procedure illustrated in Figs 5.30 and 5.31 to give the correct markings on the gear teeth.

Practical assignment – removing a clutch

When completing the task of removing and refitting the clutch, complete the following worksheet.

Questions

1. Why are the splines of the gearbox input shaft left clean and dry and not lubricated with engine oil?
2. State the type of clutch spring fitted to the vehicle.
3. Make a simple sketch of the clutch release mechanism. Label your sketch with the main components.
4. Inspect a diaphragm spring clutch assembly when attached and removed from the flywheel.

Sketch the shape (a side view will do) of the diaphragm spring in the normal run position and in the removed position.
5. State the method used to adjust the clutch and the manufacturer's recommended clearance.
6. Record the condition of the following components:
   (a) the driven plate
   (b) the pressure plate assembly
   (c) the release bearing and mechanism
   (d) the oil seals of both the engine and the gearbox
7. Make your recommendations to the customer to give adequate operation of the clutch for say the next 50 000 miles.

Practical assignment – removing and refitting a gearbox

Complete this when undertaking the removal and refitting of the gearbox.

Questions

1. Dismantle a front engine FWD gearbox. Identify wear on moving components such as bearings, shafts, gear teeth, thrust washers and oil seals. List the equipment used to measure this wear.
2. Identify the type of synchromesh unit fitted to the gearbox. From what material is it made?
3. Make a simple sketch of the method used to prevent the engagement of two gears at the same time. When does this unit operate?
4. Sketch and label the position of the selector forks, rails and location devices. Describe how the gear, once it is engaged, is held in position.
5. Identify first, second, third, fourth and where necessary fifth gear positions. State the gear ratios for each gear. The formula for calculating gear ratios is:

$$\frac{\text{driven gear}}{\text{driver gear}} \times \frac{\text{driven gear}}{\text{driver gear}}$$

6. Reassemble the gearbox and state any special precautions to be observed during reassembly and safety points to look out for.

Practical assignment – axles and final drives

Remove and dismantle the final drive and differential unit from a front engine RWD vehicle.

Questions

1. The type of final drive is:
   (a) straight bevel
   (b) spiral bevel
   (c) hypoid
   (d) worm and wheel
2. The bearings used to support the crown wheel are:
   (a) ball bearing taking axial thrust only
   (b) taper roller bearings taking axial thrust only
   (c) ball bearings taking axial and radial thrust
   (d) taper roller bearings taking axial and radial thrust
3. Why should the pinion be pre-loaded?
4. The type of bearing arrangement supporting the half shafts is:
   (a) semi-floating
   (b) three-quarter floating
   (c) fully floating
   (d) a combination of (a), (b) and (c)
5. The axle half-shaft is splined to the:
   (a) sun gear
   (b) planet gear
   (c) crown wheel
   (d) pinion gear
6. The type of axle is:
   (a) split axle
   (b) live axle
   (c) dead axle
   (d) trans-axle
7. Describe the procedure for reassembling the final drive and differential, state the following information:
   (a) the pre-load on the pinion bearing
   (b) the tolerance for backlash
   (c) the torque setting for the pinion nut
   (d) the final drive ratio
   (e) the type of oil seal used on the pinion
   (f) the type of oil seal used on the axle half-shaft
   (g) the type and grade of oil used in the axle
   (h) the service interval when the oil should be changed

Self assessment questions

1. A gearbox ratio of 4:1 means that:
   (a) the power at the gearbox output shaft is 4 times the engine power
   (b) the power at the gearbox output shaft is a quarter of the engine power
   (c) the torque at the gearbox output shaft is 4 times the engine torque
   (d) the speed of the gearbox output shaft is 4 times the speed of the engine crankshaft
2. A differential gear in the final drive:
   (a) allows the wheel on the outside of a bend to rotate slower than the one on the inside of the bend
   (b) prevents skidding during acceleration
   (c) permit's the drive to be transmitted through an angle of 90°
   (d) allows the wheel on the outside of the bend to rotate faster than the one on the inside of the bend
3. In a synchromesh gearbox:
   (a) the gears are engaged by sliding them into mesh as required
   (b) the gears are engaged by double declutching and blipping the throttle
   (c) the gears are engaged by the operation of the synchromesh units
   (d) the gears are engaged by the interlock device on the selector shaft
4. In the Hotchkiss type drive:
   (a) the distance between the driven axle and the gearbox remains constant
   (b) a sliding joint is fitted in the propeller shaft to allow for variations in the distance between the gearbox and the driven axle
   (c) only one Hooke's type coupling is required
   (d) a torque tube is always required
5. The technical details of a certain vehicle state that a road speed of 30 mile/hour is achieved when the engine speed is 2 000 rev/min and the vehicle is in top gear. On test, this vehicle achieves 25 mile/hour in top gear when the tachometer shows an engine speed of 2 300 rev/min. This is an indication that:
   (a) the clutch is slipping
   (b) the engine is producing too much power
   (c) the driver is using too much throttle
   (d) the alternator charging rate is too high and it is causing false speedometer readings

6. In a hypoid bevel final drive:
   (a) the centre line of the pinion is offset from the centre line of the crown wheel
   (b) the drive is transmitted through an angle of 100°
   (c) a differential gear is not required
   (d) the pressure between the gear teeth is very low

7. In a semi-floating live axle:
   (a) most of the load is taken by the axle casing
   (b) shearing and bending forces are carried by the axle casing
   (c) most of the load is taken by the axle shaft
   (d) a shear pin is fitted at the differential end of the half shaft

8. The stalling speed of a torque convertor is:
   (a) the speed at which the idle control valve is set to operate
   (b) the speed at which torque multiplication is at a maximum
   (c) the highest speed that the engine can reach while the vehicle is in gear and prevented from moving
   (d) lowest speed at which the vehicle can be driven

9. When moving away from a standing start, a torque convertor increases engine torque by a factor of 2:1. When the gearbox ratio is 2.5:1 and engine torque is 100 Nm:
   (a) the torque at the gearbox output shaft will be 50 Nm
   (b) the torque at the gearbox output shaft will be 250 kW
   (c) the torque at the gearbox output shaft will be 500 Nm
   (d) the torque at the gearbox output shaft will be 4.5 times the engine torque

10. In a fully floating drive axle:
    (a) shearing, bending and torsional forces are all taken by the axle shafts
    (b) a broken axle shaft may cause the wheel to fall off
    (c) shearing and bending forces are taken by the axle casing, and torsional forces only are taken by the axle shafts
    (d) the axle has to be dismantled in order to replace a half shaft

6

Suspension systems

Topics covered in this chapter

Non-independent – leaf spring suspension
Independent suspension
Leaf spring construction
Dampers (shock absorbers)

The suspension of a vehicle is present to prevent the variations in the road surface encountered by the wheels from being transmitted to the vehicle body. There are two main categories of suspension systems: **independent** and **non-independent**. An independent type has each wheel moving up and down without affecting the wheel on the opposite side of the vehicle. On non-independent systems movement of one wheel will affect the wheel on the opposite side of the vehicle. A typical system consists of:

- a wheel and pneumatic tyre;
- a spring and damper unit together with arms and links that attach the wheel or axle to the vehicle.

Many different types are used on light vehicles.

Unsprung mass

The unsprung weight (mass) is the weight of components that are situated between the suspension springs and the road surface. Unsprung weight has an undesirable effect on vehicle handling and ride comfort and suspension systems are designed to keep it to a minimum.

6.1 Non-independent suspension systems

Leaf springs

Leaf springs of the type shown in Fig. 6.1(a) are manufactured from a number of leaves (layers) of steel strip. The use of multiple leaves ensures that the bending and shearing strength of the spring is uniform along the span of the spring from the spring eye to the point where the spring is attached to the axle. At each end of the main leaf is a circular hole (eye) which is used to attach the spring to the vehicle frame. As the spring flexes in use, the surfaces of the leaves rub against each other and this provides a degree of damping, the clips on the spring serve the purpose of keeping the leaves in contact with each other. The leaf spring is able to transmit driving and braking forces between the road wheels and the chassis.

Single leaf springs of the type shown in Fig. 6.1(b) are used on the rear suspension systems of many light commercial vehicles. Modern design techniques and materials permit the use of these single leaf springs which reduce the weight and cost of springs.

Learning tasks

1. The modern trend is to use leaf springs containing a single leaf only. Why is this and what are the advantages?
2. The multi-leaf spring performs a number of functions (tasks). One is to positively locate the axle to prevent it moving sideways independent of the body. Name two other functions and give reasons for your answer.
3. What effect does the movement of the suspension have on the drive line and/or steering of the vehicle?
4. Why should a reduced 'unsprung' weight improve the ride and handling characteristics of the vehicle?

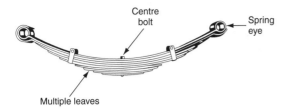

Fig. 6.1(a) A semi elliptic leaf spring

5. Remove a leaf spring assembly from a vehicle and inspect the mounting bushes and rubber for deterioration and wear. Reassemble the spring to the vehicle, replacing worn components as necessary.

Coil spring

This type of spring gives a smoother ride than the multi-leaf due to the absence of **interleaf friction**. It can be used on front and rear systems. The helical spring is normally used in conjunction with independent suspension and is now often used with beam-axle rear suspensions. Figure 6.2 illustrates this type.

The **coil spring** and the **torsion bar** suspension are alike and are superior to the leaf spring as regards the energy stored but unlike the leaf spring extra members to locate it are required which adds to the basic weight. The advantages of the coil spring are:

- a reduction in unsprung weight;
- energy storage is high;
- it can provide a softer ride;
- it allows for a greater movement of suspension;
- it is more compact.

Its main disadvantage is the location of the spring (the axle requires links and struts to hold it in place). This is shown in Fig. 6.3.

Learning tasks

1. What method would you adopt to check a suspected weak spring?

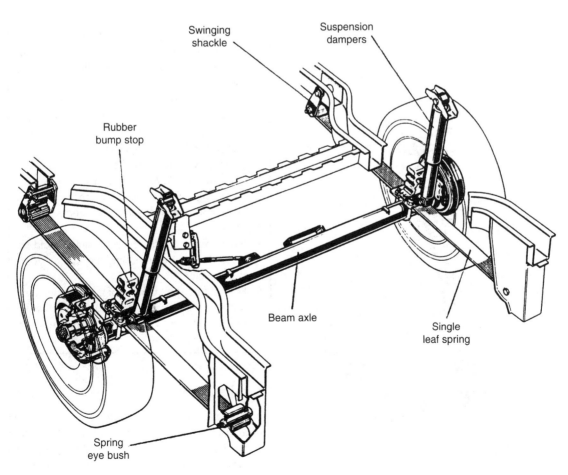

Fig. 6.1(b) Rear axle and suspension assembly

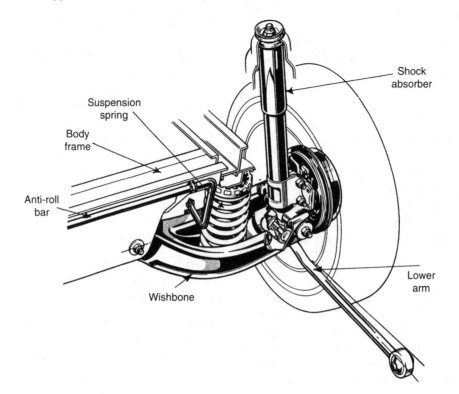

Fig. 6.2 Near side independent rear suspension

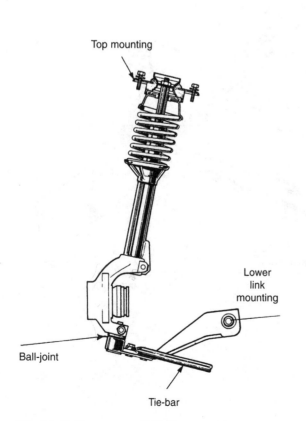

Fig. 6.3 MacPherson strut off side front suspension

2. Write out a work schedule for removing a coil spring from a MacPherson strut. Itemize the safety points that *must* be observed.
3. Using your schedule remove a MacPherson strut from a vehicle. Remove the spring from the strut. Check: the operation of the damper; oil leaks from the seals; the upper bearing for wear; rubber bump stop and damage or severe rust to the spring mountings. Reassemble the spring to the strut and the suspension assembly to the vehicle.
4. What other checks do you think should be carried out after reassembling the strut to the vehicle?
5. How could the stiffness of a coil spring suspension be increased to enable the driver to tow a caravan? What other improvements might be required?

Torsion bar

This is a straight bar which can either be round or square section and fixed at one end to the chassis. The other end is connected by a lever to the axle. At each end of the bar the section is increased and serrated or splined to connect with the lever or chassis. Adjustment is provided for at the chassis end to give the correct **ride height** for the vehicle. This type of suspension is shown in Fig. 6.4 and is used on the front of vehicles. Figure 6.5 shows the arrangement used on the rear suspension.

Since the coil spring is a form of torsion bar suspension, the rate of both types of spring is governed by the same factors.

- The length of the bar.
- The diameter of the bar. If the length is increased or the diameter decreased, the rate

of the spring will decrease, i.e. the spring will become softer.

- The material it is made from.

> *Learning tasks*
>
> 1. How is the 'ride height' adjusted on a vehicle fitted with a torsion bar front suspension? Where on the vehicle is the measurement taken?
> 2. List the main advantages and disadvantages of torsion bar suspension over the leaf and coil spring types?

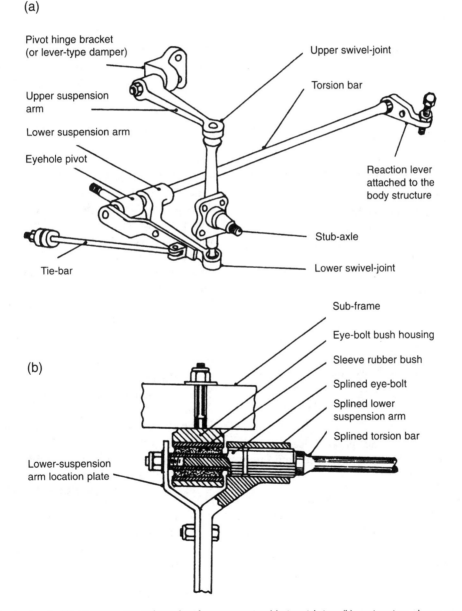

Fig. 6.4 Torsion-bar double-transverse-arm independent front suspension (a) pictorial view, (b) section view – lower-suspension-arm pivot asembly

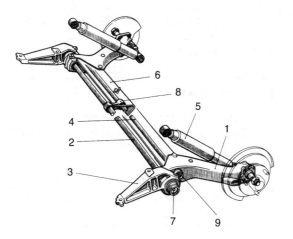

1 – Trailing link	7 – Torsion bar mounting
2 – Torsion bar (N/S)	in bracket
3 – Mounting bracket	8 – Connector
4 – Torsion bar (O/S)	9 – Serated profile mounting
5 – Shock absorber	torsion bar in trailing link
6 – Cross member	

Fig. 6.5 Transverse torsion bar rear suspension

6.2 Independent suspension

Cars and many of the lighter commercial vehicles are now fitted with some form of **independent front suspension** (IFS) and/or **independent rear suspension** (IRS). These have the following advantages over the earlier beam axle and leaf spring arrangements.

• Reduced unsprung weight.
• The steering is not affected by the 'gyro-scopic' effect of a deflected wheel being transmitted to the other wheel.
• Better steering stability due to the wider spacing of the springs.
• Better road holding as the centre of gravity is lower – due to the engine being mounted nearer the ground. This is because there is no front beam axle in the way.
• More space in the body due to the engine being lower and possible further forward.
• More comfortable ride due to the use of lower rate springs.

Where a beam axle and parallel leaf spring arrangement is used the springs are subjected to the following forces:

• suspension loads due to vehicle weight;
• driving and braking thrusts due respectively to the forward movement of the chassis and its retardation when braking;
• braking torque reaction – the spring distorting but preventing the rotating of the back plate and axle;
• twisting due to the deflection of one wheel only.

When the springs are strong enough to resist all these forces they are too stiff and heavy to provide a comfortable ride and good road holding. Independent suspension designs must provide for the control or limitation of these same forces, and their action must not interfere with the steering geometry or the operation of the braking system.

As we have already mentioned with the independent suspension when one wheel moves up or down it has little or no effect on the opposite wheel. In the case of the beam axle suspension, when one wheel rises over a bump the other wheel on the same axle is also affected. Because of this neither wheel is vertical to the road surface and so the road holding ability of the vehicle is affected. In a truly independent system each wheel is able to move without affecting any of the others. This has a number of obvious advantages:

1. a softer ride giving greater comfort for the passengers;
2. better road holding especially on rough or uneven road surfaces;
3. the engine can normally be mounted lower in the vehicle;
4. because of (1) (2) and (3) the vehicle will corner better;
5. it allows for a greater rise and fall of the wheel.

Because of the expense, greater complication and difficulty of mounting, many vehicle manufacturers of smaller cars fit independent suspension to the front and beam axle on the rear especially where the engine drives the front wheels. This arrangement gives a light, simple layout for the rear suspension using coil springs and links to support the axle to the body.

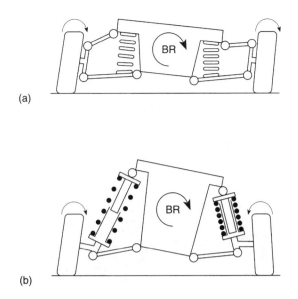

(a)

(b)

Fig. 6.6 Effects of (a) body roll and (b) irregular road surfaces on suspension geometry

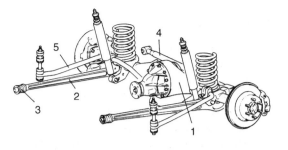

1 – Axle casing
2 – Longitudinal (down the length of the car) control arms
3 – Rubber mountings
4 – Panhard rod
5 – Anti-roll bar

Fig. 6.7 Layout of rigid line axle using coil springs and Panhard rod

The effects of uneven road surfaces and body roll on the suspension can be seen in Fig. 6.6. From the illustrations it will be seen that the **MacPherson strut** arrangement gives the most stable body and wheel alignment. This is one reason why the MacPherson strut is the most common type of suspension fitted to light vehicles.

Learning tasks

1. Explain what is meant by 'brake torque reaction'. Illustrate where necessary your answer with a simple sketch.
2. List a typical sequence of checks that should be made to find a reported knocking noise coming from the front off side suspension.
3. What effect would the lowering of the front suspension have on the steering geometry of the vehicle? Give reasons for your answer.

Axle location

As already mentioned the leaf spring locates the axle, but the coil, torsion bar, **rubber** and **air spring** suspension support the vehicle and its load and remove the unevenness of the road surface. They do not locate the wheel assembly or axle in any way. Extra arms (**radius arms**) or links are fitted to locate the wheels for both fore and aft movement and to resist the turning forces of both driving and braking. A **Panhard rod** gives sideways location between the axle and the body (see Fig. 6.7).

A number of other methods are in use that locate the axle. These include the use of **tie rods, wishbone** and **semi-trailing links** (Fig. 6.8). When an IRS is fitted (as on some of the more expensive vehicles) then the arrangement for location of the wheel assembly can become more complicated.

The method used to attach the links and arms to the body is by the use of **rubber bushes**. There is always a certain amount of 'give' in these rubber joints and this is termed compliance. This is generally allowed for in the specifications given by the manufacturer when checking the suspension for alignment.

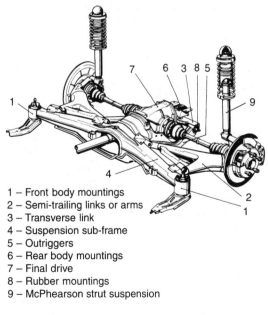

1 – Front body mountings
2 – Semi-trailing links or arms
3 – Transverse link
4 – Suspension sub-frame
5 – Outriggers
6 – Rear body mountings
7 – Final drive
8 – Rubber mountings
9 – McPhearson strut suspension

Fig. 6.8 Independent rear suspension and final drive using semi-trailing links

Rubber suspensions

Rubber can be used as the suspension medium as well as for mountings and pivots. Its main advantage is that for small wheel movements the ride is fairly soft but it becomes harder as wheel movement increases. It has the advantage of being small, light and compact and will absorb some of the energy passed to it, unlike a coil spring which gives out almost as much energy as it receives. The rubber spring is also commonly used on LGVs and trailers. A rubber suspension unit is shown in Fig. 6.9.

Hydrolastic suspension

This is a combination of rubber and fluid (which is under pressure). Each wheel is fitted with a **hydrolastic unit** which consists of a steel cylinder mounted on the body of the car. A tapered piston, complete with a rubber and nylon diaphragm which is connected to the upper suspension arm, fits in the bottom of the unit. When the wheel moves upwards the piston is also moved up and via fluid action compresses the rubber spring. The units are connected together on the same side by large bore pipes and some of the fluid is displaced down the pipe to the other unit. In this way the tendency for the vehicle to pitch (this is the movement of the body fore and aft, that is front to back) is reduced. A hydroelastic system is shown in Fig. 6.10.

Hydrogas suspension

Sometimes called **hydropneumatic** this system, shown in Fig. 6.11, is a development of the hydrolastic system. The main difference is that

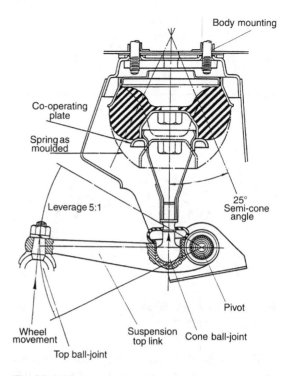

Fig. 6.9 Rubber suspension unit

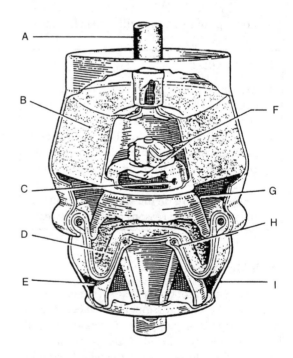

A – Fluid outlet to rear unit   F – Damper valves
B – Rubber spring   G – Dividing member
C – Fluid bleed hole   H – Reinforcing in diaphragm
D – Rubber diaphragm   I – Tapered outer cylinder
E – Tapered piston

Fig. 6.10 Hydrolastic spring displacer unit

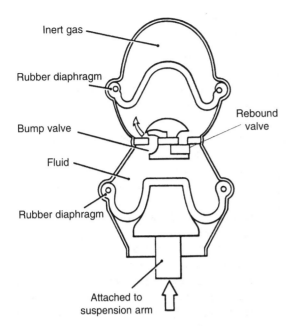

Fig. 6.11 Hydrogas suspension

the rubber is replaced by a gas (usually nitrogen), hence the term 'pneumatic'. With this type the gas remains constant irrespective of the load carried. Gas pressure will increase as volume is reduced. This means that the suspension stiffens as the load increases. The units are connected together in a similar manner to the hydrolastic suspension and the fluid used is a mixture of water, alcohol and an anti-corrosive agent.

The ride height in both these systems can be raised or lowered by the use of a hydrolastic suspension pump to give the correct ride height and ground clearance.

Active or 'live' hydropneumatic suspension

This system, shown in Fig. 6.12, allows the driver to adjust the ride height (sometimes inaccurately referred to as ground clearance) of the vehicle. It also maintains this clearance irrespective of the load being carried. First developed by Citroen it has recently been taken up by a number of other manufacturers. On the Citroen arrangement each **suspension arm** is supported by a **pneumatic spring**.

Connected between the suspension arms at both front and rear are **anti-roll bars**. These are linked to **height correctors** by means of

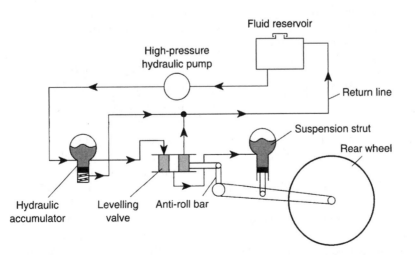

Fig. 6.12 Simplified layout of hydropneumatic suspension system

control rods. An engine driven pump supplies oil under pressure to a **hydraulic accumulator** and this is connected to the height control or **levelling valves**.

As the vehicle is loaded, the downward movement of the body causes the rotation of the anti-roll bar. This moves the slide valve in the height correctors and uncovers the port to supply oil under pressure from the accumulator to the suspension cylinders. When the body reaches the predetermined height (which can be varied by the driver moving a lever inside the vehicle), the valve moves to the 'neutral' position. Removal of the load causes the valve to vent oil from the cylinder back to the reservoir.

A delay device is incorporated to prevent rapid oil flow past the valve when the wheel contacts a bump. This prevents the valve from continuously working and giving unsatisfactory operation.

In some systems a third spring unit is fitted between the two spring units on the front axle and between the two spring units on the rear axle. This gives a variable spring rate and roll stiffness, i.e. the suspension is active. The system is controlled by an ECU (electronic control unit) which senses steering wheel movement, acceleration, speed and body movement and reacts accordingly via control valves to regulate the flow of oil to and from the suspension units. Under normal driving conditions the ECU operates the solenoid valve which directs fluid to open the regulator valves. This allows fluid to flow between the two outer spring units and the third spring units via the damper units to give a soft ride. During harder driving the solenoid valve is switched off automatically relieving the regulator valves which close, preventing fluid flow between the spring units. The third spring unit being isolated and not in use gives a firmer ride.

There are a number of benefits of this system:

- it automatically adjusts the spring and damper rate to suit road conditions and driving styles;
- it can provide a soft and comfortable ride under normal driving conditions;
- it will stiffen to give better road holding during hard driving;
- a near constant ride height can be achieved irrespective of the load on the vehicle.

Learning tasks

1. A number of safety precautions must be observed before dismantling this type of suspension system. What are these and why should they be carried out?
2. What are the main advantages/disadvantages of this type of suspension system over a mechanical system?
3. The oil in the system should be changed at regular intervals. How often is this and what is the correct procedure? State the correct type of fluid that must be used in the system.
4. Remove and refit a hydropneumatic suspension unit. Take care when undoing pipe unions that the pipes are not damaged. Observe all the safety precautions necessary. Make a note of all defects, re-pressurize the system and check for leaks.

Rear wheel 'suspension steer'

In some systems the rear suspension is arranged to produce a steering effect when cornering. As the suspension is deflected the road wheel **toes-in** due to the arc of movement of the semi-trailing link. This produces **understeer** on cornering (it gives a small degree of same direction rear-steer). Figure 6.13 shows one example that will give:

- toe-in when braking;
- understeer when cornering;
- stability in straight line running and when changing lanes.

A simpler suspension arrangement is shown in Fig. 6.14 that produces toe-in under similar operating conditions.

Air suspension

Some vehicle manufacturers fit air suspension units in place of the conventional coil springs. A typical example with electronically controlled air suspension is shown in Fig. 6.15. Three switches, positioned near the steering column, control the ride height: the upper switch raises the vehicle by approximately 40 mm for driving through deep water or over rough ground; the middle switch gives a similar ride height to the coil-sprung model; the lower switch lowers the vehicle approximately 60 mm below the standard setting to enable easier

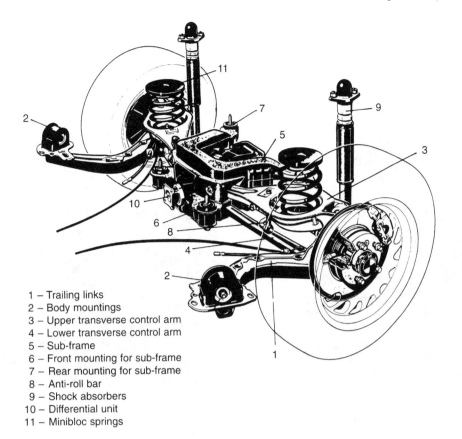

1 – Trailing links
2 – Body mountings
3 – Upper transverse control arm
4 – Lower transverse control arm
5 – Sub-frame
6 – Front mounting for sub-frame
7 – Rear mounting for sub-frame
8 – Anti-roll bar
9 – Shock absorbers
10 – Differential unit
11 – Minibloc springs

Fig. 6.13 Multi-link rear suspension

loading and for getting in and out. A number of automatic adjustments such as lowering the vehicle when driving over 50 mph to reduce wind resistance and raising the vehicle if it should become grounded are programmed into the ECU and operate independently of the driver.

Another system uses a digital controller (ECU) which reacts to suspension acceleration and ride height by varying the oil pressure. It will also adjust damper action by reference to body movement.

Some manufacturers make **ride height levellers** (Fig. 6.16), which are shock absorbers that will raise the rear of the vehicle to its normal ride height when loaded. The system is operated through a switch, electric motor and compressor; the dampers are pumped up via air pipes connecting the units together.

The **suspension damper** has to perform a number of functions. These are:

- absorb road surface variations;
- insulate the noise from the vehicle body;
- give resistance to vehicle body pitch under braking and acceleration;
- offer resistance to roll when cornering.

A passive mechanical system to tackle each of these tasks is impossible as the rate of the damper must be varied to suit differing circumstances and handling characteristics. Shock absorber manufacturers have made provision for some control over the damping rate by making them adjustable, but once set even these are not suitable for all applications and uses.

With this in mind many of the new suspension systems are **active**; in other words they will operate continuously adjusting to pre-set limits the suspension spring rate. One system uses an ECU which receives information on cruising, suspension condition, body roll and pitch. It then directs air from the compressor to the suspension units to give the correct spring rate and ride height for the load carried.

> *Learning tasks*
>
> 1. What is the purpose of the ECU and electronic sensors in an active suspension system? What attention do these units require and at what intervals should checks be carried out?

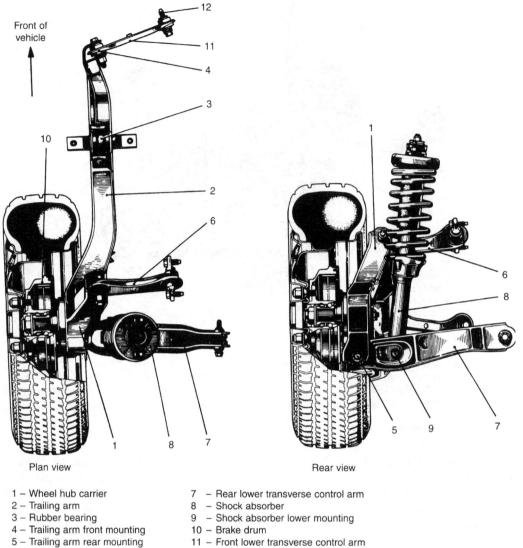

Front of vehicle

Plan view

Rear view

1 – Wheel hub carrier
2 – Trailing arm
3 – Rubber bearing
4 – Trailing arm front mounting
5 – Trailing arm rear mounting
6 – Upper transverse control arm

7 – Rear lower transverse control arm
8 – Shock absorber
9 – Shock absorber lower mounting
10 – Brake drum
11 – Front lower transverse control arm
12 – Inner mounting

Fig. 6.14 Multi-link rear suspension giving toe-in during braking and cornering

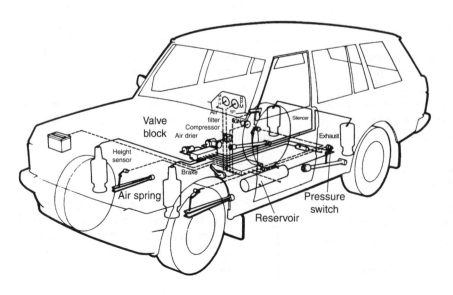

Valve block

Air filter
Ign
Compressor
Air drier

Silencer

Exhaust

Height sensor

Brake

Air spring

Reservoir

Pressure switch

Fig. 6.15 Air suspension

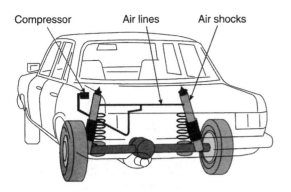

Compressor   Air lines   Air shocks

Fig. 6.16 Height levellers in the rear axle

2. On some models the system is connected to the anti-lock braking system. Explain in simple terms the reason for this and how it operates.
3. List the maintenance requirements for one of the 'active' suspension systems with which you are familiar.
4. Draw up a work schedule for removing a suspension unit from a vehicle. Special note should be made of the safety aspects of the task.
5. Remove an ECU from an active suspension system, check the terminals for corrosion or damage and clean with appropriate spray cleaner where necessary. Ensure that the mounting surface is clean and free from rust. Refit the ECU taking special care when refitting the terminal connector.

6.3 Suspension dampers

The function of the suspension damper is not to increase the resistance to the spring deflecting but to control the **oscillation** of the spring (this is the continuing up and down movement of the spring after going over a bump or hollow in the road surface). In other words it absorbs energy given to the spring, hence the more common name of **shock absorber**. It does this by forcing oil in the damper to do work by passing it through holes in the piston and converting the energy of the moving spring into heat which is passed to the atmosphere.

Two main categories of dampers are in common use: the **direct-acting** (usually telescopic) and the **lever arm**.

The twin-tube telescopic damper

This type is usually located between the chassis and the axle so that on both bump and rebound oil is forced through the holes in the piston. The reservoir is used to accommodate the excess oil that is displaced as a result of the volume in the upper cylinder being smaller. If oil is lost and air enters the damper it will affect its performance and it will become 'spongy' in operation. This will have an effect on the stability of the vehicle especially when cornering, travelling over rough ground, uneven surfaces and on braking or accelerating. When checking for correct operation disconnect one end of the damper and check the amount of resistance to movement. Figure 6.17 shows a telescopic shock absorber.

Gas-pressurized dampers

A **single tube** is used as the cylinder in which the piston operates. It is attached to the car body and suspension by rubber bushes to reduce noise and vibration and to allow for slight sideways movements as the suspension operates. A chamber in the unit, sealed by a free piston, contains an inert gas that is under pressure when the damper is filled with oil. As the suspension operates the piston moves down the cylinder and oil is forced through the 'bump' valve to the upper chamber; excess oil that cannot be accommodated in the upper chamber because of the rod moves the free piston to compress the gas. On rebound the oil is made to pass through the 'rebound' valve in the piston; by varying the size of the holes in the valves the resistance of each stroke can be altered to suit different vehicle applications. This arrangement is most commonly fitted to the MacPherson strut suspension systems. The main advantages of the single-tube damper are:

- it can displace a large volume of fluid without noise or fluid aeration;
- it is fairly consistent in service even when operating at large angles to the suspension movement;
- it has good dissipation of heat to the air flow.

Checking damper operation

Two tests are normally used to check whether the dampers are serviceable without removing them from the vehicle. The **bounce test** involves pushing down on each corner of the vehicle

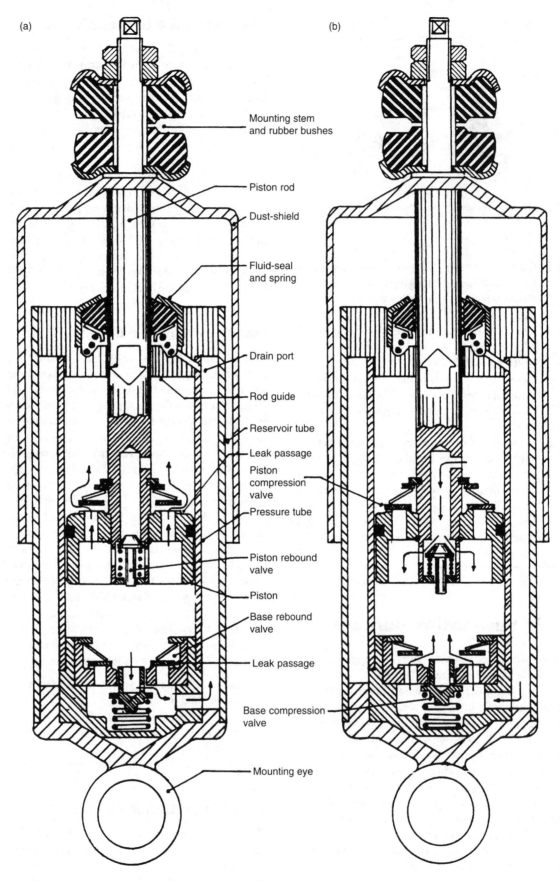

(a)   (b)

Mounting stem
and rubber bushes

Piston rod

Dust-shield

Fluid-seal
and spring

Drain port

Rod guide

Reservoir tube

Leak passage

Piston
compression
valve

Pressure tube

Piston rebound
valve

Piston

Base rebound
valve

Leak passage

Base compression
valve

Mounting eye

Fig. 6.17 Telescopic shock absorber (a) bump or compression stroke, (b) rebound or extension stroke

and observing the up and down movement of the body as it comes to its rest position. The second method involves the use of some form of **tester** of which there are several types available. One type uses an eccentric roller arrangement that the vehicle is driven onto to produce the required movement of the suspension. In another the vehicle is driven up a collapsible ramp; when set the ramp is dropped. Both types give a graphical printout of the test that shows the oscillations of the suspension.

Materials used in suspensions

- **Road springs** These are made from a silicon manganese steel as they must withstand very high stresses and fatigue yet still retain elasticity and strength.
- **Suspension links** Here a nickel steel is used which gives elasticity together with toughness.
- **Bushes** These are usually made from rubber to reduce vibration, noise and the need for lubrication.

6.4 Maintenance checks

General rules

There are a number of general rules that should be observed when carrying out maintenance checks and adjustments; and there are also methods of protecting the system against accidental damage during repair operations on the suspension system. These are listed below:

- preliminary vehicle checks – see the vehicle report sheet;
- safety – the proper use of lifting supporting and choking equipment;
- safe use of special tools such as coil spring compressor, high-pressure lubricants and high-pressure hydraulic equipment;
- care when working with and disposing of toxic and corrosive fluids;
- checking of the suspension thoroughly before road testing.

Checking suspension alignment and geometry

To check the alignment of the suspension a four wheel aligner is used. The gauges are located on each rear wheel and the light on the each front wheel; the light shines onto the gauge down each side of the vehicle. This will show any misalignment between the front and rear wheels which can then be identified, helping to prevent tyre scrub, steering pulling to one side or crabbing of the vehicle down the road. The lights also shine across the front of the vehicle to check front wheel alignment. When all four gauges give the correct readings, the vehicle will drive down the road in a straight line.

Learning tasks

1. What are the symptoms of faulty dampers? How will the driver notice this problem? How should they be checked for correct operation?
2. Draw up a work schedule for stripping and rebuilding a MacPherson strut-type suspension. Name any special tools that should be used and identify any safety precautions that should be observed?
3. For each of the following suspension faults: describe the fault (what is wrong?); identify the symptom (how will the driver notice?); state the probable cause (what has caused the fault?); and give the preventative or corrective action that should be taken.
   (a) Excessive uneven tyre wear
   (b) Excessive component wear
   (c) Premature failure of component
   (d) Vibration and/or noise from suspension
   (e) Uneven braking
   (f) Steering pulling to one side
   (g) Incorrect trim height
   (h) Axle misalignment
   (i) Excessive pitch or roll
   (j) Vehicle instability over rough road surfaces
   (k) Poor handling and ride quality
   (l) Noisy suspension

Practical assignment – suspension system

Introduction

At the end of this assignment you should be able to:

- carry out a visual inspection
- remove and replace a
  - MacPherson strut
  - coil spring

– shock absorber, checking for correct operation
– suspension bush
- check alignment of suspension
- make a report together with recommendations

Tools and equipment

- A vehicle fitted with suitable suspension
- A vehicle lift or jack and axle stands
- Workshop manual
- Selection of tools to include specialist equipment such as a spring compressor

Objective

- To check operation of damper
- To replace suspension bushes
- To check 'free height' of coil spring
- To investigate oil leaks
- To replace worn/damaged components

Activity

1. After suitably raising and supporting the vehicle remove the wheels (*do not* support the vehicle under the suspension to be removed).
2. Observe and note the type of suspension, e.g. coil spring, leaf spring, torsion bar, rubber, etc.
3. Make a simple sketch of the layout.
4. Before cleaning around the mounting points look for signs of:
   (a) insecure or loose components
   (b) places where dirt may be rubbed off by something catching
   (c) bright or rusty streak marks where body or chassis may be cracked
   (d) rust where thickness of material may be reduced to a failure level
   (e) excessively worn components
   (f) accident damage to body or components
5. Remove suspension and dismantle where necessary.
6. Check operation of damper.
7. Reassemble suspension (replacing any worn/broken components) according to manufacturer's manual.
8. Check for correct assembly and tightness of all mounting bolts before fitting wheel.
9. Complete an inspection report and recommendations for the customer.

Checklist
Vehicle
Removal and refitting

MacPherson strut
Coil spring
Leaf spring
Shock absorber
Rubber suspension bush

Dismantle and reassemble

MacPherson strut
Shock absorber

Checking and adjusting

Castor angle
Camber angle
King pin inclination
Toe-out on turns
Wheel alignment
Axle alignment
Body alignment
Ride height

Investigating and reporting

Rubber
Hydrogas
Hydrolastic
Hydropneumatic
Coil spring
Leaf spring
Air

Student's signature

Supervisor's signature

Self assessment questions

1. The rear suspension system shown in Fig. 6.13:
   (a) is a trailing arm type
   (b) is a torque tube transaxle type
   (c) has no unsprung weight
   (d) is a dead axle
2. A principal task of the suspension dampers:
   (a) is to dampen out spring vibrations
   (b) is to allow the use of weaker springs
   (c) is to keep the vehicle upright on corners
   (d) is to eliminate axle tramp under braking

3. Coil spring suspension systems:
   (a) are lighter than leaf spring suspension systems
   (b) require the use of radius arms to transmit driving and braking forces from the wheels to the vehicle frame
   (c) do not require the use of suspension dampers
   (d) are only suitable for small vehicles

4. With independent front suspension:
   (a) the engine can be set lower in the frame because there is no axle beam
   (b) vertical movement of a wheel on one side of the vehicle will cause the wheel on the opposite side to move in a similar way
   (c) double acting dampers are essential
   (d) tyre wear is lower than it is on beam axle systems

5. In leaf spring type suspension systems:
   (a) extra tie rods are needed in order to transmit driving and braking forces from the wheels to the vehicle body
   (b) both ends of the spring are rigidly fixed to the chassis
   (c) driving and braking forces are transmitted to the chassis through the spring leaves

   (d) extra strong dampers are required to cope with the greater tendency for this type of suspension to vibrate

6. Figure 6.7 shows a Panhard rod fitted to a coil spring suspension. The purpose of this rod is to:
   (a) limit vertical movement of the vehicle body relative to the road wheels
   (b) transmit cornering forces between the vehicle body and the axle
   (c) absorb braking torque reaction
   (d) secure the brake caliper to the vehicle frame

7. In a hydro pneumatic suspension unit:
   (a) ride height is fixed at a constant level
   (b) ride height is varied by introducing or removing hydraulic fluid from the chamber below the gas diaphragm
   (c) gas is pumped in and out of the chamber above the gas diaphragm
   (d) the equivalent spring rate is constant

8. An active suspension system:
   (a) uses an ECU to adjust suspension characteristics to suit driving conditions
   (b) is only suitable for use on racing cars
   (c) has fewer working parts than a non-active type suspension system
   (d) can only work on torsion bar type suspension

7

Braking systems

Topics covered in this chapter

Hydraulic braking systems
Drum brakes
Disc brakes
Master cylinder
Wheel cylinders
Brake fluid
Braking efficiency

7.1 Vehicle braking systems

The purpose of the braking system is to slow down or stop the vehicle and, when the vehicle is stationary, to hold the vehicle in the chosen position. When a vehicle is moving it contains energy of motion (kinetic energy) and the function of the braking system is to convert this kinetic energy into heat energy. It does so through the friction at the brake linings and the brake drum, or the brake pads and the disc.

Some large vehicles are fitted with secondary braking systems that are known as retarders. Examples of retarders are exhaust brakes and electric brakes. In all cases, the factor that ultimately determines how much braking can be applied is the grip of the tyres on the driving surface.

7.2 Types of brakes

Two basic types of friction brakes are in common use on vehicles; these are:

1. the drum brake
2. the disc brake

The drum brake

Figure 7.1 shows a drum brake as used on a large vehicle. This cut-away view shows that the linings on the shoes are pressed into contact with the inside of the drum by the action of the cam. In this case the cam is partially rotated by the action of a compressed air cylinder. The road wheel is attached to the brake drum by means of the wheel studs and nuts.

A brake of this type has a leading shoe and a trailing shoe. The leading shoe is the one whose leading edge comes into contact with the drum first, in the direction of rotation. A leading shoe is more powerful than a trailing shoe and this shows up in the wear pattern because a leading shoe generally wears more than a trailing shoe owing to the extra work that it does.

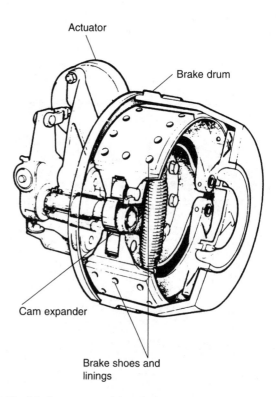

Actuator

Brake drum

Cam expander

Brake shoes and linings

Fig. 7.1 A cam operated drum brake

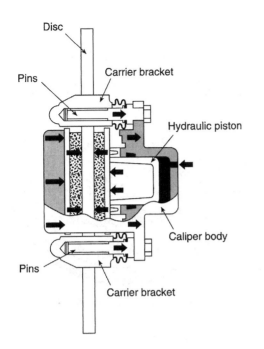

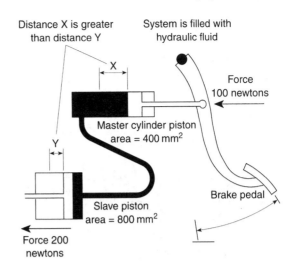

Fig. 7.3 The principle of a hydraulic braking system. The hydraulic pressure is equal at all parts of the system

Fig. 7.2 A disc brake

The disc brake

Figure 7.2 shows the principle of the disc brake. The road wheel is attached to the disc and the slowing down or stopping action is achieved by the clamping action of the brake pads on the disc.

In this brake the disc is gripped by the two friction pads. When hydraulic pressure is applied to the hydraulic cylinder in the caliper body, the pressure acts on the piston and pushes the brake pad into contact with the disc. This creates a reaction force which causes the pins to slide in the carrier bracket and this action pulls the other pad into contact with the disc so that the disc is tightly clamped by both pads.

7.3 Hydraulic operation of brakes

The main braking systems on cars and most light commercial vehicles are operated by hydraulic systems. At the heart of a hydraulic braking system is the master cylinder as this is where the pressure that operates the brakes is generated.

Principle of the hydraulic system

The small diameter master cylinder is connected to the large diameter actuating cylinder by a

strong metal pipe. The cylinders and the pipe are filled with hydraulic fluid. When a force is applied to the master cylinder piston a pressure is created that is the same at all parts of the interior of the system. Because pressure is the amount of force acting on each square millimetre of surface, the force exerted on the larger piston will be greater than the force applied to the small piston. In the example shown in Fig. 7.3 the force of 100 Newtons on an area of 400 square millimetres of the master cylinder piston creates a pressure of 0.25 Newtons per square millimetre. The piston of the actuating cylinder has an area of 800 square millimetres and this gives a force of 200 Newtons at this cylinder.

The master cylinder

The part of the hydraulic braking system where the hydraulic operating pressure is generated is the master cylinder. Force is applied to the master cylinder piston by the action of the driver's foot on the brake pedal.

In the example shown in Fig. 7.4 the action is as follows. When force is applied to the push rod the piston moves along the bore of the master cylinder to take up slack. As soon as the lip of the main rubber seal has covered the by-pass hole, the fluid in the cylinder, and the system to which it is connected, is pressurised. When the force on the brake pedal and the master cylinder push rod is released, the return spring pushes the piston back and the hydraulic operating pressure is removed. The action of the main piston seal ensures that the master cylinder remains filled with fluid.

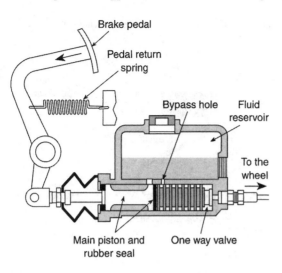

Fig. 7.4 A simple type of hydraulic master cylinder

7.4 Wheel cylinders

The hydraulic cylinders that push the drum brake shoes apart, or apply the clamping force in the disc brake, are the wheel cylinders. There are two principal types of wheel cylinders, a single acting cylinder and a double acting cylinder.

The single acting wheel cylinder

The space in the wheel cylinder, behind the rubber seal and piston, is filled with brake fluid. Pressure from the master cylinder is applied to the wheel cylinders through pipes. Increased fluid pressure pushes the piston out and this force is applied to the brake shoe or brake pad.

Double acting wheel cylinder

Figure 7.6 shows that the double acting wheel cylinder has two pistons and rubber seals. Hydraulic pressure applied between the pistons pushes them apart. The pistons then act on the brake shoes and moves the linings into contact with the inside of the brake drum.

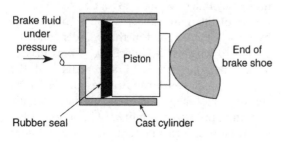

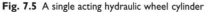

Fig. 7.5 A single acting hydraulic wheel cylinder

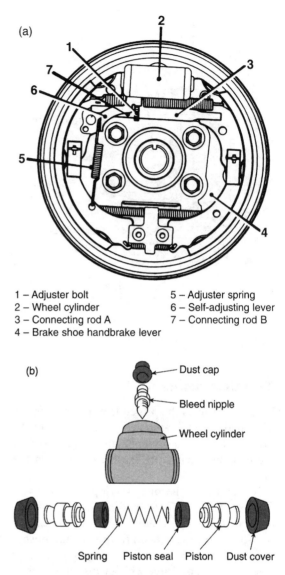

1 – Adjuster bolt
2 – Wheel cylinder
3 – Connecting rod A
4 – Brake shoe handbrake lever
5 – Adjuster spring
6 – Self-adjusting lever
7 – Connecting rod B

Fig. 7.6 A double acting hydraulic wheel cylinder (a) Rover 200 series drum brake, (b) a double acting wheel cylinder

7.5 The hand-brake

The handbrake (parking brake) is required to hold the vehicle in any chosen position when the vehicle is stationary. In addition to its function as a parking brake, the handbrake is also used when making hill starts and similar manoeuvres. The handbrake also serves as an emergency brake in the event of failure of the main braking system. Figure 7.7 shows the layout and main features of a handbrake for a car or light van. The vehicle has trailing arm rear suspension and the swivel sector pivots (4) are needed to guide the cable on these suspension arms. The purpose of the compensator is ensure that equal braking force is applied to each side of the vehicle. The

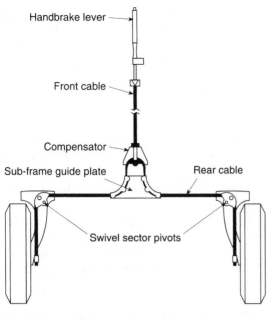

Fig. 7.7 A handbrake system

7.6 Split braking systems

The split braking circuit provides for emergencies such as a leak in one area of the braking system. There are various methods of providing split braking circuits and some of these are shown in Figs 7.9(a, b & c).

Front/rear split

Fig. 7.9(a) shows how the front brakes are operated by one part of the master cylinder and the

handbrake normally operates through the brakes at the rear of the vehicle.

With normal use, the brake linings will wear and it is also possible that the handbrake cable may stretch a little. In order to keep the handbrake working properly it is provided with an adjustment, as shown in Fig. 7.8.

The nut (2) is the adjusting nut and the nut (1) is the lock nut. In order to adjust the handbrake, the rear of the vehicle must be lifted so that the wheels are clear of the ground. The normal safety precautions must be observed and the wheels should be checked for freedom of rotation after the cable has been adjusted.

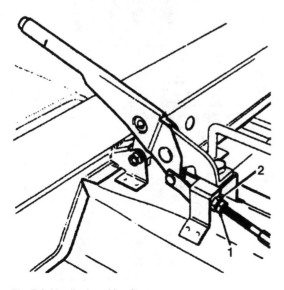

Fig. 7.8 Handbrake cable adjustment

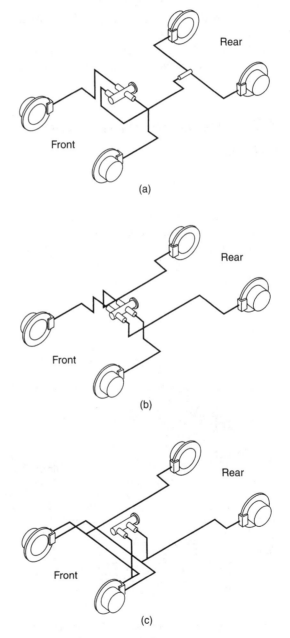

Fig. 7.9 (a) Front/Rear split of braking system, (b) diagonal split of braking system, (c) front axle and one rear wheel split

rear brakes by the other part. This system is used for rear wheel drive cars.

Diagonal split

One part of the tandem master cylinder is connected to the offside front and the nearside rear brakes and the other part to the nearside front and the offside rear, as shown in Fig. 7.9(b).

Front axle and one rear wheel split

This is known as the 'L' split. One part of the master cylinder is connected to the front brakes and the offside rear wheel brake and the other part to the front brakes and the near side rear wheel brake. This arrangement requires four piston calipers for the front disc brakes or two pistons, if they are floating calipers.

7.7 The tandem master cylinder

In each split system a tandem master cylinder is used. Figure 7.10 shows a tandem master cylinder.

The master cylinder is designed to ensure that the brakes are applied evenly. There are two pistons, a primary piston and a secondary piston. The spring (1) is part of the primary piston and it is stronger than the spring (2) that

is fitted to the secondary piston. Application of force to the piston causes the spring (1) to apply force to the secondary piston so that both pistons move along the cylinder together. The two piston recuperation seals cover their respective cut-off ports at the same time and this ensures simultaneous build up of pressure in the primary and secondary circuits.

When the brake pedal force is released, the primary and secondary pistons are pushed back by the secondary spring and both cut-off ports are opened at the same time, thus releasing the brakes simultaneously.

In the event of leakage from one part of the master cylinder, the other cylinder remains operative. This is achieved through the design of the stops and other features of the master cylinder.

7.8 The brake fluid reservoir

The fluid reservoir is usually mounted directly onto the master cylinder. Figure 7.11 shows a

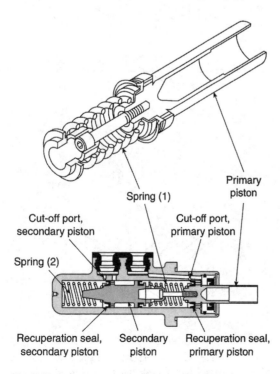

Fig. 7.10 A tandem master cylinder

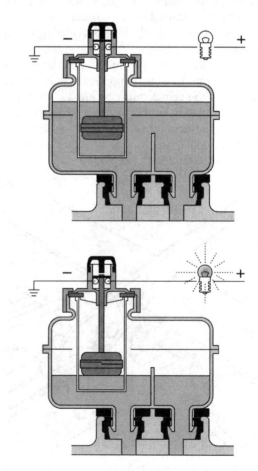

Fig. 7.11 The fluid level indicator

fluid reservoir equipped with a fluid level indicator. The fluid reservoir is normally made from translucent plastic and, provided it is kept clean, it is possible to view the fluid level by under bonnet inspection. The reservoir is fitted with an internal dividing panel ensuring that one section remains operative should the other section develop a loss of fluid. The fluid level indicator operates a warning light. Should the fluid level drop below the required level, the switch contacts will close and the warning light on the dash panel will be illuminated.

7.9 Brake pipes and hoses

The hydraulic pressure created at the master cylinder is conveyed to the wheel brakes through strong metal pipes, where they can be clipped firmly to the vehicle, and through flexible hoses where there is relative movement between the parts, for example axles and steered wheels and the vehicle frame.

7.10 The brake servo

It is common practice to provide some means of increasing the force that the driver applies to the brake pedal. The servo is the device which allows the driver to apply a large braking force by the application of relatively light force from the foot. The amount of increased force that is produced by the servo is dependent on the driver's effort that is applied to the brake pedal. This ensures that braking effort is proportional to the force applied to the brake pedal.

On petrol engined vehicles, manifold vacuum is used to provide the boost that the servo generates. On diesel engined vehicles there is often no appreciable manifold vacuum, owing to the way in which the engine is governed. In these cases, the engine is equipped with a vacuum pump that is known as an exhauster.

The master cylinder (1) is firmly bolted to the servo unit on one side and, on the other side, the servo unit is firmly bolted to the vehicle bulkhead (4). The brake pedal effort is applied to the servo input shaft and this in turn pushes directly on the master cylinder input piston. Figure 7.13 shows the servo in more detail.

At this stage you should concentrate on the flexible diaphragm, the sealed container, the two chambers (A) and (B) and the vacuum connection. The valve body and the control piston are designed so that when the brakes are off, the manifold vacuum will draw out air and create a partial vacuum in chambers (A) and (B) on both sides of the diaphragm.

When force is applied to the brake pedal, the control piston closes the vacuum port and effectively shuts off chamber (B) from chamber (A). At this stage, the control piston moves away from the control valve and atmospheric pressure air is admitted into chamber (B). The greater pressure in chamber (B), compared with the partial vacuum in chamber (A), creates a force that adds to the pedal effort applied by the driver. The servo output rod pushes direct on to the master cylinder piston, as shown in Fig. 7.14, so that there is no lost motion and the resulting force applied to the master cylinder is proportional to the effort that the driver applies to the brake pedal.

It is common practice to fit a non-return valve in the flexible pipe, between the manifold and the servo unit, as shown in Fig. 7.15. This valve serves to retain vacuum in the servo after the engine is stopped and also prevents petrol engine vapours and 'backfire' gases from entering the servo.

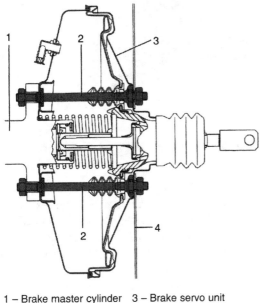

1 – Brake master cylinder 3 – Brake servo unit
2 – Tie bolts   4 – Vehicle bulkhead

Fig. 7.12 A vacuum servo

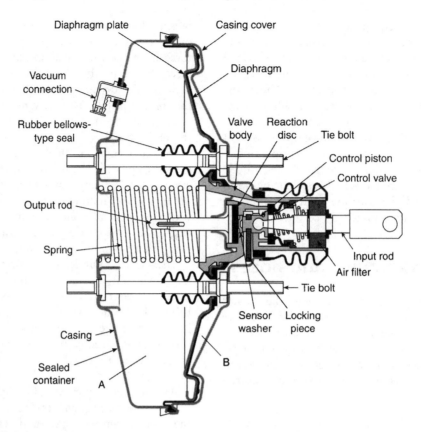

Fig. 7.13 Details of a vacuum servo

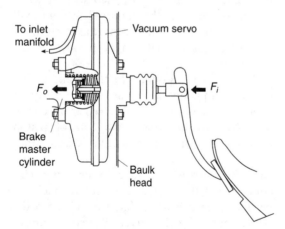

Fig. 7.14 Vacuum servo and master cylinder

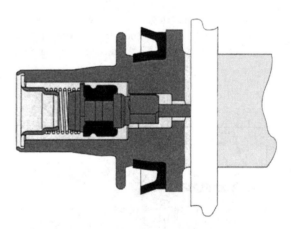

Fig. 7.15 The non-return valve in the vacuum pipe

7.11 Brake adjustment

During normal usage, the brake friction surfaces wear. In disc brakes this means that the actuating pistons push the pads closer to the disc and the light rubbing contact between friction material (pads) and the discs is maintained; disc brakes are thus 'self adjusting'. There is some displacement of brake fluid from the reservoir to the wheel cylinders and this will be noticed by the lowering of fluid level in the reservoir. When new brake pads are fitted there may be an excess of fluid in the system, if the fluid was topped up when the brakes were in a worn condition.

When the friction linings on brake shoes wear, the gap between the lining and the inside of the brake drum increases. To compensate for this wear, drum brakes are provided with adjusters. Adjusters take two forms; 1) manual adjusters, and 2) automatic adjusters.

Manual brake adjuster

Figure 7.16 shows a type of manual brake adjuster that has been in use for many years. The threaded portion is provided with a slotted part that permits it to be rotated by means of a lever. The adjuster is accessed through a hole in the brake drum and the lever, probably a screwdriver, is applied through this hole. The two ends of the adjuster are located onto the brake shoes and 'screwing out' the threaded portion pushes the shoes apart until the correct lining to drum clearance is obtained.

Automatic brake adjuster

This adjustment mechanism relies on the movement of the operating mechanism to operate a ratchet and pawl. The mechanism together with part of the two brake shoes is shown in Fig. 7.17.

The ratchet is a small toothed wheel that is fixed to the adjusting bolt. The pawl is on the end of the adjusting lever and it engages with the ratchet. The connecting rod which is in two parts, (A) and (B), and fits between the two brake shoes. The pawl is pulled lightly into contact with the ratchet by the spring. Operation of the brake shoes pushes the shoes apart and this creates a clearance at the end of the connecting rod. When there is sufficient clearance, the ratchet will rotate by one notch, which increases the length of the connecting

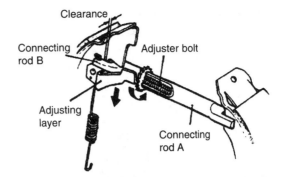

Fig. 7.17 An automatic brake adjuster

rod and takes up the excess clearance between the shoes and the drum.

These mechanisms work well when they are properly maintained and service schedules must be properly observed to ensure that the brakes are maintained in good order.

7.12 Wear indicators

The purpose of the friction pad wear indicator is to alert the driver to the fact that the pads have worn thin. The warning light on the dash panel is illuminated when the pads have worn by a certain amount. Figure 7.18 shows a pair of brake pads. One of the pads is equipped with a pair of wires whose ends are embedded in the friction material of the pad. When the pad wears

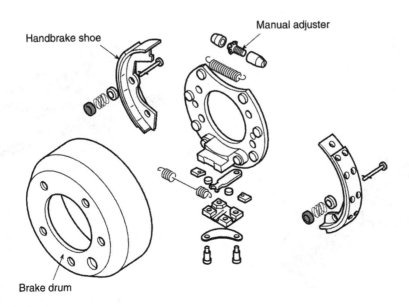

Fig. 7.16 A manual brake adjuster

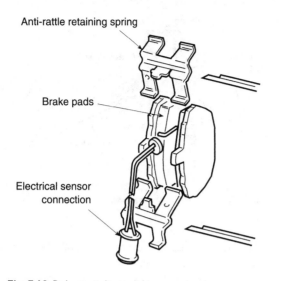

Fig. 7.18 Pad wear indicator light

down to the level of the ends of the wires, the wires are bridged electrically by the metal of the brake disc. This completes a circuit and illuminates the warning light.

7.13 Stop lamp switch

The purpose of the stop lamp switch is to alert following road users to the fact that the driver of the vehicle in front of them is applying the brakes. Figure 7.19 shows an arrangement that is frequently used to operate the brake-light switch. As the foot is applied to the pedal, a spring inside the switch closes the switch contacts to switch on the brake lights.

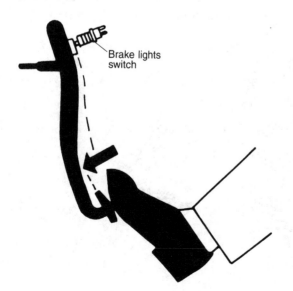

Fig. 7.19 A brake light switch

7.14 Brake pressure control valve

A pressure control valve is fitted between the front and rear brakes to prevent the rear wheels from 'locking up' before the front wheels. This arrangement contributes to safer braking in emergency stops. Figure 7.20 shows how such a valve is mounted on a vehicle.

The internal details of the valve are shown in Fig. 7.21.

'Normal' braking fluid under pressure passes through the valve, from port (B) to port (E) and the rear brakes. If the deceleration of the vehicle reaches the critical level, the ball (D) will move and seal off the fluid path to the bore

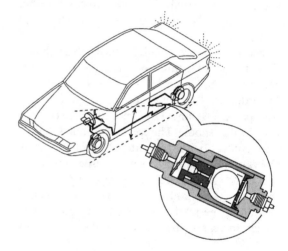

Fig. 7.20 The pressure limiting valve on the vehicle

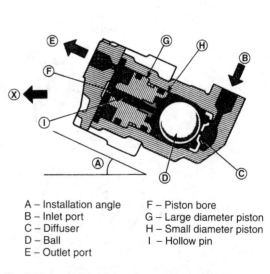

A – Installation angle F – Piston bore
B – Inlet port G – Large diameter piston
C – Diffuser H – Small diameter piston
D – Ball I – Hollow pin
E – Outlet port

Fig. 7.21 Internal detail (cross-sectional view) of inertia-type limiting brake pressure control valve – diagram simplified to show schematic flow of fluid

of the valve (F). At this point the rear brakes are effectively 'cut off' from the front brakes and the pressure on the rear brakes is held at its original level. Further pressure on the brake pedal increases the front brake pressure without increasing the rear brake pressure. Dependent on the design of the valve, further brake pedal pressure will cause the piston (G–H) to move and increase the pressure in the rear brake line.

When the deceleration falls, the ball will 'roll' back to its seat on the diffuser (C) and the pressure throughout the braking system is stabilised.

7.15 Brake fluid

Brake fluid must have a boiling temperature of not less than 190 °C, and a freezing temperature not higher than −40 °C. Brake fluid is hygroscopic, which means that it absorbs water from the atmosphere. Water in brake fluid affects its boiling and freezing temperatures, which is one of the reasons why brake fluid needs to be changed at the recommended intervals.

Brake fluid is normally based on vegetable oil and its composition is carefully controlled to ensure that it is compatible with the rubber seals and not corrosive to the metal parts. Some manufacturers use mineral oil as a base for brake fluids and their systems are designed to work with this fluid. It is important always to use only the type of fluid that a vehicle manufacturer recommends for use in their vehicles.

7.16 Braking efficiency

The concept of braking efficiency is based on the 'idea' that the maximum retardation (rate of slowing down) that can be obtained from a vehicle braking system is gravitational acceleration $g = 9.8 \, \text{m/s}^2$. The actual retardation

Fig. 7.22 Roller-type brake tester measuring the braking force at each wheel

obtained from a vehicle is expressed as a percentage of 'g' and this is the braking efficiency. For example, suppose that a vehicle braking system produces a retardation of $7 \, \text{m/s}^2$. The braking efficiency $= (7/9.8) \times 100 = 71\%$. Because of the physics of vehicle dynamics the braking efficiency can be obtained without actually measuring deceleration. This is so because the total weight of the vehicle divided by the sum of the braking forces applied between the tyres and the driving surface produces the same result. Brake testing equipment, as used in garages, uses this principle to measure braking efficiency.

The left (F1) and right (F2) front braking forces are measured and recorded. The vehicle is moved forward so the that the rear wheels are on the rollers. The test procedure is repeated and the two rear wheel braking forces (F3) and (F4) are recorded. The four forces are added together and divided by the weight of the vehicle.

For example, in a certain brake test on a vehicle weighing 1 700 kg, the four braking forces add up to 1 070 kg. This gives a braking efficiency of $(1\,070/1\,700) \times 100 = 70\%$. Often the data relating to a particular vehicle is contained on a chart kept near the test bay.

The regulations regarding braking efficiency and the permitted differences in braking, from side to side of the vehicle, together with other data are contained in the tester's manual. Any technician carrying out brake tests must familiarise themselves with the current regulations.

7.17 Anti lock braking system (ABS)

The term ABS covers a range of electronically controlled systems that are designed to provide optimum braking in difficult conditions. ABS systems are used on many cars, commercial vehicles and trailers.

The purpose of anti-skid braking systems is to provide safer vehicle handling in difficult conditions. If wheels are skidding it is not possible to steer the vehicle correctly and a tyre that is still rolling, not sliding, on the surface will provide a better braking performance. ABS does not usually operate under normal braking. It comes into play in poor road surface conditions, such as ice, snow, water, etc., or during emergency stops.

Figure 7.23 shows a simplified diagram of an ABS system, which gives an insight into the way that such systems operate.

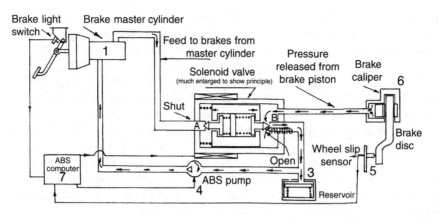

Fig. 7.23 A simplified version of an ABS

The master cylinder (1) is operated via the brake pedal. During normal braking, manually developed hydraulic pressure operates the brakes and, should an ABS defect develop, the system reverts to normal pedal operated braking.

The solenoid operated shuttle valve (2) contains two valves, A and B. When the wheel sensor (5) signals the ABS computer (ECU) (7) that driving conditions require ABS control, a procedure is initiated which energises the shuttle valve solenoid. The valve (A) blocks off the fluid inlet from the master cylinder and the valve (B) opens to release brake line pressure at the wheel cylinder (6) into the reservoir (3) and the pump (4) where it is returned to the master cylinder.

In this simplified diagram, the shuttle valve is enlarged in relation to the other components. In practice, the movement of the shuttle valve is small and movements of the valve occur in fractions of a second.

In practical systems, the solenoids, pump and valves, etc. are incorporated into a single unit, as shown in Fig. 7.24. This unit is known as a modulator.

This brief overview shows that an anti-lock braking system has sensors, an actuator, an ECU and interconnecting circuits. In order that the whole system functions correctly, each of the separate elements needs to be working correctly.

When deciding whether or not a vehicle wheel is skidding, or on the point of doing so, it is necessary to compare the rotational movement of the wheel and brake disc, or drum, with some part such as the brake back plate which is fixed to the vehicle. This task is performed by the wheel speed sensing system and the procedure for doing this is reasonably similar in all ABS systems, so the wheel speed sensor is a good point at which to delve a little deeper into the operating principles of ABS.

The wheel speed sensor

Figures 7.25 and 7.26 shows a typical wheel sensor and reluctor ring installation. The sensor contains a coil and a permanent magnet. The reluctor ring has teeth and when the ring rotates past the sensor pick up the lines of magnetic force in the sensor coil vary. This variation of magnetic force causes a varying voltage (emf) to be induced in the coil and it is this varying voltage that is used as the basic signal for the wheel sensor. The particular application is for a Toyota but its principle of operation is typical of most ABS wheel speed sensors.

The raw output voltage waveform from the sensor is approximately of the form shown in Fig. 7.27. It will be seen that the voltage and frequency increase as the wheel speed, relative to the brake back plate, increases. This property means that the sensor output is a good representation of the wheel behaviour relative to the back plate and, thus, to provide a signal that indicates whether or not the wheel is about to skid.

In most cases, this raw curved waveform is not used directly in the controlling process and it has first to be shaped to a rectangular waveform, and tidied up before being encoded for control purposes.

If the brake is applied and the reluctor (rotor) starts to decelerate rapidly, relative to the sensor pick-up, it is an indication that the wheel rotation is slowing down. If the road surface is dry and the tyre is gripping well, the retardation of the wheel will match that of the vehicle and normal braking will occur. However, if the road surface is slippery, a sudden braking application will cause the reluctor rotor and road wheel to decelerate at a greater rate than the vehicle, indicating that a skid is about to happen. This condition is interpreted by the electronic control

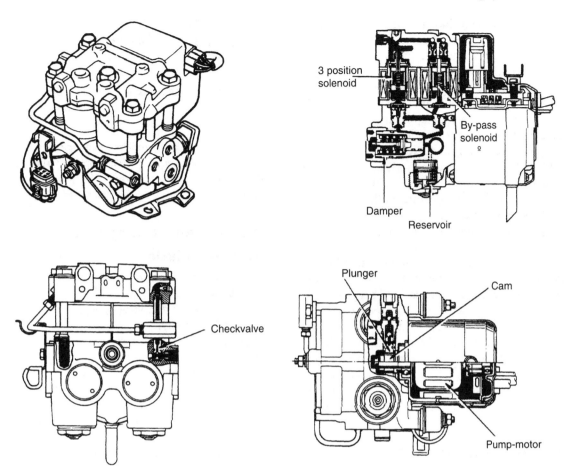

Fig. 7.24 ABS modulator

unit. Comparisons are made with the signals from the other wheel sensors and the brake line pressure will be released automatically, for sufficient time (a fraction of a second) to prevent the wheel from locking.

In hydraulic brakes, on cars the pressure release and re-application is achieved by solenoid valves, a pump and a hydraulic accumulator and

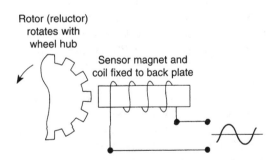

Fig. 7.26 The basic principle of the ABS sensor

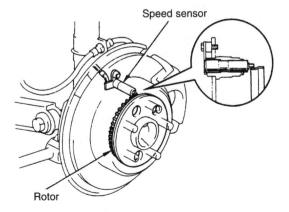

Fig. 7.25 ABS wheel sensor

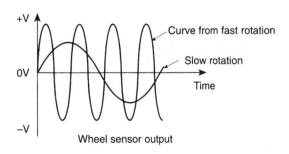

Fig. 7.27 The voltage waveform of the ABS sensor output

these are normally incorporated into one unit called the modulator. The frequency of 'pulsing' of the brakes is a few times per second, depending on conditions, and the pressure pulsations can normally be felt at the brake pedal.

With air brakes on heavy vehicles, the principle is much the same, except that the pressure is derived from the air braking system and the actuator is called a modulator. The valves that release the brakes during anti-lock operation are solenoid operated on the basis of ECU signals, and the wheel sensors operate on the same principle as those on cars.

As for the strategy that is deployed to determine when to initiate ABS operation, there appears to be some debate. Some systems operate what is known as 'select low', which means that brake release is initiated by the signal from the wheel with the least grip, irrespective of what the grip is at other wheels. An alternative strategy is to use individual wheel control. Whichever strategy is deployed, the aim is to provide better vehicle control in difficult driving conditions and it may be that the stopping distance is greater than it would be with expert manual braking.

The ABS warning light

ABS systems are equipped with a warning light. This lamp is illuminated when the system is not operating. When the vehicle is first started, the ABS warning light is illuminated and the system runs through a self check procedure. As the vehicle moves away, the ABS warning lamp will remain 'on' until a speed of 3 mile/h (7 km/h) is reached. If the ABS system is functioning properly, the warning lamp will not 'come on' again until the vehicle stops. The system is constantly monitoring itself when the vehicle is in motion. Should a fault occur, the warning light will again come on. Should this happen, the system reverts to normal braking operation and the vehicle should receive urgent attention to ascertain the cause of the problem.

7.18 Bleeding the brakes

During repair work, such as replacing hoses and wheel cylinders, air will probably enter the hydraulic system. This air must be removed before the vehicle is returned to use and the process of removing the air is called 'bleeding the brakes'.

Practical assignment

1. With the aid of sketches, describe the equipment and procedures for bleeding brakes on the types of vehicle that you work on.
2. Braking System Check.

This check should be completed in a logical manner and in the correct sequence. Make up a tick sheet to indicate pass/fail for each item. The following list should be used to make up the 'tick sheet'.

Checklist for braking system
Inside the vehicle
1. Check the operation of the brake pedal.
   (i) brake pedal travel,
   (ii) pedal feel, e.g., soft or spongy,
   (iii) pedal security,
   (iv) operation of brake light switch,
   (v) security of switch and cables.
2. Check operation of the handbrake.
   (i) distance travelled when applied (no more than 3 to 7 clicks),
   (ii) handbrake resistance (does it feel right?),
   (iii) security of handbrake lever,
   (iv) operation of ratchet and release button,
   (v) operation of warning lamp switch,
   (vi) security of switch and cables.

Under the bonnet
3. Master cylinder.
   (i) check the fluid level in the reservoir,
   (ii) check operation of fluid level warning device,
   (iii) inspect for security of mounting bolts, brackets, etc.,
   (iv) check operation of the servo unit,
   (v) look for evidence of hydraulic fluid leaks.
4. Pipes and hoses.
   (i) inspect brake pipes for signs of corrosion,
   (ii) check security of pipes (are the retaining clips in position?)
   (iii) look for evidence of leaks and check the security of connections and unions,
   (iv) test flexible hoses for signs of aging and damage,
   (v) check the condition of the servo vacuum hose.
5. Front brakes.
   (i) check operation of calipers/brakes,
   (ii) measure thickness of brake pads,
   (iii) measure thickness of brake discs and check the condition of the disc surfaces,

(iv)   movement of caliper on the slide (where fitted)

(v)   check for splits and cracks in flexible hoses,

(vi)   check for corrosion and damage in metal pipes and check the unions for security and tightness,

(vii)   check the security and operation of the pad wear indicator light and cables,

(viii)   check security and condition of anti-rattle devices,

(ix)   check for evidence of leaks,

(x)   check for movement and excessive free play in wheel bearings.

6. Rear Brakes.

(i)   check condition of brake drums and look for evidence of scoring (ensure that dust removal is done with a vacuum cleaner whilst wearing a mask)

(ii)   measure thickness of brake shoe linings.

(iii)   check operation of wheel cylinders (will require an assistant),

(iv)   check to see that all parts are in place and that brakes are correctly assembled,

(v)   check for fluid leaks,

(vi)   check security and condition of brake pipes and unions.

7. Under the Car.

(i)   security of pipes and flexible hoses, check hat all retaining clips are properly fitted,

(ii)   look for any evidence of fluid leaks,

(iii)   check the operation of handbrake cables, compensator, etc.,

(iv)   lubrication of cables and clevis pins,

(v)   security of handbrake linkages.

Self assessment questions

1. Brake fluid is said to be hygroscopic because:
   (a) it is like water
   (b) it absorbs water
   (c) it is heavier than water
   (d) it freezes at a higher temperature than water

2. Brake fluid absorbs water over a period of time. In order to prevent problems that may arise from this it is recommended that:
   (a) brakes are not allowed to overheat
   (b) brake shoes are changed at regular intervals
   (c) brake fluid is replaced at regular intervals
   (d) vehicles are kept under cover in bad weather

3. A diagonal split braking system requires:
   (a) two master cylinders
   (b) a tandem master cylinder
   (c) two calipers on each front brake
   (d) disc brakes at the front and drum brakes at the rear

4. In a test to determine braking efficiency a vehicle weighing 1 200 kg is placed on a brake testing machine, the brake tester showing the following readings. Front right 2 120 N, front left 2 080 N, rear right 1 490 N, rear left 1 510 N. The braking efficiency is:
   (a) 60%
   (b) 17%
   (c) 25%
   (d) 75%

5. On diesel engined vehicles the vacuum for the brake servo is provided by:
   (a) manifold vacuum
   (b) a compressor
   (c) a vacuum pump known as an exhauster
   (d) a venturi fitted in the engine air intake

6. Anti lock braking systems:
   (a) decrease stopping distance in all conditions
   (b) assist the driver to retain control during emergency stops
   (c) are not used on heavy vehicles
   (d) are a form of re-generative braking system

7. Disc brakes are said to be self adjusting because:
   (a) they use a ratchet and pawl mechanism
   (b) the pads are kept in light rubbing contact with the disc
   (c) the discs expand when heated
   (d) the brake servo compensates for wear

8

Steering, wheels and tyres

Topics covered in this chapter

Ackerman principle
Steering angles
Steering geometry
Track adjustment
Steering boxes
Toe-out on turns

The steering mechanism has two main purposes. It must enable the driver to: easily maintain the straight ahead direction of the vehicle even when bumps are encountered at high speeds; and to change the direction of the vehicle with the minimum amount of effort at the steering wheel. One of the simplest layouts is the **beam axle** arrangement (Fig. 8.1) as used on large commercial vehicles. This is where the hub pivots or swivels on a king pin (in the case of a car a top and bottom ball-joint) to give the steering action.

As can be seen the two stub axles are connected together by two steering arms and a track rod with ball-joints at each end. The **steering gearbox** converts the rotary movement of the steering wheel into a straight line movement of the steering linkage; it also makes it easier for the driver to steer by giving a **gear reduction**. The drop arm and drag link connect the steering gear box to the stub axle. The steering arms, track rod and ball joints connect the stub axles together and allow the movement to be transferred from one side of the vehicle to the other as well as providing for the movement of the linkage as the suspension operates.

8.1 Light vehicle steering layouts

To provide means of turning the front wheels of a vehicle left or right would not be too difficult were it not also necessary to make provision for their movement up and down with the suspension. Most modern cars now

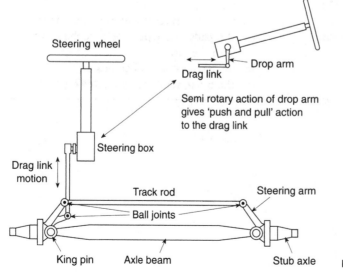

Fig. 8.1 Layout for beam axle steering system

have a fully independent front suspension. This creates serious problems when one wheel moves upwards or downwards independent of the opposite wheel. If a single track rod were used the tracking would alter every time the wheels moved causing the vehicle to wander from the straight ahead position. This problem has been overcome by the use of two or three part track rods. As can be seen from Fig. 8.2 on a vehicle fitted with a rack and pinion type steering the centre track rod has been replaced by the rack.

Learning tasks

1. Inspect several different types of vehicles and make a simple line diagram of the layout of the steering system identifying the following components: steering wheel and column, gearbox, linkage and ball-joints, stub axles and swivel joints.
2. Identify how the tracking is adjusted, check the manufacturer's setting for the tracking and state the tolerance given (the difference between the upper and lower readings). What equipment is used to check the measurements and how should it be set up on the vehicle?
3. When undertaking an MOT the steering system should be checked. Look in the MOT tester's manual and list the areas that are subject to testing. What in particular is the tester looking for? Your list should include the following areas:
(a) inside the vehicle
   - security of steering column mountings
   - play in upper steering column bush
   - amount of rotary movement in steering wheel before movement of the steered wheels
   - amount of lift in steering column
   - security of steering wheel to column
   - any undue noise or stiffness when operating the steering from lock to lock
(b) under the vehicle
   - security of steering box to chassis
   - signs of rust in the chassis around steering box mounting
   - excessive wear in steering column universal joints (these may be fitted inside the vehicle)
   - excessive play in steering box

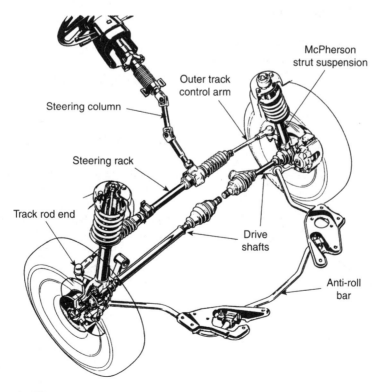

Fig. 8.2 Steering layout for IFS system

- amount of play in inner and outer track rod ends
- splits or holes in rubber boots (steering rack arrangement)
- loss of oil from steering box
- play in suspension mountings and pivots that may affect the operation of the steering
- signs of uneven tread pattern wear that may indicate a fault in the steering mechanism

8.2 Steering geometry

The subject of steering geometry is a very complex one in its own right and because of this many manufacturers import the knowledge and skills of specialists. However, the basic principles are relatively simple to understand and apply to most vehicles. The development of the steering and suspension (to which it is very closely linked) is based on past experience; much of the work has been through an evolutionary process learning from mistakes and modifying systems to suit varying applications. Many of the terms used are applied only to the steering and have no alternative; hopefully most of these are explained in the text.

The geometry of steering may best be understood by looking at Fig. 8.3. A swinging beam mounted on a turntable frame turns the wheels. This keeps all the wheels at right angles to the

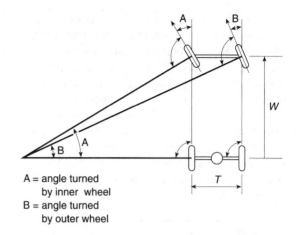

A = angle turned by inner wheel
B = angle turned by outer wheel

Fig. 8.4 Differing angles of front wheels about centre of turn

centre of the turn which reduces tyre wear especially when turning a corner.

Rudolph Ackermann took out a patent in 1818 in England which is now widely used and is known as the **Ackermann layout**. The angles of the front wheels about the turning point depend upon the wheel base (W) and the width of the track (T). In 1878 Jeantaud showed that the layout should conform to Fig. 8.4. In this arrangement the inner wheel (A) turns through a larger angle than the outer wheel (B) to give true rolling motion. The Ackermann layout does not fully achieve these conditions in all wheel positions; normally it is only accurate when the wheels are straight ahead and in one position on each left and right turn wheel setting. This system gives as near true rolling motion as possible together with simplicity.

8.3 Ackermann principle

The Ackermann layout is obtained by arranging for the stub axles to swivel on king pins or ball-joints to give the steering action of the wheels. The track rod ball-joints are positioned on an imaginary line drawn between the king pins and the centre line of the vehicle. When the track rod is positioned behind the swivel pins it is made shorter and has the protection of the axle, but the rod must be made stronger as it is in compression when the vehicle is being driven. This is shown in Fig. 8.5(a). When it is positioned in front of the swivel pins (Fig. 8.5(b)), it is made longer and can be out of the way of the engine and is made thinner as it is in tension when the vehicle is being driven.

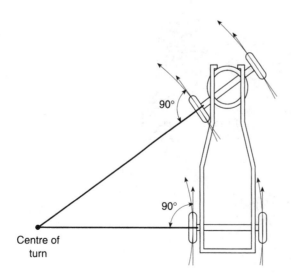

Fig. 8.3 Swing beam steering

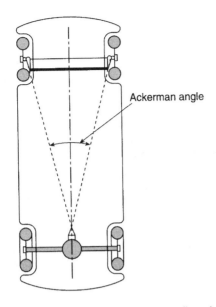

Fig. 8.5(a) Track rod behind swivel pins still conform to Ackerman principle

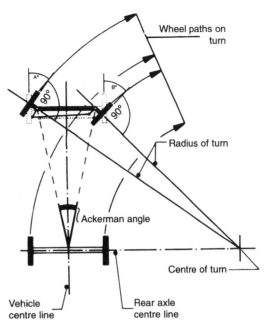

Fig. 8.6 Operation of Ackerman principle right-hand wheel turns more than left-hand wheel

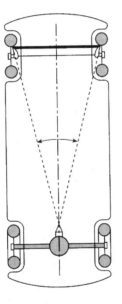

Fig. 8.5(b) Track rod in front of swivel pins still conform to Ackerman principle

Figure 8.6 shows the arrangements for an independent front suspension where the top and bottom ball-joints are placed in a line which forms the king pin inclination and also allows for the movement of the steered wheels both up and down over the irregularities of the road surface and as the driver turns the wheels to negotiate corners.

8.4 Centre-point steering

The stub axle arrangement shown in Fig. 8.7 has certain disadvantages due to the **off-set** (x). These are:

- there is a large force generated, due to the resistance at the wheel (R) from the road surface trying to turn the steering about the swivel pin (F) especially when the brakes are applied;
- the forces generated produce large bending stresses in the stub axle and steering linkages;
- heavy steering as the steered wheel has to rotate around the king pin. When this off-set is eliminated and the centre line of the wheel and the centre line of the king pin coincide at the road surface then **centre-point steering** is produced. This condition is achieved by the use of **camber**, swivel axis inclination (often referred to as **king pin inclination** or KPI for short) and **dishing** of the wheel.

Camber

Camber (Fig. 8.8) is the amount the wheel slopes in or out at the top relative to the imaginary vertical line when viewed from the front of the vehicle. This reduces the **bending** and **'splaying out' stresses** on the stub axle and steering linkages. The amount of camber angle is usually quite small as large angles

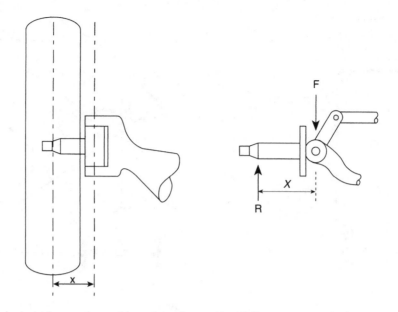

Fig. 8.7 Diagram shows the forces acting on the steering without centre point

will produce rapid tyre wear on the shoulder of the tyre tread as the inside tread of the tyre (*R*) travels a greater distance than the outside tread (*r*), even when travelling in a straight line. A cambered wheel tends to roll in the direction in which it is leaning which has the effect of producing a side force. Two beneficial effects of this are that it reduces any small sideways forces imposed on the wheel by ridges in the road surface, and that it also produces a small lateral pre-load in the steering linkage. The actual angle varies depending on the suspension system used, but normally it is no more than 2°.

King pin or swivel axis inclination (KPI)

When the king pin is tilted inwards at the top the resulting angle between the vertical line and the king pin centre line is called the king pin

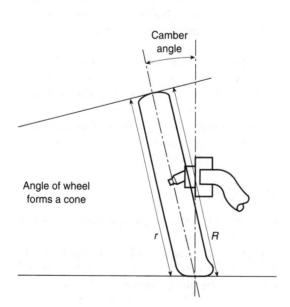

Fig. 8.8 Camber angle

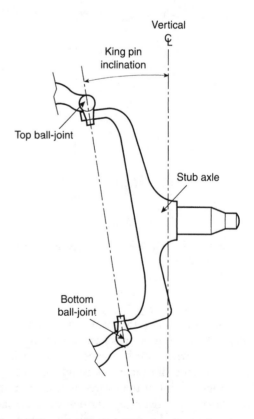

Fig. 8.9 King pin inclination (KPI) sometimes referred to as swivel axis

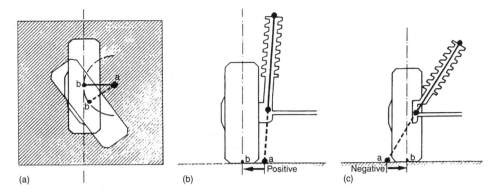

Fig. 8.10 Steering roll radius (a) no steering roll radius, (b) positive roll radius, (c) negative roll radius

inclination (Fig. 8.9). Normally this will be between 5° and 15° to produce **positive off-set** and to accommodate the brake, wheel bearings and drive shaft joint. When the king pin is set outwards at the bottom then **negative off-set** is produced.

As the wheel is steered through an angle it will pivot around the king pin. This will have the effect of lifting the front of the vehicle helping to give a self-centring action to the steering.

8.5 Steering roll radius

If the steering wheel is turned, the front wheels move along an arc around (a) in Fig. 8.10; this is where the extension of the king pin centre line meets the ground. The point (b) is where the tyre centre line meets the ground; radius (a–b) is the steering roll radius.

Whether the steering roll radius is positive or negative depends on the position of the KPI. It is positive if the extension of the KPI axis meets the ground on the inside of the **tyre contact centre**. It is negative if the extension of the KPI meets the ground outside the tyre contact centre. The advantage of having **negative steering roll radius** is increased directional stability in the case of uneven braking forces on the front wheels, or if a tyre is suddenly deflated.

As can be seen in Fig. 8.11 if the brake force on the right-hand front wheel is greater, then the vehicle tends to slide in an arc around that wheel, which means that the rear of the vehicle veers out to the left. In a vehicle with negative steering roll radius, the force of the car in motion will turn the wheel with the stronger braking force around the lower arm formed by the steering roll radius.

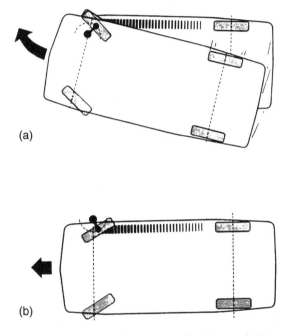

Fig. 8.11 (a) With positive steering roll and uneven braking on one front wheel, (b) with negative roll radius extra braking effort equals direction of steered wheels giving greater stability

8.6 Castor

The action of the castor may best be understood by looking at the castor wheel on a shopping trolley. When the trolley is pushed forwards the castor wheels always follow behind the **pivot points** where they are attached to the trolley. The same action applies to the front wheels of the motor vehicle. When the pivot point is in front of the steered wheel the wheel will always follow behind. This is called **positive castor** and is used on rear wheel drive vehicles; most front wheel drive and four wheel drive vehicles have **negative castor**.

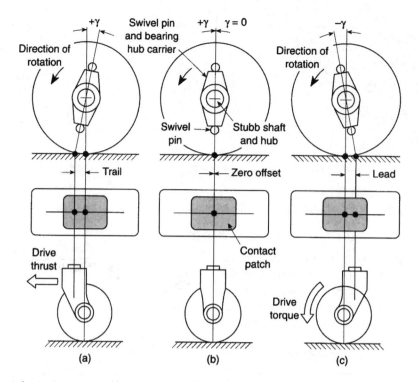

Fig. 8.12 Castor angle steering geometry (a) positive castor, (b) zero castor, (c) negative castor

This angle viewed from the side of the vehicle gives some 'driver feel' to the steering. It enables the steering to **self-centre** and a force must be exerted on the steering wheel to overcome the castor action of the steered wheels. It is produced by tilting the king pin forwards at the bottom approximately 2° to 5° so that when the line of the swivel axis of the king pin is extended it lies in front of the tyre contact point on the road surface; too much castor gives heavy steering and too little causes the vehicle to 'wander'. Castor angle steering geometry is shown in Fig. 8.12.

Tolerances may vary for all the steering angles so it is important to make reference to the manufacturer's specification as even small variations from this can lead to rapid tyre wear and poor handling characteristics. The checking of camber, king pin inclination and castor is done by using a set of gauges such as the Dunlop castor, camber and king pin inclination gauges together with a set of turn tables. The vehicle is driven onto the turn tables and the camber is measured by placing the camber gauge on the side of the wheel. The bubble in the level is adjusted to the central position to give the reading. The castor and KPI gauges are attached to the wheel and set approximately horizontal and the turn tables are adjusted to zero. Steer the wheel to

be checked 20° in (i.e. the right-hand wheel steered to the left and the left-hand wheel steered to the right). Set castor and KPI dials to zero at the pointer and centre the bubbles in the spirit levels by turning the lower knurled screws. Steer the wheel to be checked 20° out and centre the bubbles in the spirit levels by turning the castor and KPI dials and take the readings off the dials.

8.7 Wheel alignment

Adjusting the track

Often referred to as **toe-in/toe-out** or **tracking**, wheel alignment is a plan view of the wheels and means 'are the wheels correctly aligned with each other?' When the vehicle is travelling in a straight line all the wheels must be parallel, especially the steered wheels.

Arrangement for changing or adjusting the track is made in the track rod or outer track control arms and, when correctly aligned for toe-in, the distance across the front of the steered wheels measures less than the distance across the rear. These measurements are taken using very accurate measuring gauges of which several types are available (e.g. the **Dunlop optical alignment gauge** (Fig. 8.13)). The gauges

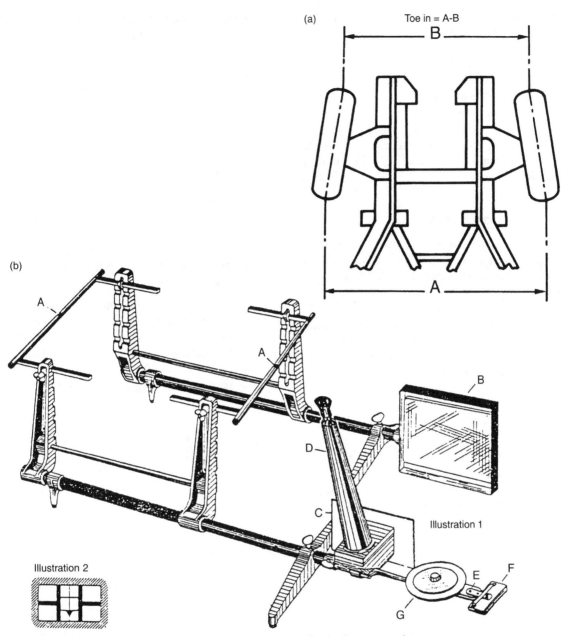

To assemble gauge

Assemble gauge as shown in illustration 1 with the periscope (D) fixed on the left hand unit and the mirror (B) on the right hand unit.

The contact bars may be fitted at any of five different height positions to suit the radius of the tyre and wheel assembly being checked. The height of all the bars must be the same and should be selected to bring the bars as near hub centre height as possible.

Each bar may be fitted into the support arms in either of two directions providing a range of width sufficient to cover all tyres on 9" to 24" diameter rims. The contact bars may be both inboard of the support arms, both outboard of the support arms, or one inboard and one outboard according to need.

To check accuracy of gauge

1. Stand the complete gauge on a level, clean floor with contact bars touching as shown in (A) illustration 1.

2. Adjust mirror and periscope until the reflection of target plate (C) is visible through periscope.

3. Sighting through periscope move pointer (E) until the image reflects the hair-line in the centre of the triangle between the vertical lines as in illustration 2.

The pointer should now be at zero on graduated scale (F). If not, slacken the two wing nuts holding the scale, adjust the scale to zero and retighten wing nuts, The gauge is now ready for use.

Fig. 8.13 (a) Toe-in, (b) Dunlop AGO/40 optical alignment gauge

are set to zero and are positioned on the wheels. The alignment is adjusted to give a reading on the scale, which is then checked against the manufacturer's specifications.

A more accurate method would be to use four **wheel alignment gauges**. This would give the mechanic more precise information as to which wheel needs adjusting, whether the front axle lines up with the rear axle and if the vehicle suspension has been misaligned in an accident.

Toe-out on turns

When turning a corner, the inner wheel turns through a greater angle than the outer wheel. This difference in angles is called 'toe-out on turns' (Fig. 8.14). To check these angles the wheels are placed on turn tables and the outer wheel is turned on its axis through 20°. The inner wheel should now read a larger angle (typically approximately 22°). This is because it is rolling around a smaller radius.

When checking steering alignment such as toe-in or toe-out, a number of factors must be attended to, for example:

1. The vehicle must be on a level surface.
2. The vehicle must be loaded to the makers specification, such as may be the kerbside unladen weight figure.
3. The tyres must be inflated to the correct pressure.
4. The front wheels should be in the straight ahead position.
5. The vehicle should be rocked from side to side to ease any stresses in the steering system.
6. Prior to checking the track, the vehicle should be rolled forward by about a vehicle's length in order to take up small amounts of slack in the linkage.

If the track requires adjustment (toe-in or toe-out) it is important to ensure that the track rods are maintained at the correct lengths, because this will affect the toe-out on turns. In the example shown in Fig. 8.14(b), the track rods are of equal length. This may not always apply and

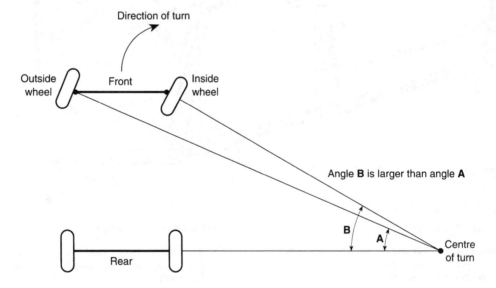

Fig. 8.14(a) Toe-out angle on turns. Inner wheel turns through a greater angle

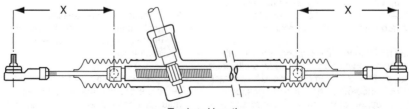

Fig. 8.14(b) Track rod length

such details must be checked prior to making any adjustments.

Learning tasks

1. Draw a simple diagram of the steering layout to conform to the Ackermann principle. Show the position of the front and rear wheels, point of turn, approximate angles of the front wheels and the position of the track rod. Indicate on your diagram where adjustments may be made.
2. Make a list of reasons why the steering mechanism would require adjusting.
3. Draw up a workshop schedule that you could use when checking a steering system for signs of wear and serviceability.
4. What important safety factors would you consider when working under a vehicle on the steering system.
5. Name two steering faults that would cause uneven tyre wear. How would these faults be put right?
6. If a customer came into the workshop complaining that his vehicle was 'pulling to one side' when being driven in a straight line, what would you suspect the fault to be? Describe the tests you would undertake to confirm your diagnosis.
7. Check the tracking on a vehicle using both optical and four wheel alignment gauges. Adjust the steering to the middle of the limits set by the manufacturer.

8.8 Steering gearboxes

The steering gearbox is incorporated into the steering mechanism for two main reasons:

- to change the rotary movement of the steering wheel into the straight line movement of the **drag link**;
- to provide a gear reduction and therefore a torque increase, thus reducing the effort required by the driver at the steering wheel.

Quite a number of different types of steering gearboxes have been used over the years, the most common ones being:

- worm and roller
- cam and peg
- recirculating ball (half nut)
- rack and pinion

The gearbox should have a degree of reversibility to provide some driver 'feel' and at the same time keep the transmission of road shocks through to the steering wheel to a minimum. The arrangement should be positive, i.e. it should have a minimum amount of backlash. Most steering gearboxes are designed with the following adjustments:

- end float of the steering column, usually by shims;
- end float of the rocker shaft, either by shims or adjusting screw and lock nut;
- backlash between the gears – these can be moved closer together, again by the use of shims or adjusting screw and lock nut.

It is essential to keep the 'backlash' to a minimum to provide a positive operation of the steering mechanism and to give directional stability to the path of the vehicle. The components inside the gearbox are lubricated by gear oil and the level plug serves as the topping up and level position for the oil.

Worm and roller

In this arrangement, shown in Fig. 8.15, the **worm** (in the shape of an hour glass) is formed on the inner steering column. Meshing with the worm is a **roller** which is attached to the **rocker shaft**. As the steering is operated the roller rotates in an arc about the rocker shaft giving the minimum amount of backlash together with a large gear reduction. This means that for a large number of turns on the steering wheel there is a very small number of turns on the rocker shaft, with very little free play. Because of the specialized machining geometry and shape, the hour glass worm is really in the form of a cam rather than a gear. This is why the arrangement is sometimes known as a cam-and-roller steering gearbox. It will mainly be found on LGVs as it provides a large gear reduction and can transmit heavy loads.

Cam and peg

In the cam and peg steering box a **tapered** peg is used in place of the roller. This engages with a special **cam** formed on the end of the inner steering column. The peg may be made to rotate on needle roller bearings in the rocker arm to reduce friction as the steering column is rotated and the peg moves up and down the cam.

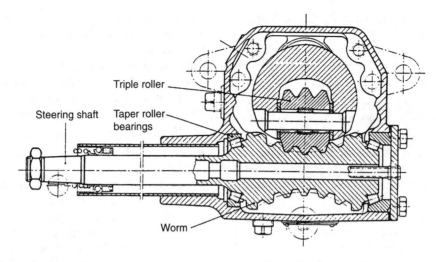

Fig. 8.15 Worm and roller steering gearbox

Recirculating ball

In this arrangement the worm is in the form of a thread machined on the inner steering column. A nut with **steel ball bearings** acting as the thread operates inside the nut; as the worm rotates the balls reduce the friction to a minimum. In many cases a half nut is used and a transfer tube returns the balls back to the other side of the nut. A peg on the nut locates in the rocker arm which transfers a rocking motion to the rocker shaft. In Fig. 8.16 a sector gear is used to transfer the movement to the rocker shaft.

Rack and pinion

This type, shown in Fig. 8.17, is now probably the most common type in use on cars and light vehicles. It has a rack which takes the place of the middle track rod and outer track rods (sometimes called tie rods) which connect to the steering arms at the hub. The pinion is

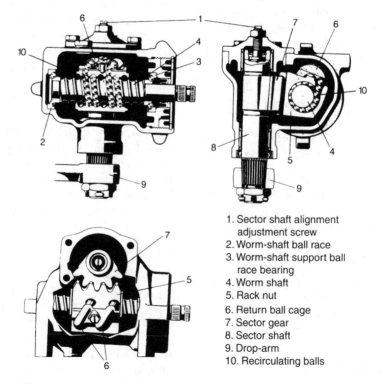

1. Sector shaft alignment adjustment screw
2. Worm-shaft ball race
3. Worm-shaft support ball race bearing
4. Worm shaft
5. Rack nut
6. Return ball cage
7. Sector gear
8. Sector shaft
9. Drop-arm
10. Recirculating balls

Fig. 8.16 Recirculating ball rack and sector steering box

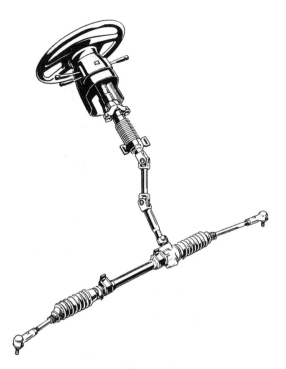

Fig. 8.17 The complete layout of a rack and pinion steering system showing the steering column, with universal joints (UJs)

mounted to the steering column by a universal joint as often the steering column is not in line with the input to the steering rack. This gives ease of mounting and operation of the gearbox.

On each end of the rack is a ball joint to which the track rod is mounted; these are spring loaded to allow for movement together with the minimum of play in the joint. The system is arranged so that in the event of an accident the column, because it is out of alignment (not a solid straight shaft) will tend to bend at the joints. This helps to prevent the steering wheel from hitting the driver's chest and causing serious injury. A spring loaded rubbing pad (called a slipper, Fig. 8.18) presses on the underside of the rack to reduce backlash to a minimum and also to act as a damper absorbing road shocks that are passed back through the steering mechanism from the road surface.

Adjustment of the steering rack

The inner ball-joints of the outer track rod are adjusted to give the correct pre-load and the locking ring of the ball-joint housing is tightened. In some cases it is locked to the rack by locking pins which must be drilled out to dismantle the joint as shown in Fig. 8.19.

Shims are used to adjust the slipper/rubbing block to give the correct torque on rotating the pinion (Fig. 8.20). Shims are also used to adjust the bearings on the pinion. It is important that the correct data is obtained and used as these may vary from model to model.

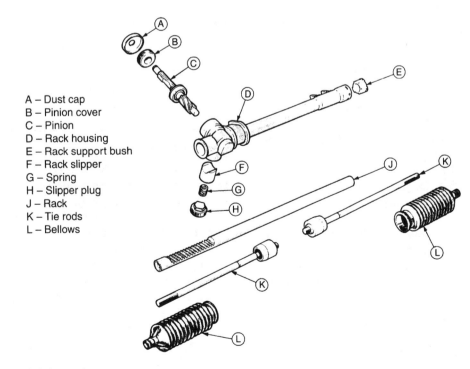

A – Dust cap
B – Pinion cover
C – Pinion
D – Rack housing
E – Rack support bush
F – Rack slipper
G – Spring
H – Slipper plug
J – Rack
K – Tie rods
L – Bellows

Fig. 8.18 Exploded view of steering gear

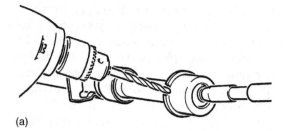

(a)

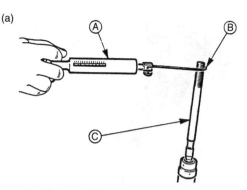

(a)

A – Piston pull side or spring balance
B – Wire hook 6mm (0.25") from end of tie rod
C – Tie rod

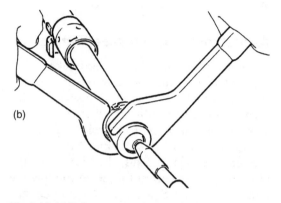

(b)

Fig. 8.19 (a) Drilling out tie rod inner ball-joint housing locking pins, (b) unscrewing tie rod inner ball-joint housings

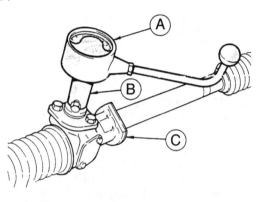

(b)

A – Torque gauge, tool number 15-041
B – Adaptor, tool number 13-008
C – Steering gear

Fig. 8.20 (a) Checking tie rod articulation, (b) checking pinion turning torque

Ball-joints

There are a number of requirements that must be fulfilled by steering ball-joints. These are:

- they must be free from excess movement (backlash) to give accurate control of the steering over the service life of the vehicle;
- they must be able to accommodate the angular movement of the suspension and the rotational movement of the steering levers;
- they must have some degree of damping on the steering to give better control of the wobble of the wheels especially at low speeds;
- where possible eliminate the need for lubrication at regular intervals – this is achieved through the use of good bearing materials and sealing the bearing with a good-quality grease at manufacture.

Ball-joints that are used on modern cars do not need lubricating as they are sealed for life, although on some medium to heavy vehicles they may require greasing at regular service intervals. Track rod and ball-joint assemblies are shown in Fig. 8.21.

> *Learning tasks*
>
> 1. What is the difference between angular movement and rotational movement? Which component in the steering mechanism uses angular movement and which uses rotational movement?
> 2. Using workshop manuals identify the method for adjusting the types of steering gearboxes identified in this chapter.
> 3. Dismantle each of the different types of steering gearboxes. Fill in a job sheet for each type. Note any faults or defects and make recommendations for their serviceability.

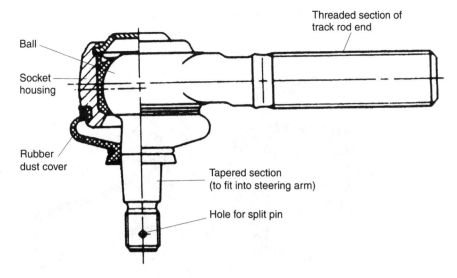

Fig. 8.21 Track rod end ball-joint assembly

8.9 Front hub assemblies

Non-driving hubs

Figure 8.22 shows one type of front hub assembly used on a front engine rear wheel drive vehicle. As can be seen the hub rotates on a pair of bearings that are either taper roller or angular-contact ball bearings. The inner bearing is usually slightly bigger in diameter than the outer, as this carries a larger part of the load. Various methods of adjustment are used depending on the type of bearing and the load to be carried. Some manufacturers specify a torque setting for the hub nut (usually where a spacer or shims are positioned between the bearings); others provide a castellated nut that has to be tightened to a specified torque and then released before the split pin is located. The hubs are lubricated with a grease that has a high melting point; this is because the heat from the brakes can be transferred to the hub, melt the grease and cause it to leak out onto the brakes. A synthetic lip-type seal is fitted on the inside of the hub to prevent the grease escaping. It also prevents water, dust and dirt from entering into the bearing. The arrangement in Fig. 8.23 is used mainly on larger vehicles where a beam axle is fitted instead of independent suspension.

Driving hubs

Figure 8.24 shows one arrangement of hub and drive shaft location using thrust-type ball bearings, in this case a single bearing with a double row of ball bearings. When the nut is tensioned/tightened to the correct torque it also adjusts the bearing to the correct clearance.

On a front wheel drive vehicle the drive shaft passes through the hub to engage with the flange that drives the wheel. In this way the hub becomes a carrier for the bearings (normally taper roller – Fig. 8.25) which are mounted close together. Where angular-contact bearings are used they are of the double-row type which have a split inner ring, in this way large-diameter balls can be used to give greater load carrying capacities; they also have the added advantage that

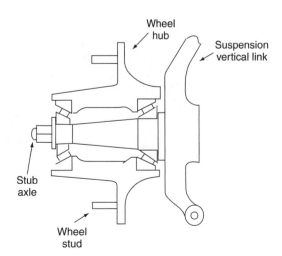

Fig. 8.22 Layout of non-driving hub assembly

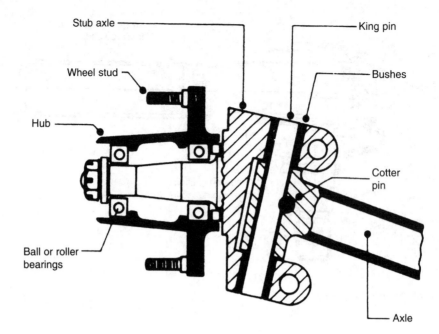

Fig. 8.23 Axle-beam and stub-axle assembly

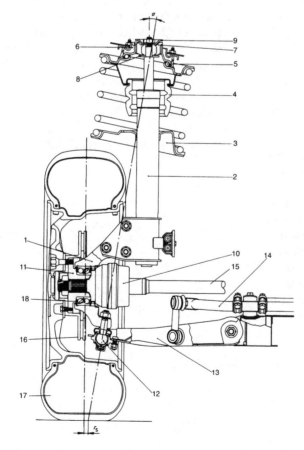

1 – Upper transverse arm
2 – Telescopic damper
3 – Spring mount
4 – Rubber bump stop
5 – Upper bearing
6 – Upper mounting
7 – Rubber mounting
8 – Upper swivel ball-joint
9 – Retaining nut
10 – Constant-velocity joint
11 – Wheel hub
12 – Lower swivel-pin ball-joint
13 – Lower wishbone
14 – Anti-roll bar
15 – Drive shaft
16 – Brake disc (air cooled)
17 – Road wheel
18 – Double row ball wheel bearing

Fig. 8.24 Independent suspension front wheel drive stub-shaft and swivel-pin carrier assembly

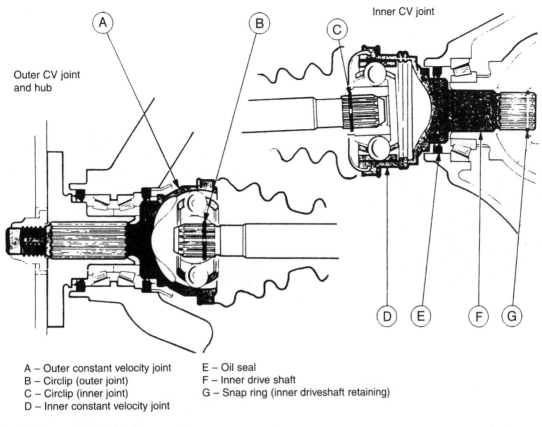

A – Outer constant velocity joint
B – Circlip (outer joint)
C – Circlip (inner joint)
D – Inner constant velocity joint
E – Oil seal
F – Inner drive shaft
G – Snap ring (inner driveshaft retaining)

Fig. 8.25 Layout of a driving hub arrangement

they require no adjustment and the torque setting for the drive nut can be high to give good hub-to-shaft security.

8.10 Light vehicle wheels

Most cars use wheels with a well base rim of the types shown in Fig. 8.26. The standard type of rim has the form shown in Fig. 8.26(a). The rim has a slight taper and the internal pressure in the tyre forces the bead of the tyre into tight frictional contact with the wheel. This tight contact is required so as to maintain an air-tight seal and also to provide the grip that will transmit driving and braking forces between the tyre and the wheel. The grip is dependent on the correct tyre pressure being maintained and, in the event of a puncture, the tyre bead is likely to break away from the wheel rim. The double hump rim shown in Fig. 8.26(b) is intended to provide a more secure fixing for the tyre in the sense that, should a puncture occur, the bead of the tyre will be held in place by the humps

and this should provide a degree of control in the event of a sudden, rapid puncture.

8.11 Types of car wheels

Most car wheels are fabricated from **steel pressings** and their appearance is often altered by the addition of an **embellisher**. Figures 8.27 and 8.28 show a range of wheels and wheel trims as used on Ford Escort vehicles. When weight and possibly 'sporty' appearance is a consideration **cast alloy wheels** are used. The metals used are normally **aluminium** alloy, or **magnesium** based alloy, and the wheels are constructed by casting. An alloy wheel is shown in Fig. 8.27(e).

Figures 8.27(f) and 8.27(g) shows the different types of wheel bolts and balance weights that are required. The steel wheel uses a bolt with a taper under the head. This taper screws into a corresponding taper on the wheel and, when the correct torque is applied, the wheel is secured to the vehicle hub. In order to prevent damage to the alloy wheel its bolts are provided with a

(a) (b)

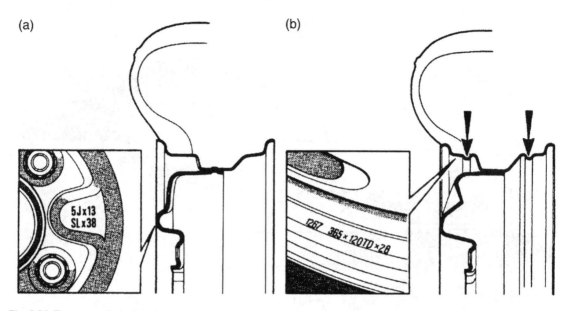

Fig. 8.26 Two types of wheel rim in common use

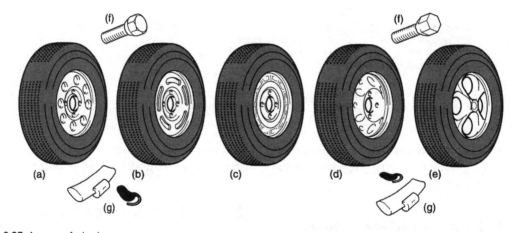

Fig. 8.27 A range of wheels

tapered washer. The taper on the washer sits in the taper on the wheel and the bolt can be tightened without 'tearing' the alloy of the wheel. Figure 8.27 also shows the two different types of wheel balancing weights that are required. The obvious point to make here is that the steel wheel weight is not suitable for use on the alloy wheel, and vice versa.

8.12 Wire wheels

These are wheels that are built up in a similar way to spoked bicycle wheels. Figure 8.29 shows a typical wire wheel. They are normally to be found on sports cars.

8.13 Wheel nut torque

In order to ensure that wheels are properly secure it is advisable to use a torque wrench for the final tightening of the wheel nuts. This is shown in Fig. 8.30.

8.14 Tyres

Tyres play an important part in the steering, braking, traction, suspension and general control of the vehicle. Two types of tyre construction are in use. These are:

- **cross-ply** tyres
- **radial-ply** tyres

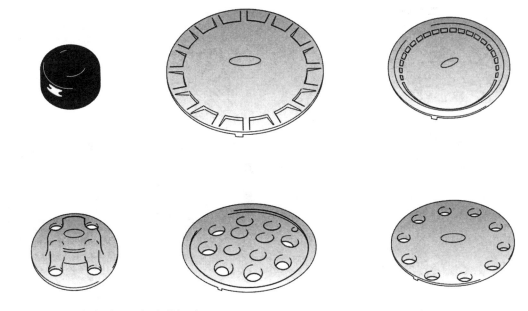

Fig. 8.28 A range of wheel trims (embellishers)

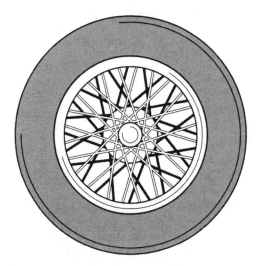

Fig. 8.29 A wire (spoked) wheel

Fig. 8.30 Using a torque wrench for the final tightening

The term **ply** refers to the layers of material from which the tyre casing is constructed.

Cross-ply tyre

Figure 8.31 shows part of a cross-ply tyre. The plies are placed one upon the other and each adjoining ply has the bias angle of the cords running in opposite directions. The angle between the cords is approximately 100° and the cords in each ply make an angle of approximately 40° with the tyre bead and wheel rim.

Radial-ply tyre

Figure 8.32 shows part of a radial-ply tyre. The plies are constructed so that the cords of the tyre wall run at right angles to the tyre bead and wheel rim.

Cross-ply versus radial-ply tyre

Radial-ply tyres have a more flexible side wall and this, together with the **braced tread**, ensures that a greater area of tread remains in contact with the road when the vehicle is cornering. Figure 8.33 shows the difference between cross-ply and radial-ply tyres when the vehicle is cornering. In effect, radial-ply tyres produce a better grip.

It is this remarkable difference between the performance of cross-ply and radial-ply tyres

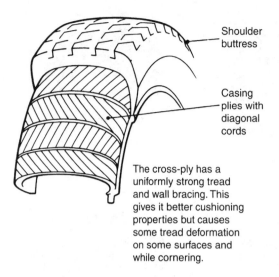

Shoulder buttress

Casing plies with diagonal cords

The cross-ply has a uniformly strong tread and wall bracing. This gives it better cushioning properties but causes some tread deformation on some surfaces and while cornering.

Fig. 8.31 Cross-ply tyre construction

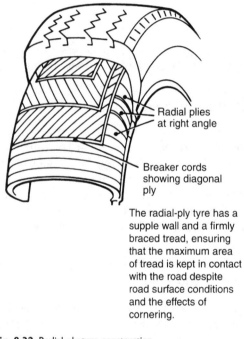

Radial plies at right angle

Breaker cords showing diagonal ply

The radial-ply tyre has a supple wall and a firmly braced tread, ensuring that the maximum area of tread is kept in contact with the road despite road surface conditions and the effects of cornering.

Fig. 8.32 Radial-ply tyre construction

that leads to the rules about mixing of cross-ply and radial-ply tyres on a vehicle. Cross-ply tyres on the front axle and radial-ply tyres on the rear axle are the only combination that can be used.

Tyre sizes

In general tyres have two size markings, one indicating the width of the tyre and other giving the diameter of the wheel rim on which the tyre fits. Thus, a 5.20 by 10 marking indicates a tyre which is normally 5.20 inches wide which fits a

The radial-ply tyre has a great number of advantages over the cross-ply tyre. Cornering stability is greatly improved; as you can see in the Figure the flexible side walls of the radial together with the braced tread ensure that the contact area is held firmly on the road when cornering. For this reason also, tyre life is extended.

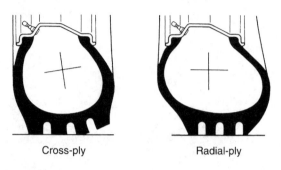

Cross-ply   Radial-ply

Fig. 8.33 Cross-ply and radial-ply tyres under cornering conditions

10 inch diameter wheel rim. Radial-ply tyres have a marking which gives the nominal width in millimetres and a wheel rim diameter in inches; for example, a 145–14 tyre is a tyre with a nominal width of 145 mm with a wheel rim diameter of 14 inches.

Originally vehicle tyres were of circular cross-section and the width of the tyre section was equal to its height. Over the years, developments have led to a considerable change in the shape of the tyre cross-section and it is now common to see tyres where the width of the cross-section is greater than its height. These tyres are known as 'low-profile' tyres. Modern tyres now have an additional number in the size; for example, 185/70–13. The 185 is the tyre width in millimetres, the 13 is the rim diameter in inches, and the 70 refers to the fact that the height of the tyre section is 70% of the width. This 70% figure is also known as the **aspect ratio**. The aspect ratio = (tyre section height/tyre section width) × 100 per cent. Tyre size and other information is carried on the side wall and Fig. 8.34 shows how they appear on a typical tyre.

Load and speed ratings

The tyre load and speed ratings are the figures and letters that refer to the maximum load and speed rating of a tyre. For example 82S marked on a tyre means a load index of 82 and a speed index of S. In this case the 82 means a load capacity of 475 kg per tyre

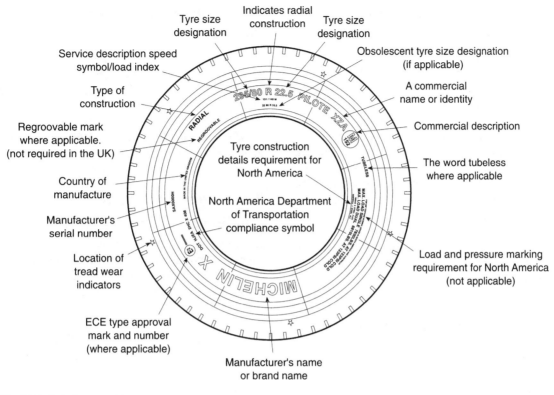

Fig. 8.34 Typical tyre markings

and the S relates to a maximum speed of 180 km/h or 113 mph. One of the main points about these markings is that they assist in ensuring that only tyres of the correct specification are fitted to a vehicle.

Tread wear indicators (TWI)

Tyres carry lateral ridges (bars) 1.6 mm high in the grooves between the treads at intervals around the tyre. The purpose of these **tread wear indicators** is to show when the tyre tread is reaching its minimum depth. Their position on the tyre is marked by the letters **TWI** on the tyre wall, near the tread.

Tyre regulations

Tyres must be free from any defect which might cause damage to any person or to the surface of the road. There are strict laws about worn tyres. They vary from time to time and it is important that you should be aware of current regulations. The following list is not complete but it does contain some of the more widely known ones. It is illegal to use a tyre which:

- is not inflated to the correct pressure;
- does not have a tread depth of at least 1.6 mm in the grooves of the tread pattern throughout a continuous band measuring at least 3/4 of the breadth of tread and round the entire circumference of the tyre;
- has a cut in excess of 25 mm or 10% of the section width of the tyre, whichever is the greater, which is deep enough to reach the ply or the cord.

There are several other regulations and details can be found in tyre manufacturers' data and Construction and Use Regulations.

In addition to the rules about condition of tyres there are also strict regulations about mixing of cross-ply and radial-ply tyres. For example, it is illegal to have a mixture of cross-ply and radial-ply tyres on the same axle. It is also illegal to have cross-ply tyres on the rear axle of a vehicle that has radial-ply tyres on the front axle. If radial-ply and cross-ply tyres are to be used on the same car, the rule is *radials on the rear*.

8.15 Wheel and tyre balance

Unbalanced wheels give rise to vibrations that affect the steering and suspension, they can be dangerous and, if not rectified, can lead to wear and damage to components. Wheel balance can be affected in many ways; for example, a damaged wheel or tyre caused by hitting the kerb. Such damage must be rectified immediately and when the repair has been made the wheel and tyre assembly must be re-balanced.

A **wheel balancer** is a standard item of garage equipment. When new tyres are fitted, or if wheels are being moved around on the vehicle to balance tyre wear, the wheel and tyre assembly should be checked and, if necessary, rebalanced.

8.16 Tyre maintenance

Tyre pressures

Tyre pressures have an effect on the steering, braking and general control of a vehicle and it is important to ensure that pressures are maintained at the correct level for the conditions that the vehicle is operating in. In addition to affecting vehicle handling, tyre pressures have an effect on the useful life of a tyre. The bar chart in Fig. 8.35 shows how various degrees of under inflation affect tyre life.

Inspecting tyres for wear and damage

Tyres can be damaged in a number of ways, such as running over sharp objects, impact with a kerb, or collision with another vehicle. Tyre damage can also be caused by incorrect tyre pressures, badly adjusted steering track and accidental damage that has affected steering geometry angles, such as castor and camber.

Tread wear patterns
Inflation pressures

In addition to checking tyre tread depth, the walls and casing of the tyre should be thoroughly examined. Tread wear patterns are a guide to the probable cause of a problem. For example, under-inflated tyres produce more wear at the outer edges because the centre of the tread is pushed up into the tyre. Over-inflated tyres wear at the centre of the tread because that is where most of the load is carried.

Steering track, camber
Toe-in

Figure 8.36 shows an axle and wheel arrangement where the front wheels have toe-in. In this diagram, the outer edges of the tyre are referred to as the shoulders. If the amount of toe-in is

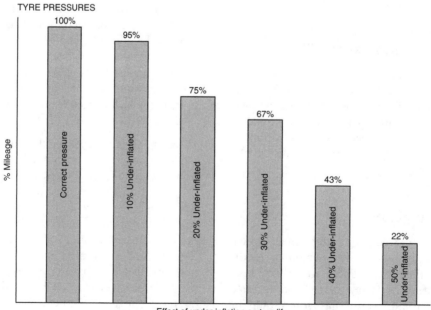

Fig. 8.35 Tyre wear and effect of pressures

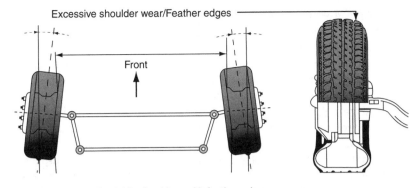

Excessive shoulder wear/Feather edges

Front

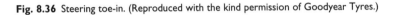

Toe-in (positive): faster wear of outside shoulders with feather edges outside to inside.
Toe-out (negative): faster wear of inside shoulders with feather edges inside to outside

Fig. 8.36 Steering toe-in. (Reproduced with the kind permission of Goodyear Tyres.)

too great, the tyre tread will wear in a feathered edge at the outer shoulder.

Toe-out

If the wheels have excessive toe-out, instead of toe-in, the tyre tread will wear in a feathered edge at the inner shoulder. (A feathered edge is a thin layer of rubber that projects outwards away from the tread.)

Camber

Excessive positive camber, where the wheel tilts out at the top, causes the tyre to wear at the outer shoulder. Excessive negative camber causes the tyre to wear at the inner shoulder.

Incorrect camber angle also affects the steering, because the vehicle will pull in the direction of the wheel that has the greatest camber angle.

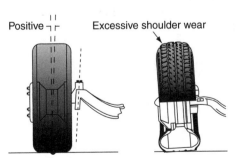

Positive ┐┌   Excessive shoulder wear

Positive camber (upper wheel part tilted outwards): faster wear of outside shoulder.
Negative camber (upper wheel end tilted inwards): faster wear of inside shoulder.

Fig. 8.37 Steering camber angle. (Reproduced with the kind permission of Goodyear Tyres.)

Learning tasks

1. Remove the wheels and tyres complete from a vehicle. Check the following and record your results on the job sheet for the customer:
   (a) tread depth
   (b) cuts and bulges
   (c) tyre pressures
   (d) signs of abnormal wear
   Complete you report with a set of recommendations.
2. Remove a tyre from a wheel, remove all the balance weights, clean the rim and inspect the tyre bead for damage. Check the inside of the tyre for intrusions and signs of entry of any foreign body such as a nail. Lubricate the bead of the tyre and replace on the wheel, inflate to the correct pressure, rebalance the wheel and refit to vehicle.
3. Produce a list of the current MOT requirements for wheels and tyres.
4. Describe a procedure that you have used for fitting a new tyre.
5. Make a list of the precautions that must be taken when removing a wheel in order to do some work on the vehicle. Make special note of the steps taken to prevent the vehicle from moving and also the steps taken to ensure that the vehicle cannot slip on the jack.
6. Examine a selection of wheel nuts. State why the conical part of the nut, or set bolt, is necessary. State why some vehicles, especially trucks, have left-hand threaded wheel nuts, on the near side. (Left-hand side of the vehicle when sitting in the driving seat).
7. Make a note of the type of wheel-balancing machine that is used in your workshop.

State the safety precautions that must be taken when using it and make a list of the major points that you need to know about when balancing a wheel and tyre assembly.

8. State the type of tyre-thread depth gauge that you use. Describe how to use it and state the minimum legal tread depth in the UK.

9. Examine a number of tyres and locate the tyre-wear indicator bars. State which mark on the tyre wall helps you to locate these wear indicators.

Practical assignment – steering worksheet

Introduction

After this practical exercise you should be able to:

- check and adjust steering alignment
- assess for wear in steering components
- remove and replace track rod ends
- check tyres for misalignment
- make a written report on your findings

Tools and equipment

- vehicle with appropriate steering mechanism
- appropriate workshop manual
- selection of tools and specialist equipment, e.g. alignment gauges
- vehicle lift or jacks and stands

Activity

1. Inspect the steering for wear and security, e.g.
   (a) uneven wear on tyres
   (b) play in track-rod ends, steering rack and column
   (c) play in wheel bearings
   (d) security of shock absorbers/dampers, check for leaks
   (e) wear/play in suspension joints
   (f) security/wear in anti-roll bar mountings
2. Remove and refit at least one track rod end.
3. Using appropriate equipment check tracking of vehicle's steered wheels.
4. Check steering angles – castor, camber, king pin inclination, toe-out on turns.
5. Make any adjustments necessary.
6. Investigate where possible the following types of steering gearboxes:
   (a) worm and roller
   (b) cam and peg
   (c) recirculating ball
   (d) rack and pinion
7. Make any adjustments on the above types of steering gearboxes after stripping and rebuilding them.
8. Produce a report on the condition of the steering including the gearboxes, indicating wear and serviceability.
9. State the meaning of the following terms:
   (a) tracking
   (b) toe-out on turns
   (c) steering angles – castor, camber, KPI, Ackerman angle
   (d) roll radius

Checklist

Vehicle

Exercise	Date started	Date finished	Serviceable	Unserviceable
Rack and pinion				
Recirculating ball				
Worm and peg				
Worm and roller (hour glass)				
Castor				
Camber				
KPI				
Toe-out on turns				
Optical alignment gauge				
Four wheel aignment				
Body alignment				

Student's signature

Supervisor's signature

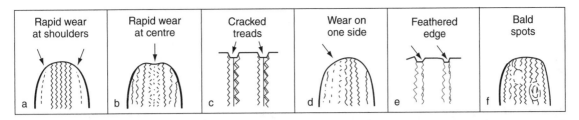

Rapid wear at shoulders	Rapid wear at centre	Cracked treads	Wear on one side	Feathered edge	Bald spots
a	b	c	d	e	f

Fig. 8.38 Tyre tread wear patterns

Self assessment questions

1. Figure 8.38 shows six examples of tyre wear problems. State a probable cause for each case.
2. Tread wear indicators on tyres:
   (a) are located in the grooves between the tyre treads
   (b) only become visible when the tread is worn to the legal limit
   (c) have their positions indicated by an arrow on the wheel rim
   (d) indicate that the tyre is completely legal
3. A car is to be fitted with two radial ply tyres and two cross ply tyres. In this case:
   (a) both radial ply tyres should be fitted to one side of the vehicle
   (b) the radial ply tyres should be fitted to the rear wheels
   (c) the tyres should be mixed diagonally
   (d) it does not matter where they are fitted
4. In a tyre size of 185/70 – 13, the aspect ratio is:
   (a) 185
   (b) 70%
   (c) 13
   (d) 198
5. In a steering system that uses a cam and peg steering box, the drop arm:
   (a) controls back lash in the mechanism
   (b) transmits a pull-push motion to the drag link
   (c) connects the steering arms to the stub axle
   (d) transmits steering wheel motion to the cam
6. When checking the steering geometry on a certain vehicle, it is found that the camber angle on the nearside is 2.6 degrees and on the offside it is only 1.2 degrees. This difference:
   (a) may have caused the vehicle to pull to the right
   (b) may have caused the left front tyre to wear on the inside edge
   (c) may have caused the vehicle to pull to the left
   (d) is not significant
7. With centre point steering:
   (a) the centre of tyre contact meets the road surface at the same spot as the centre line of the king pin
   (b) the vehicle's centre of turn is on a line that passes through the centre of gravity of the vehicle
   (c) the steering gear box in the centre of the vehicle
   (d) no king pin inclination or camber angle is required

Learning tasks (wheels and tyres)

1. Describe a procedure that you have used for fitting a new tyre.
2. Make a list of the precautions that must be taking when removing a wheel in order to do some work on the vehicle. Make special note of the steps taken to prevent the vehicle from moving and also the steps taken to ensure that the vehicle cannot slip on the jack.
3. Examine a selection of wheel nuts. State why the conical part of the nut, or set bolt, is necessary.
   State why some vehicles, especially trucks, have left-hand threaded wheel nuts on the near side. (Left-hand side of the vehicle when sitting in the driving seat.)
4. Make a note of the type of wheel balancing machine that is used in your workshop. State the safety precautions that must be taken when using it and make a list of the major points that you need to know about when balancing a wheel and tyre assembly.
5. State the type of tyre tread depth gauge that you use. Describe how to use it and state the minimum legal tread depth in the UK.
6. Examine a number of tyres and locate the Tyre Wear Indicator bars. State which mark on the tyre wall helps you to locate these wear indicators.

9

Electrical and electronic principles

Topics covered in this chapter

Principles of magnetism and electricity
Introductory electronics
Computer controlled systems
Networking and CAN
Multiplexing
Meters

9.1 Electricity and magnetism

Permaent magnets

Permanent magnets are found in nature. Magnetite, which is also known as lodestone, is naturally magnetic and ancient mariners used it as an aid to navigation. Modern permanent magnets are made from alloys. One particular permanent (hard) magnetic material named 'alnico', is an alloy of iron, aluminium, nickel, cobalt and a small amount of copper. The magnetism that it contains arises from the structure of the atoms of the metals that are alloyed together.

Magnetic force is a natural phenomenon. To deal with magnetism, rules about its behaviour have been established. The rules that are important to our study are related to the behaviour of magnets and the main rules that concern us are:

- Magnets have North and South poles;
- Magnets have magnetic fields;
- Magnetic fields are made up from lines of magnetic force;
- Magnetic fields flow from North to South.

If two bar magnets are placed close to each other, so that the north pole of one is close to the south pole of the other, the magnets will be drawn together. If the north pole of one magnet is placed next to the north pole of the other, the magnets will be pushed apart. This tells us that like poles repel each other and unlike poles attract each other.

Electromagnetism
The magnetic effect of an electric current

Figure 9.2 shows how a circular magnetic field is set up around a wire (conductor) which is carrying electric current.

Direction of the magnetic field due to an electric current in a straight conductor.

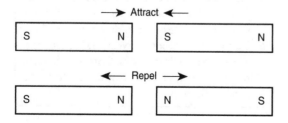

Fig. 9.1 Attraction and repulsion of magnets

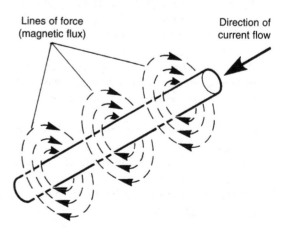

Fig. 9.2 Magnetic field around a straight conductor

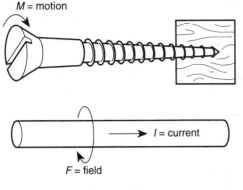

Fig. 9.3 Direction of magnetic field

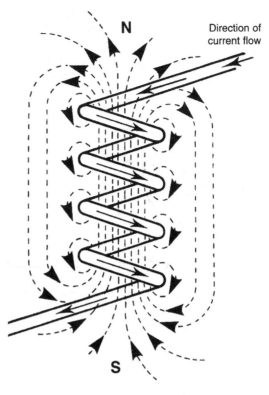

Current in   Current out

Fig. 9.4 Convention for representing current flow and magnetic field

Think of screwing a right-hand threaded bolt into a nut, or a screw into a piece of wood. The screw is rotated clockwise (this corresponds to the direction, north to south, of the magnetic field). The screw enters into the wood and this corresponds to the direction of the electric current in the conductor. The field runs in a clockwise direction and the current flows in towards the wood, as shown in Fig. 9.3.

This leads to a convention for representing current flowing into a conductor and current flowing out. At the end of the conductor, where current is flowing into the wire, a + is placed, this representing an arrow head. At the opposite end of the wire, where current is flowing out, a dot is placed, this representing the tip of the arrow.

Magnetic field caused by a coil of wire

When a conductor (wire) is made into a coil, the magnetic field created is of the form shown in Fig. 9.5; a coil such as this is the basis of a solenoid.

9.2 Electromagnetic induction

Figure 9.6 shows a length of wire that has a voltmeter connected to its ends. The small arrows pointing from the North pole to the South pole of the magnet, represent lines of magnetic force (flux) that make a magnetic field.

The arrow in the wire (conductor) shows the direction of current flow and the larger arrow,

Fig. 9.5 Magnetic field of a coil

with the curved tail, shows the direction in which the wire is being moved. Movement of the wire through the magnetic field, so that it cuts across the lines of magnetic force, causes an electromotive force EMF (voltage) to be produced in the wire. In Fig. 9.6(a), where the wire is being moved upwards, the electric current flows in towards the page. In Fig. 9.6(b), the direction of motion of the wire is downwards, across the lines of magnetic force, and the current in the wire is in the opposite direction. Current is also produced in the wire if the magnet is moved up and down while the wire is held stationary. Mechanical energy is being converted into electrical energy, and this is the principle that is used in generators, such as alternators and dynamos.

9.3 The electric motor effect

If the length of wire shown in Fig. 9.6(a) is made into a loop, as shown in Fig. 9.7, and electric current is fed into the loop, opposing magnetic fields are set up.

In Fig. 9.7, the current is fed into the loop via brushes and a split ring. This split ring is a

(a)

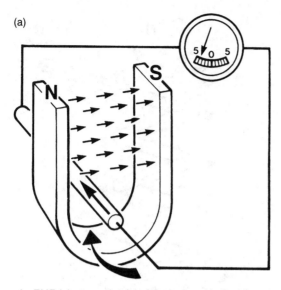

An EMF (electromotive force) is produced in the wire whenever it moves across the lines of magnetic flux.

Note: The reversal of current takes place when direction of movement is reversed.

(b)

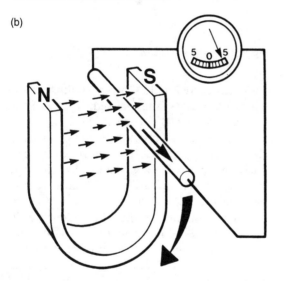

Fig. 9.6 Current flow, magnetic field and motion

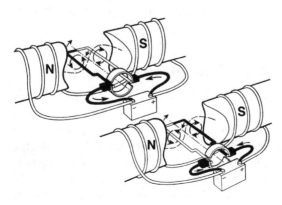

Fig. 9.7 Current flowing through the loop

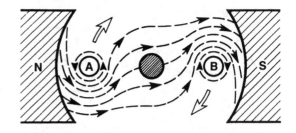

Fig. 9.8 Opposing magnetic fields

simple commutator. One half of the split ring is connected to one end of the loop and the other half is connected to the other end. The effect of this is that the current in the loop flows in one direction. The coil of wire that is mounted on the pole pieces, marked N and S, creates the magnetic field. These opposing fields create forces, which force (push) against each other, as shown in Fig. 9.8, and cause the loop to rotate.

In Fig. 9.8 the ends of the loop are marked A and B. B is the end that the current is flowing into, and A is the end of the loop where the current is leaving the loop.

By these processes, electrical energy is converted into mechanical energy and it is the principle of operation of electric motors, such as the starter motor. Just to convince yourself, try using Fleming's rule to work out the direction of the current, the field and the motion.

9.4 Fleming's rule

Consider the thumb and first two fingers on each hand. Hold them in the manner shown in Fig. 9.9.

- **Motion** This is represented by the direction that the thumb (M for motion) is pointing.

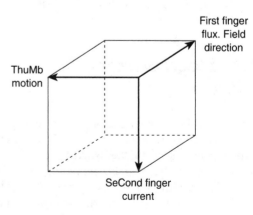

Fig. 9.9 Fleming's rule

- **Field** The direction (North to South) of the magnetic field is represented by the first finger (F for field).
- **Current** Think of the hard **c** in second. The second finger represents the direction of the electric current.

An aid to memory

We think that it is helpful for vehicle technicians to think of the MG car badge. If you imagine yourself standing in front of an MG, looking towards the front of the vehicle, the badge will present you with an M on your left and a G on your right. So you can remember M for MOTORS and G for GENERATORS, left-hand rule for motors, right-hand rule for generators.

9.5 Electrical circuit principles

Materials can usefully be divided into three categories:

1. conductors
2. semi-conductors
3. insulators

Electrical conductance and resistance

The ability of a material to conduct electricity is known as its conductivity. The opposite (inverse) of conductivity is resistivity.

Conductors

Examples of commonly used conductors are copper, aluminium and, in electronics, silver and gold.

Semi-conductors

Materials such as silicon do not have good insulating properties, and are not good conductors. These materials are known as semi-conductors.

Silicon is a semi-conductor material that is widely used in electronics and its conductivity can be varied by doping.

Insulators

These are materials where large electrical potential (applied voltage) causes only very small current to flow. Rubber, most plastics and ceramics, such as porcelain, are commonly used insulators.

Temperature coefficient of resistance

The resistance of most materials changes with temperature. In the case of conductors, the resistance of the material increases as the temperature increases. The relationship between resistance and temperature is affected by a factor known as the temperature coefficient of resistance.

Negative temperature coefficient of resistance (NTC)

When the temperature of a semi-conductor rises, the resistance falls and is said to have a negative temperature coefficient. A material with a negative temperature coefficient is used in the thermistor sensor that detects engine coolant temperature.

9.6 Electric circuits

Electricity will only perform its task if it has circuits to flow through. It is, therefore, necessary to have some knowledge of electric circuits. To begin with, we need to consider three electrical units. These are:

(i) **The volt** This is the unit of electrical pressure.
(ii) **The ampere, or amp** This is the unit of electric current.
(iii) **The ohm** This is the unit of electrical resistance.

Ohm's law

Voltage, current and resistance are related by Ohm's law. Ohm's law is often taken to mean that if a voltage is applied to a circuit that contains resistance, the current that flows in the circuit will be proportional to the voltage, provided that the temperature remains constant. Ohm's law is normally expressed mathematically as

VOLTAGE = CURRENT × RESISTANCE

The symbol V is used for volts. I = current in amperes. R = resistance in Ohms. Ohm's law is expressed algebraically as

$$V = IR$$

A common practice in teaching is to build circuits and by using low voltage (2 volts), direct current supplies and resistors of known value, to take meter readings of current and voltage. These readings are noted down and compared

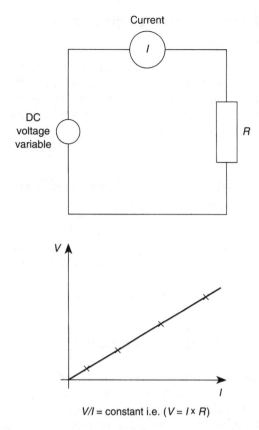

Fig. 9.10 Demonstrating Ohm's law

with those that are calculated by using Ohm's law. The examples shown here should make the point adequately. If you are a complete beginner you will need to take a short course in principles of electricity, preferably a course that offers plenty of practical work with circuits.

An Ohm's law circuit

Figure 9.10 shows a circuit that may be used to check the relation between voltage, current and resistance. The graph obtained by varying the voltage and measuring the current, for a given resistance, shows that $V = I \times R$.

9.7 Circuit testing and meters

Electrical systems are designed so that each element of a circuit will have a known resistance value. These resistance values must be maintained throughout the life of the system. The operation of vehicle electrical/electronic systems can be upset if resistance values, for any part of a circuit, are incorrect. Checking of electrical resistance in vehicle circuits is therefore an important part of a

vehicle technician's skill. Whilst it is possible to perform resistance checks on vehicle circuits without a great deal of background knowledge, it is beneficial to have some understanding of principles because a little deeper understanding often assists technicians to work out reasons why some device is not working correctly.

Meters

The instrument panel on most vehicles carries a great deal of driver information, for example, vehicle speed (speedometer), coolant temperature, oil pressure and so on. Meters are also used for fault diagnosis and it is useful to have an insight into the operating principles of two test meters as this will give an understanding that should assist in understanding how other meters work. Figure 9.11 shows two meters that can be used for

(a)

(b)

Fig. 9.11 Two types of test meters (a) moving coil, (b) digital

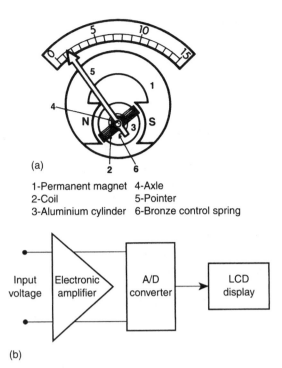

(a)

1-Permanent magnet 4-Axle
2-Coil 5-Pointer
3-Aluminium cylinder 6-Bronze control spring

(b)

Fig. 9.12 Detail of the test meters (a) moving coil, (b) digital voltmeter circuit

circuit testing. The one on the left is a moving coil meter and the one the right is a digital meter.

A moving coil meter

The moving coil meter takes the sample current into the moving coil. This causes the moving coil magnetic field to react with the field of the permanent magnet. Partial rotation of the coil and its pointer then occurs, against the force of the control spring. The amount of movement is related to the size of the variable being measured, for example, volts, amperes and ohms. A good-quality moving coil meter with a resistance of 20 000 ohms per volt is suitable for most circuit testing on vehicles.

A digital meter

A digital meter samples the variable being measured, such as volts, amperes and ohms and processes them in the electronic circuits of the meter. The resulting reading is displayed on a liquid crystal (LCD) panel. Digital meters have a high input impedance (resistance), which makes them suitable for diagnostic checks on electronic systems.

9.8 Electronics

Electronics is a branch of electrical engineering and it is concerned with the 'the science and technology

of the conduction of electricity in a vacuum, a gas or a semi-conductor'. It is the semi-conductor that concerns us most, at this stage, so it is necessary to explain what the term means.

Semi-conductors

Metals such as copper, aluminium and gold are good conductors of electricity and other materials like rubber and PVC, are bad conductors and are known as insulators. Materials such as silicon and germanium have conductivity which lies between that of good conductors and insulators. These materials are semi-conductors. Semi-conductors allow an electric current to flow only under certain circumstances.

Doping

Pure silicon, when treated (doped) with substances such as aluminium or gallium, is widely used in the manufacture of electronic components. The doping agents, which are added in tiny concentrations (probably less than one part per million) cause the silicon to become more conductive. The doping agents are divided into two groups, called 1) acceptors and 2) donors.

Typical acceptors are boron, gallium, indium and aluminium. When an acceptor is added to the silicon it forms p-type silicon. The acceptor forms holes, which behave like a mobile positive charge, and so this silicon is called p-type.

Typical donors are arsenic, phosphorous and antimony. When a donor is added to the silicon, it forms n-type silicon. It is called n-type because electric currents with in it are carried by (negatively charged) electrons.

The p-n junction

A p-n junction is a junction, between p-type silicon and n-type silicon, within a single crystal. If wires are attached to the material on either sided of the p-n junction, it is found that current will pass through the device in one direction, but very little in the opposite direction. Such a device is a junction rectifier, which is commonly known as a junction diode, or just a 'diode'.

Thus we have a device that acts like a one-way valve, which will pass current in one direction but not the other.

Figure 9.13 shows a representation of a p-n junction, in silicon, connected to a source of electricity. When it is connected in the way shown, with the positive battery terminal connected to the p-type material, the junction is

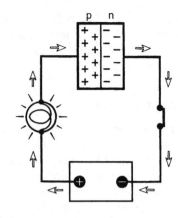

Fig. 9.13 A p-n junction

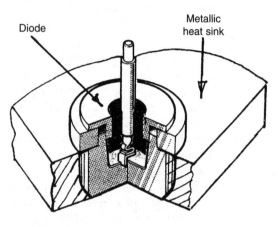

Fig. 9.14 A diode

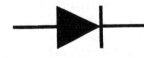

Fig. 9.15 Circuit symbol for a diode

said to be forward biased. When the forward voltage reaches approximately 0.7 volts, a continuous current will flow in the forward direction. Diodes are made for many applications, such as rectification of a.c. to d.c., switching circuits and so on. Figure 9.14 shows a diode in a metallic heat sink. The second terminal of the diode is formed from the metal case into which it is built and the heat sink conducts heat away from the diode.

Figure 9.15 shows the symbol for a diode. The arrow indicates the direction in which current will flow. Diodes are used in circuits to permit current flow in one direction only. A common use of a diode is in the conversion (rectifying) of alternating current (a.c.) to direct current (d.c.), as shown in the section on alternators.

Testing a junction diode

The fact that a diode conducts freely in the forward direction and virtually not at all in the reverse direction, means that an ohm-meter can be used to check for correct operation of the diode.

Other types of diodes

In addition to the junction diode, extensive use is made of other types of diodes. We shall have a brief look at two of these, namely the Zener diode and a light emitting diode (LED).

Zener diode

The Zener effect occurs when a heavily doped diode is reverse biassed. The Zener effect permits a low reverse voltage to produce breakdown and the diode conducts in the reverse direction. Diodes that are made to work in this way are called 'Zener' diodes. The Zener diode may be thought of as a voltage conscious switch and it is often used as a voltage reference in devices such as voltage regulators.

Figure 9.17 shows a simple circuit with a Zener diode in series with a bulb. If the potential divider slider is placed in the zero volts position and then moved towards the right, the Zener diode will conduct when the Zener voltage is reached

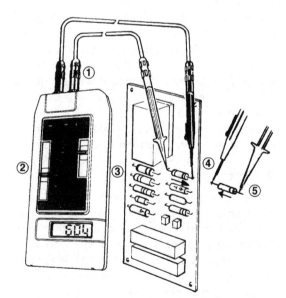

1 – Test leads
2 – Multimeter scale setting (continuity)
3 – Electronic components on a circuit board
4–5 – Test probes placed on the diode to check for correct operation

Fig. 9.16 Diode test using Avo 2002 (Thorn EMI)

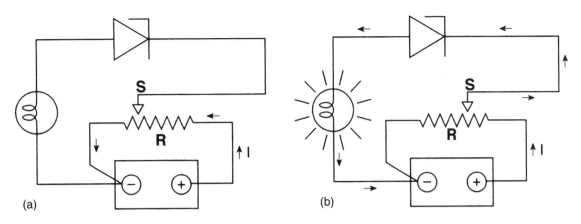

Fig. 9.17 Action of the Zener diode (a) low voltage – no light, (b) higher voltage – lamp lights

Fig. 9.18 An LED and its circuit symbol

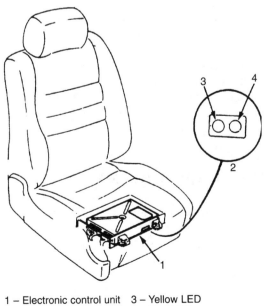

1 – Electronic control unit   3 – Yellow LED
2 – LED display   4 – Red LED

Fig. 9.19 LEDs for fault code display

and the bulb will light up. Note the direction of current flow through the Zener diode.

A light emitting diode (LED)

Silicon is an opaque material which blocks the passage of light and the energy which derives from the action of the diode is given off as heat. LEDs are different because they use other semi-conductor materials, such as gallium, arsenic and phosphorous and the energy deriving from their operation is given out as light. The material from which the LED is made determines the colour of the light given out.

It may be considered that the LED is a semi-conductor device which gives out light when current flows through it. LEDs are used as indicator lights and for fault indicator codes on some electronic systems.

Transistors

The transistor is another basic building block of electronic circuits. It may be thought of as two p-n junction diodes, connected back to back, as shown in Fig. 9.20 (this makes a bi-polar transistor). Transistors are used as switches or current/voltage amplifiers, and in many other applications.

Transistors have 3 connections, named as follows:

● base – (b)
● collector – (c)
● emitter – (e)

Transistors are non-conductive until a very small current flows through the base/emitter circuit. This very small current, in the base/emitter circuit, 'turns on' the collector/emitter circuit and permits a larger current to flow through the collector/emitter circuit. When

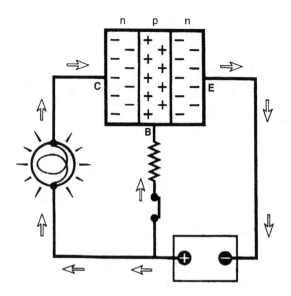

Fig. 9.20 Two p-n junctions back to back

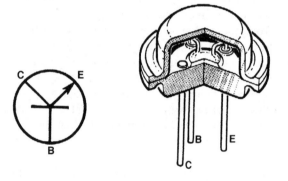

Fig. 9.21 A transistor

the base/emitter current is switched off the collector emitter immediately ceases to conduct and the transistor is effectively switched off. This happens in a fraction of a second (a few nano-seconds, usually) and makes the transistor suitable for high-speed switching operations, as required in many vehicle applications. The use of switching transistors is covered in the chapter on ignition systems.

Integrated circuits (I/Cs)

Many diodes, resistors and transistors can be made on a single piece of silicon (a chip). When these are connected together in specific ways, to make circuits on one piece of silicon, the resulting device is known as an integrated circuit (I/C). When a small number of diodes, transistors and resistors are connected together (about 12) the result is called small-scale integration (SSI). When the number is greater than 100, it is called

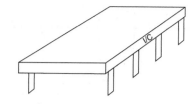

Fig. 9.22 A typical I/C

large-scale integration (LSI). From the outside, an I/C looks like Fig. 9.22.

The 'legs' along the sides are metal tags which are used to connect the I/C to other components. The number of tags is related to the complexity of the I/C. When I/CS are packaged as shown, they are known as DIL type, which means that the 'legs' are dual in line. Figure 9.23 shows an I/C built into an ECU.

Field effect transistors (FETs)

This type of transistor is faster in operation and uses less power than a bi-polar transistor. They are used in integrated circuits and many other applications.

The FET shown in Fig. 9.24 is a CMOS device. CMOS stands for complementary, metal, oxide, semi-conductor. These words relate to the way that the transistor works and the various layers of material that make up the transistor.

When the gate voltage reaches a certain level, the channel becomes conductive and an electrical connection is made between the source and the drain. The source and drain can thus be connected and disconnected electrically, by the application and removal of the gate voltage. There are several different types of FETs and the circuit symbol shown in Fig. 9.25 is typical of the type of symbol used to represent them on circuit diagrams.

9.9 Logic devices

A common use of transistors is to make logic devices, such as the NOR gate.

Figure 9.26 shows how a 'logic' gate is built up from an arrangement of resistors and a transistor. There are three inputs, A, B and C. If one or more of these inputs is high (logic 1), the output will be low (logic 0). The output is shown as A + B + C with a line or bar over the top; the + sign means OR. Thus, the A + B + C

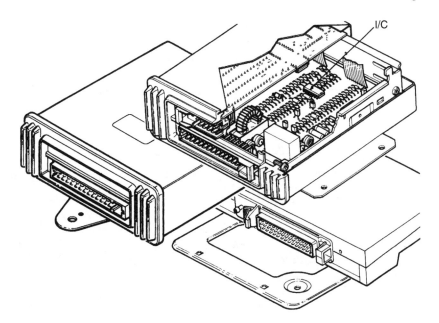

Fig. 9.23 An I/C in an ECU

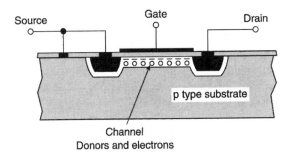

Fig. 9.24 A CMOS field effect transistor

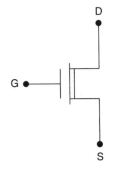

Fig. 9.25 Circuit symbol for FET

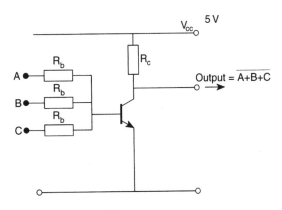

Fig. 9.26 The RTL NOR gate

Truth tables

Logic circuits operate on the basis of Boolean logic and terms such as NOT, NOR, NAND, etc. derive from Boolean algebra. This need not concern us here but it is necessary to know that the input–output behaviour of logic devices is expressed in the form of a 'truth table'.

The truth table for the NOR gate is given in Fig. 9.27.

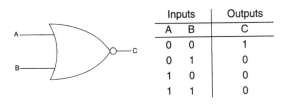

Inputs		Outputs
A	B	C
0	0	1
0	1	0
1	0	0
1	1	0

Fig. 9.27 NOR gate symbol and truth table

with the line above means 'not A or B or C' (NOR : NOT OR).

The base resistors R_b have a value that ensures that the base current, even when only one input is high (logic 1), will drive the transistor into saturation to make the output low (logic 0). (RTL stands for resistor transistor logic).

In computing and control systems, a system known as TTL (transistor to transistor logic) is used. In TTL logic 0 is a voltage between 0 and 0.8 volts. Logic 1 is a voltage between 2.0 and 5.0 volts.

In the NOR truth table, when the inputs A and B are both 0 the gate output, C, is 1. The other 3 input combinations each gives an output C = 1.

A range of other commonly used logic gates and their truth tables is given in Fig. 9.28.

Type of logic gate	USA symbol	UK symbol	Truth table
AND	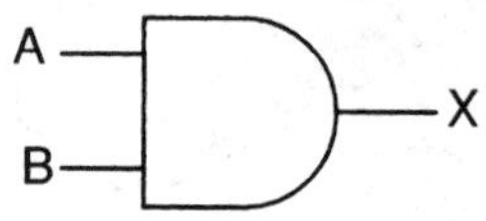	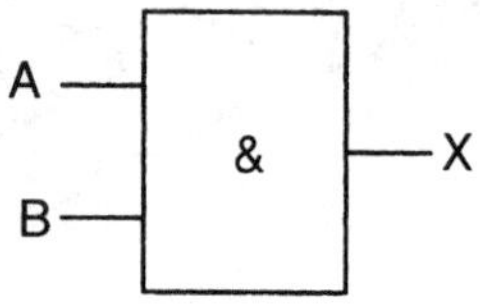	Inputs: A B; Outputs: X 0 0 0 0 1 0 1 0 0 1 1 1
OR	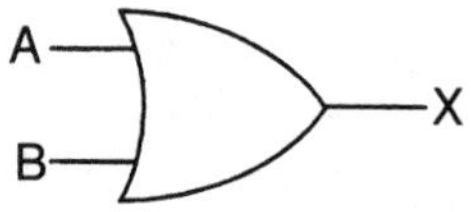	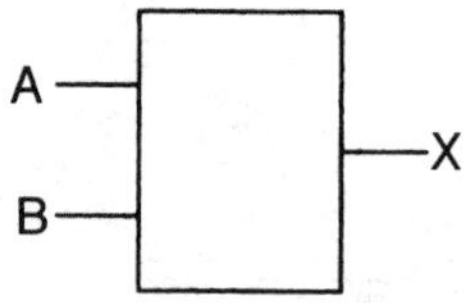	Inputs: A B; Outputs: X 0 0 0 0 1 1 1 0 1 1 1 1
(NOT AND) NAND	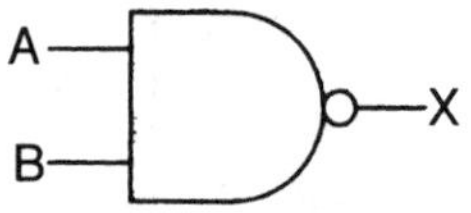	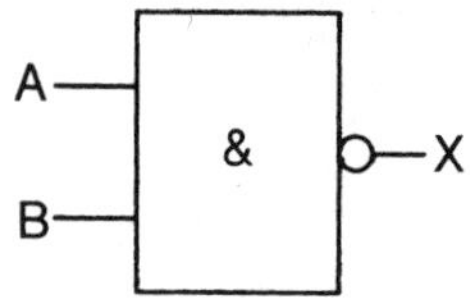	Inputs: A B; Outputs: X 0 0 1 0 1 1 1 0 1 1 1 0
NOT inverter	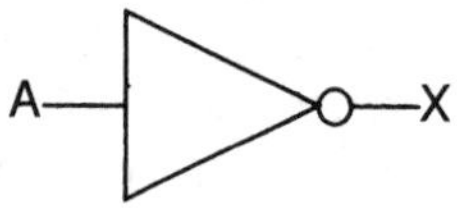	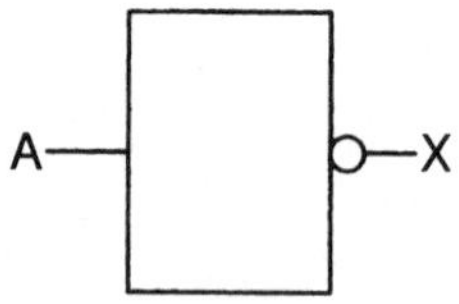	Inputs: A; Outputs: X 0 1 1 0
(NOT OR) NOR	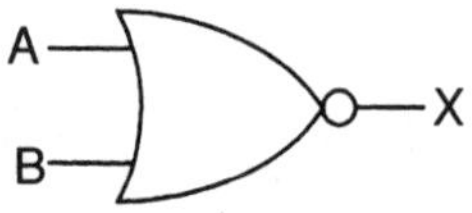	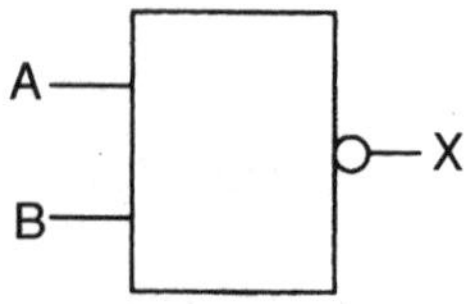	Inputs: A B; Outputs: X 0 0 1 0 1 0 1 0 0 1 1 0

Fig. 9.28 A table of logic gates and symbols

9.10 Computer (ECM) controlled systems

Whilst vehicle computers (ECMs) are not made to be repaired in garage workshops, there are certain factors that require technicians to have an appreciation of computer technology. For example, diagnostic trouble codes (DTCs) are an important part of fault finding and DTCs are stored in computer memory. The means by which these codes are read out varies from vehicle to vehicle and it is helpful for technicians to understand why a procedure for reading out DTCs on one vehicle may not work on another vehicle. It is also the case that technicians in some main dealer workshops are required to use special equipment to amend the computer operating program. Increasingly, use is being made of 'freeze frame' data. This is 'live' data that is captured while the system is in operation and it is useful in helping to determine the causes of a system fault. Whilst these operations are normally performed through the use of 'user friendly' diagnostic equipment, it is still the case that an understanding of what can and what cannot be done via the ECM is useful.

The fundamental parts of a computer

Figure 9.29 shows the general form of a computer that consists of the following parts:

1. A central processing unit (CPU)
2. Input and output devices (I/O)
3. Memory
4. A program
5. A clock for timing purposes.

Data processing is one of the main functions that computers perform. Data, in computer terms, is the representation of facts or ideas in a special way that allows it to be used by the computer. In the case of digital computers, this usually means binary data where numbers and letters are represented by codes made up from 0s and 1s. The input and output interfaces enable the computer to read inputs and to make the required outputs. Processing is the manipulation and movement of data and this is controlled by the clock. Memory is required to hold the main operating program and to temporarily hold data while it is being worked on.

Computer memory

Read only memory (ROM)

The ROM is the place where the operating program for the computer is placed. It consists of an electronic circuit which gives certain outputs for predetermined input values. ROMs have large storage capacity.

Read and write, or random access memory (RAM)

The RAM is the place where data is held temporarily while it is being worked on by the processing unit. Placing data in memory is 'writing', and the process of using this data is called 'reading'.

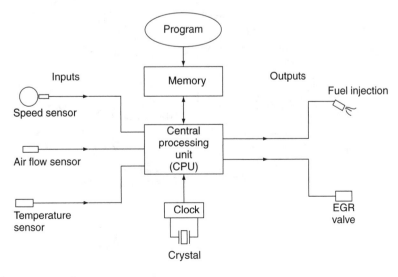

Fig. 9.29 The basic components of a computer system

Keep alive memory (KAM)

The term keep alive memory (KAM) refers to the systems where the ECM has a permanent fused supply of electricity. Here the fault codes are preserved, but only while there is battery power. Figure 9.30 shows a circuit for a KAM system.

Electrically erasable programmable read only memories (EEPROMs) are sometimes used for the storage of fault codes and other data relating to events connected with the vehicle system. This type of memory is sustained even when power is removed.

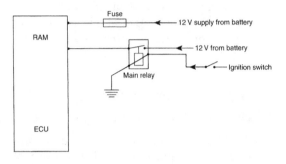

Fig. 9.30 A KAM system

The clock

The clock is an electronic circuit that utilises the piezo-electric effect of a quartz crystal to produce accurately timed electrical pulses that are use to control the actions of the computer. Clock speeds are measured in the number of electrical pulses generated in one second. One pulse per second is 1 Hertz and most computer clocks operate in millions of pulses per second. One million pulses per second is 1 mega Hertz (1 Mz).

A practical automotive computer system

Figure 9.31 shows a computer controlled transmission system.

At the heart of the system is an electronic module, or ECU. This particular module is a self-contained computer which is also known as a microcontroller. Microcontrollers are available in many sizes, for example 4, 8, 16 and 32 bit which refers to the length of the binary code words that they work on. The microcontroller in this transmission is an 8 bit one.

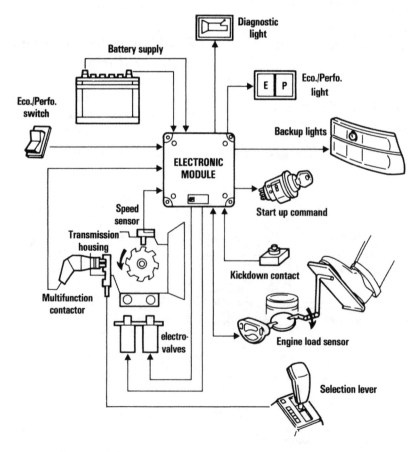

Fig. 9.31 A computer controlled transmission system

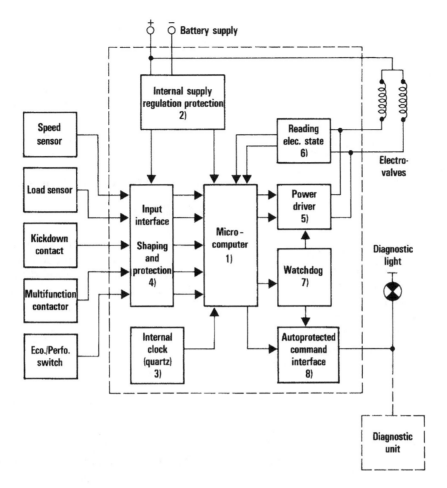

Fig. 9.32 Internal details of the computer

Figure 9.32 shows some of the internal detail of the computer and the following description gives an insight into the way that it operates:

1. **The microcomputer** This is an 8 bit microcontroller. In computer language a bit is a 0 or a 1. The 0 normally represents zero, or low voltage, and the 1 normally represents a higher voltage, probably 1.9 volts. The microcontroller integrated circuit (chip) has a ROM capacity of 2 049 bytes. There are 8 bits to one byte. There is also a RAM that holds 64 bytes.

   The microcontroller also has an on-chip capacity to convert 4 analogue inputs into 9 bit digital codes.

2. **The power supply** is a circuit that takes its supply from the vehicle battery. It then provides a regulated d.c. supply of 5 volts to the microcontroller and this is its working voltage. The power supply also includes protection against over voltage and low voltage. The low voltage protection is required if battery voltage is low and often takes the form of a capacitor.

3. **The clock circuit** In this particular application, the clock operates at 4 MHz. The clock controls the actions of the computer, such as counting sensor pulses to determine speed and timing the output pulses to the electrovalves so that gear changes take place smoothly and at the required time.

4. **The input interface** The input interface at 4) contains the electronic circuits that provide the electrical power for the sensors and switches that are connected to it. Some of these inputs are in an electrical form (analogue) that cannot be read directly into the computer and these inputs must be converted into computer (digital) form at the interface.

5. **The output (power) interface** The power driver at 5) consists of power transistors that are switched electronically to operate electrovalves that operate the gear change hydraulics.

6. **Feedback** At 6) on the diagram the inscription reads 'Reading electrical state'. This means

that the computer is being made aware of the positions (on or off) of the electro-valves.

7. **The watchdog** The watch dog circuit at 7) is a timer circuit that prevents the computer from going into an endless loop that can sometimes happen if false readings occur.

8. **The diagnostic interface** The diagnostic interface at 9) is a circuit that causes a warning lamp to be illuminated in case of a system malfunction and it can also be used to connect to the diagnostic kit.

Fault codes

When a microcontroller (computer) is controlling the operation of an automotive system, such as engine management, it is constantly taking readings from a range of sensors. These sensor readings are compared with readings held in the operating program and if the sensor reading accords with the program value in the ROM, the microcontroller will make decisions about the required output to actuators, such as injectors. If the sensor reading is not within limits it will be read again and if it continues to be 'out of limits', a fault code will be stored in a section of RAM. It is also probable that the designer will have written the main program so that the microcontroller will cause the system to operate on different criteria until a repair can be made, or until the fault has cleared. The fault codes, or diagnostic trouble codes DTCs as they are sometimes called, are of great importance to service technicians and the procedures for gaining access to them need to be understood. It should be clear that, if they are held in ordinary RAM, they will be erased when the ECM power is removed and that is why various methods of preserving them, such as KAMs and EEPROMS, are used.

Adaptive operating strategy of the ECM

During the normal lifetime of a vehicle, often compression pressures and other operating factors change. To minimise the effect of these changes, many computer controlled systems are programmed to generate new settings that are used as references by the computer when it is controlling the system. These new (learned) settings are stored in a section of memory, normally RAM. This means that such 'temporary' operating settings can be lost if electrical power

is removed from the ECM. In general, when a part is replaced or the electrical power is removed for some reason, the vehicle must be test driven for a specified period in order to permit the ECM to 'learn' the new settings. It is always necessary to refer to the repair instructions for the vehicle in question, because procedures do vary from vehicle to vehicle.

9.11 Networking and bus systems

When a number of computer controlled systems on a vehicle are required to communicate with each other, as happens with cruise control, traction control, diagnostics and other systems, the computers are connected to each other for communication purposes. Linking computers together, for communications purposes, is called networking and the lines (wires) that are used for this purpose are known as buses. The bus systems that are commonly used on vehicles are described in the following sections.

CAN Controller Area Network

CAN is the commercial name (Robert Bosch) of a system that is used to provide a means by which different computer controlled systems on a vehicle can communicate with each other.

Communication between systems is necessary in cases such as traction control systems, where the purpose of the system is to improve the vehicle's performance by eliminating wheel spin. Because the engine computer, the ABS computer and the instrument panel computer must communicate (effectively talk to each other) at high speed (in real time) they are linked together, for communication purposes, by a pair of wires. These wires form a communications system known as a bus – in this case a CAN bus. The communication buses carry messages only and operate at low electrical current values. The electrical power that operates devices such as ABS actuators, fuel valves, etc., is supplied by separate cables.

CAN details

The speed at which computers communicate along the bus is known as the baud rate. A data transmission rate of one binary bit (0 or 1)

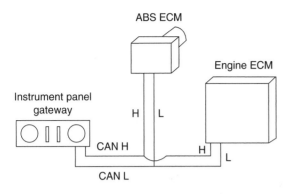

ABS ECM

Engine ECM

Instrument panel
gateway

H L

CAN H

H

L

CAN L

Fig. 9.33 Elements of a CAN bus system

per second is one baud. The CAN Systems used on vehicles operate in real time, which is taken to mean up to 500 kbaud. 1 kbaud = 1 000 bits/second.

CAN is a two wire system and the wires are twisted together in order to reduce electromagnetic interference that may corrupt the data that is being transmitted. The wires are known as CAN H and CAN L and the voltages used for high speed (high baud rate) systems are 3.5 volts, 2.5 volts and 1.5 volts.

The computing messages that are transmitted in serial form, 'one bit at a time' along the CAN bus are constructed to prevent errors occurring. The way in which data messages are constructed and used by the computers are known as **protocols**.

ISO 9141 Diagnostic bus system

The International Standards Organisation (ISO) develops standards for vehicle bus systems and a significant one that applies to vehicle systems is known as ISO 9141. The ISO 9141 diagnostic bus system is a system that networks vehicle computer control units for diagnostic purposes. The ISO 9141 system may be a single wire which is known as the K line, or it may be a two wire system when the wires are known as the K line and L line. The K line is connected to a diagnostic socket which permits certain scan tools and manufactures' dedicated test equipment to access fault codes and other diagnostic information. The K line bus operates at a slow speed of 10.4 kbaud.

Under European on Board Diagnostics (EOBD) and OBDII regulations, all systems that contribute to vehicle emissions control must be accessible, for diagnostic purposes, through the diagnostic connector.

Other bus systems used on vehicles

There are several areas of vehicle control where data-buses can be used to advantage. Some of these, such as lighting and instrumentation systems, can operate at fairly low speeds of data transfer, say 1 000 bits per second, while others, such as engine and transmission control, require much higher speeds, probably 250 000 bits per second and these are said to operate in 'real time'. To cater for these differing requirements, the Society of Automotive Engineers (SAE) recommends 3 classes, known as Class A, Class B and Class C.

- **Class A** Low speed data transmission. Up to 10 k bits per second. Used for body wiring such as exterior lamps, etc.
- **Class B** Medium speed data transmission. 10 k bits per second, up to 125 k bits per second. Used for vehicle speed controls, instrumentation, emission control, etc.
- **Class C** High speed (real time) data transmission. 125 k bits per second, up to 1 M bits (or more) per second. Used for brake by wire, traction and stability control, etc.

Gateway

The instrument pack acts as a central point, or **gateway**, to which all bus systems are connected. The instrument pack is equipped with sufficient computing power and interfaces to allow the various systems to share sensor data and other information. This means that the bus systems such as CAN and the ISO 9141 diagnostic system, that operate at different **baud rates** and use different **protocols**, can communicate effectively for control and diagnostic purposes.

Power supply and drivers

The electrical power supply for the various actuators, such as ABS modulators, fuel injectors, ignition coils and similar devices, comes from the battery and alternator; on cars this is 12 volts supply. The actuators are operated by heavy current transistorised circuits (switches) that are operated by the computers (ECUs) that control the respective systems. These transistor switches are known as drivers and a simplified example is shown in Fig. 9.35.

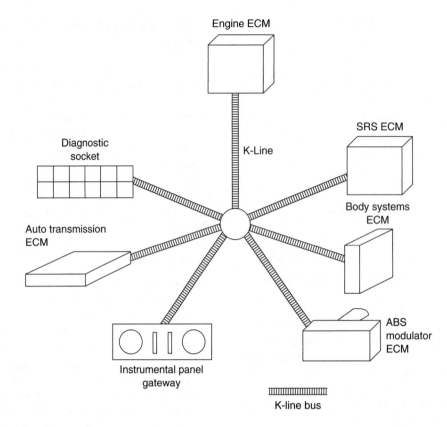

Fig. 9.34 The K line diagnostic bus

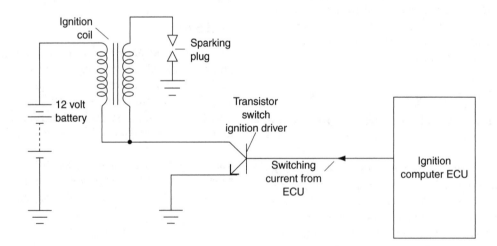

Fig. 9.35 Ignition circuit drive in computerised system

9.12 Multiplexing

The 12 volt supply to each injector or other electrically operated unit is carried through a separate cable which means that there are many cables on a single vehicle.

In order to overcome some of the problems associated with multiple cables, a system known as multiplexing is used. A multiplexer is an electronic device that has several inputs and a single output and it is used to reduce the number of electrical cables that are needed on some vehicle systems. Some estimates suggest that as much as 15 kg of wire can be eliminated by the use of networking and multiplexing on a single vehicle.

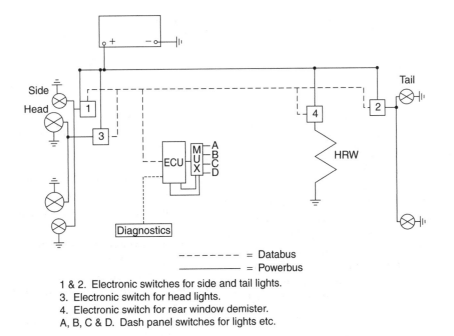

1 & 2. Electronic switches for side and tail lights.
3. Electronic switch for head lights.
4. Electronic switch for rear window demister.
A, B, C & D. Dash panel switches for lights etc.

Fig. 9.36 The multiplexed wiring concept

Figure 9.36 shows the basic concept of multiplexed vehicle wiring; in this case, lights and heated rear window. In order to keep it as simple as possible, fuses, etc. have been omitted.

As the legend for the diagram states, the broken line represents the data-bus. This is the electrical conductor (wire) which conveys messages along the data-bus to the respective remote control units. These messages are composed of digital data (0s and 1s). The rectangles numbered 1, 2, 3 and 4 represent the electronic interfaces that permit two-way communication between the ECU and the lamps, or the heated rear window. The dash panel switches are connected to a multiplexer (MUX) which permits binary codes to represent different combinations of switch positions to be transmitted via the ECU onto the data-bus. For example, side and tail lamps on and the other switches off, could result in a binary code of 1000, plus the other bits (0s&1s) required by the protocol, to be placed on the data-bus so that the side lamps are energised. Operating other switches, for example, switching on the HRW would result in a different code and this would be transmitted by the ECU to the data-bus, in a similar way. A process known as time division multiplexing is used and this allows several systems to use the data-bus. In effect, the devices are switched on and off many times

each second with the result that there is no effect that is visible to the human eye.

European On Board Diagnostics (EOBD) and OBD2

The US version of on board diagnostics, known as OBD2, for electronic systems, has been in use since the mid-1990s. The European version is much the same as OBD2 and it has been in use since 2000. The regulations for both of these systems require that vehicles should be equipped with a standardised means of gaining access to defects that may affect the performance of emissions control systems.

European On Board Diagnostics (EOBD)

In order to effectively maintain and repair networked and multiplexed systems, it is necessary for diagnostic access to be gained to all systems. Since January 2001, all petrol engined vehicles sold in Europe must be equipped with a self diagnostic system that alerts the driver to any fault that may lead to excessive emissions. To comply with EOBD rules, vehicles must be equipped with a malfunction indicator lamp (MIL). The purpose of the MIL is to alert drivers to the presence of a problem and it operates as follows. When the ignition is switched on, the MIL illuminates and goes out as soon as the engine is started. If a fault is detected while the engine is running, the MIL operates in one of two ways as follows:

1. The MIL comes on and stays on – this indicates a fault that may affect emissions and indicates that the vehicle must be taken to an approved repairer to have the fault investigated, as soon as possible.
2. The MIL flashes at regular intervals, once per second. This indicates an emergency situation and requires the driver to reduce speed because the control computer (ECM) has detected a misfire that has the potential to harm the catalytic convertor. Continued flashing of the MIL requires the vehicle to be taken to an approved repairer for early attention.

Additionally, vehicles must be equipped with a standardised diagnostic port that permits approved repairers to gain access to fault codes and other diagnostic data.

Self assessment questions

1. The unit of electro motive force is:
   (a) an ampere
   (b) an Ohm
   (c) a volt
   (d) a watt
2. The unit of electrical resistance is:
   (a) a kilovolt
   (b) an Ohm
   (c) an ampere
   (d) a mho
3. The unit of electric current is:
   (a) a kilowatt hour
   (b) a millivolt
   (c) an ampere
   (d) a Sievert
4. Circuit protection is usually provided by:
   (a) a bypass circuit
   (b) heavy insulation on earth cables
   (c) fuses
   (d) field effect transistors

5. Semi conductors are:
   (a) wires that conduct half the amount of current of ordinary conductors
   (b) materials that have electrical conductivity that falls between that of good conductors and that of insulators
   (c) materials with conductivity better than silver
   (d) the materials that are used for communication buses
6. Multiplexed systems:
   (a) use separate wires and earth return wires for each component
   (b) reduce the amount of wire used in vehicle systems
   (c) have several alternative methods of switching on vehicle lights
   (d) are not suitable for use on heavy vehicles
7. When considering the operation of an electric motor:
   (a) it useful to use Fleming's left-hand rule
   (b) Ohm's law
   (c) Fleming's right-hand rule
   (d) Lenz's law
8. A short circuit occurs when:
   (a) a conductor makes contact with earth and provides an alternative path for electric current
   (b) the circuit is not complete
   (c) the battery voltage is low
   (d) the alternator charging rate is too high
9. In an EOBD system:
   (a) the malfunction indicator lamp lights up whenever there is a fault that may interfere with the correct working of emission control systems
   (b) when the ABS has failed
   (c) a complete answer to all problems is contained in the fault code memory
   (d) the engine is shut down if the downstream oxygen sensor fails
10. On a vehicle that uses CAN for some systems and an SAE K bus for others:
    (a) the systems cannot communicate with each other
    (b) the systems communicate with each other through the GATEWAY
    (c) the systems use the same protocols
    (d) the systems use the same logic levels

10

Electrical and electronic systems

Topics covered in this chapter

Alternator – generates electricity when the engine is running, recharges the battery and supplies most of the electricity to operate systems on the vehicle when the engine is running.

Battery – supplies the electricity to operate the starter system and other electrical systems when the engine is not running.

Charging system – controls the alternator output to ensure that the supply of electricity meets demand and to display a warning light when this is not happening.

Starting systems – normally a high torque electric motor that is temporarily engaged with the stationary engine flywheel in order to rotate the engine at sufficient speed to get it started under a range of conditions.

Electrical principles. Vehicle circuits and wiring systems – distributes electrical energy to all systems on the vehicle through an organised and overload protected system of cables.

Modern vehicles are equipped with many electrical and electronic systems, as shown by the list given in Table 10.1.

We think that it is a good idea to start with the battery. This is where the energy is stored that operates the starter motor and, as vehicle journeys usually begin by starting up the vehicle engine, this seems like a good point at which to start this study of vehicle electrical and electronic systems.

10.1 Vehicle batteries

Chemical effect of electric current

If two suitably supported lead plates are placed in a weak solution of sulphuric acid in water (electrolyte) and the plates connected to a suitable low voltage electricity supply as shown in Fig. 10.1, an electric current will flow through the electrolyte from one plate to the other.

After a few minutes the lead plate P will appear a brownish colour and the plate Q will appear unchanged. The cell has caused the electric current to have a chemical effect.

This electro-chemical effect can be reversed, i.e., removing the battery and replacing it with a lamp (bulb) will allow electricity to flow out of the cell and 'light up' the lamp. Electric current will continue to flow out of the battery until the

Table 10.1 Electrical and Electronic Equipment, the main items of electrical/electronic equipment on vehicles

Vehicle system	Comment
Batteries and charging systems	*Alternators, voltage and current control*
Starter motors	**Axial engagement**
Ignition systems	Distributor type, direct type, distributor-less
Engine management	
Fuel systems	Single point, multi-point, direct injection
Emission control	Oxygen sensor, catalyst, purge canister
Diagnostics	EOBD
Body electrics	Remote central locking, electrically operated windows, heated screens, screen wash and wipe
Security	Immobiliser, anti theft alarm
Anti lock brakes, traction control, stability control	CAN systems
Lights and signals	Head lights, side and tail, stop lights, hazard lights, interior lights

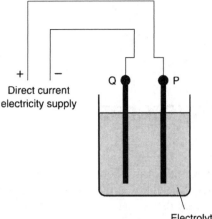

Fig. 10.1 A simple cell

cell has returned to its original state, or until the lamp is disconnected.

As a result of the electro-chemical action, the two lead plates develop a difference in electrical potential. One plate becomes electrically positive and the other becomes electrically negative. The difference in electrical potential causes an electric force. This force is known as an electro-motive-force (emf). EMFs are measured in volts and the emf generated by the two plates, as shown in Fig. 10.1 can be measured by connecting a voltmeter across the metal plates. This simple cell is known as a secondary cell. It can be charged and recharged.

Dry batteries, similar to those used in a hand torch, exhaust their active materials in the process of making an electric current and they cannot be recharged. Such dry batteries are primary cells. These are not to be confused with modern secondary cells of similar appearance.

Because secondary cells can be charged and recharged they are used to make vehicle batteries. There are two types of vehicle batteries: one is lead acid and the other is nickel-iron alkaline. The lead acid battery is the one that is most widely used and it is this type that we propose to concentrate on.

The lead acid vehicle battery

Safety note

Before we proceed with this work on batteries, there are several points that need to be emphasised.

Batteries are heavy. Proper lifting practices must be used when lifting them.

Sulphuric acid is corrosive and it can cause serious injury to the person and damage to materials. It must be handled correctly. You must be aware of the actions to take in case of contact with acid. Every workshop must have proper safety provision and the rules that are set must be observed. In the event of spillage of acid on to a person, the area affected should be washed off with plenty of clean water. Any contaminated clothing should be removed. If acid comes into contact with eyes, they should be carefully washed out with clean water and urgent medical attention obtained.

Batteries give off explosive gases, e.g. hydrogen. There should be no smoking or use of naked flame near a battery. Sparks must be avoided when connecting or disconnecting batteries. This applies to work on the battery when it is on the vehicle and also when the battery has been removed from the vehicle for recharging.

When disconnecting the battery on the vehicle, make sure everything is switched off. Remove the earth terminal FIRST.

When reconnecting the battery, make sure everything is switched off. Connect the earth terminal LAST.

In the case of batteries 'on charge', the battery charger must be switched off before disconnecting the battery from the charger leads.

A typical vehicle battery

Figure 10.2 shows a battery in the engine compartment of a light vehicle. The notes on the diagram indicate factors that require attention.

The type of battery shown in Fig. 10.2 is typical of batteries used on modern vehicles. Most light vehicles use 12 volt batteries. The 12 volts is obtained by connecting 6 cells in series. Each single lead acid cell produces 2 volts.

Some larger trucks and buses use a 24 volt electrical system. In this case, two 12 volt batteries are connected in series to give the 24 volts.

Battery construction

The case

The battery casing needs to be acid resistant and as light and as strong as possible. Plastic material, such as polypropylene is often used for the purpose. Figure 10.3 shows the construction of a fairly typical battery case.

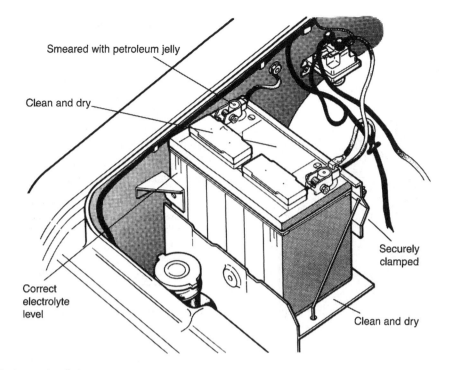

Smeared with petroleum jelly

Clean and dry

Correct electrolyte level

Securely clamped

Clean and dry

Fig. 10.2 The battery installation

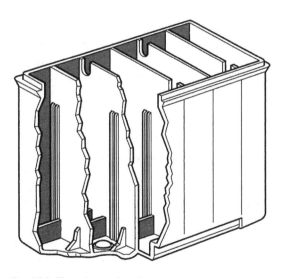

Fig. 10.3 The polypropylene battery case

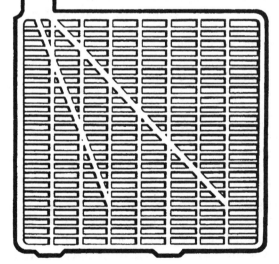

Fig. 10.4 A battery plate

Each cell of the battery is housed in a separate container and there are ribs at the bottom of the cell compartment which support the plates so that they are clear of the container bottom. Sediment can accumulate in this bottom space without short circuiting the plates.

The plates are made in the form of a grid which is cast in lead. In order to facilitate the casting process a small amount of antimony is added to the lead. In maintenance-free batteries, calcium is added to the lead because this reduces

losses caused by gassing. Figure 10.4 shows a battery plate.

During the manufacturing process, other materials are pressed into the plate grid. On the positive plate the material added is lead peroxide (a browny red substance) and on the negative plate it is spongey lead, which is grey. The completed plates are assembled, as shown in Fig. 10.5.

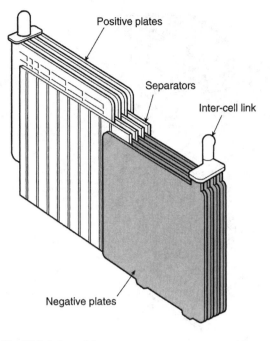

Fig. 10.5 Battery plates

The positive plates are connected together with a space in between. The negative plates are also joined together, with a space in between. The negative plates are placed in the gaps between the positive plates and the plates are prevented from touching each other by the separators which slide in between them. The separators are made from porous material which allows the electrolyte to make maximum contact with the surfaces of the plates.

Battery capacity. The ampere-hour (A–H)

The unit of quantity of electricity is the Coulomb. One coulomb is the equivalent of a current of 1 ampere flowing for 1 second. This is a very small amount of electricity in comparison with the amount that is required to drive a starter motor. It is usual, therefore, to use a larger unit of quantity of electricity to describe the electrical energy capacity of a vehicle battery. The unit that is used is the ampere-hour. One ampere-hour is equivalent to a current of 1 ampere flowing for 1 hour. The size of the battery plates and the number of plates per cell are factors that have a bearing on battery capacity. In general, a 75 ampere-hour battery will be larger and heavier than a 50 ampere-hour battery.

Relative density of electrolyte (specific gravity)

One of the chemical changes that takes place in the battery, during charging and discharging, is that the relative density of the electrolyte (dilute sulphuric acid) changes. There is a link between relative density of the electrolyte and the cell voltage. Relative density of the electrolyte is a good guide to the state of charge of a battery. The following figures will apply to most lead acid batteries:

State of charge	Relative density
Fully charged	1.290
Half charged	1.200
Completely discharged	1.100

The relative density (specific gravity) varies with temperature, as shown in Fig. 10.6.

Examination of these figures show that the relative density figure falls by 0.007 for each 10 °C rise in temperature. This should be taken into account when using a hydrometer.

If the electrolyte (battery acid) is accessible, for example, not a maintenance-free battery, a hydrometer can be used to check battery condition. Figure 10.7 shows a hydrometer together with the readings showing the state of charge.

Battery charging

A properly maintained battery should be kept in a charged condition by the vehicle alternator, during normal operation of the vehicle. If it becomes necessary to remove the battery from the vehicle to recharge it, with a battery charger, the ampere-hour capacity of the battery gives a good guide to the current that should be used and the length of time that the battery should stay on charge.

SG (corrected to 15 °C)	Density (kg/m³) (corrected to 25 °C)	State of charge %
1.273	1280	100
1.253	1260	90
1.233	1240	80
1.213	1220	70
1.193	1200	60
1.173	1180	50
1.133	1140	30
1.093	1100	10
1.073	1080	0 (flat)

Fig. 10.6 Variation of electrolyte relative density with temperature

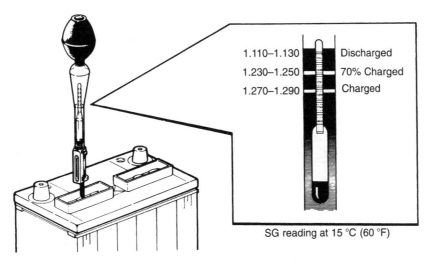

SG reading at 15 °C (60 °F)

Fig. 10.7 The hydrometer

Battery charging rate

The test for battery capacity can be a 10 hour test or a 20 hour test. If, as a guide, we take the 10 hour rating of 50 A–H battery, in theory, this means that the battery will give 5 A for 10 hours, i.e. $5 \times 10 = 50$. If this 50 A–H battery was being completely recharged, it would require a charging current of 5 A to be applied for 10 hours.

Reserve capacity

In recent years it has become the practice to give the reserve capacity of a battery rather than the A–H capacity, because reserve capacity is considered to be a better guide to how long the battery will provide sufficient current to operate vehicle systems. Reserve capacity is the time in minutes that it will take a current of 25 A to cause the cell voltage to fall to 1.75 V. That is to say, the number of minutes that it takes for a 12 volt battery voltage to drop to 10.5 volts, when it is supplying a current of 25 A.

Battery suppliers will provide the data for the batteries that they supply. Manufacturers' charging rates must be used.

Testing batteries

Batteries are normally reliable for a number of years. However, batteries do sometimes fail to provide sufficient voltage to operate the vehicle starting system. Such failure can arise for reasons such as:

● lights, or some other systems has been left switched on, and the battery has become discharged, etc.;

● the battery is no longer capable of 'holding' a charge;

● the battery is discharged because the alternator charging rate is too low.

If it is a), and the lights have been left on and caused the battery to 'run down', a quick check of switch positions (i.e., are they in the 'on' position?) will reveal the most likely cause.

A hydrometer test, or a voltage check at the battery terminals, will confirm that the battery is discharged. The remedy will be to recharge the battery, probably by removing it and replacing it, temporarily, with a 'service' battery.

If it is b) or c), it will be necessary to check the battery more thoroughly. A high rate discharge tester that tests the battery's ability to provide a high current for a short period is very effective, but it can only be used if the battery is at least 70% charged. The state of charge of the battery can be determined by using a hydrometer. If the specific gravity (relative density) reading shows that the battery is less than 70% charged, the battery should be recharged at the recommended rate. After charging, and leaving time for the battery to settle and the gases to disperse, the high rate discharge test can be applied. Many different forms of high rate discharge testers are available and it is important to carefully read the instructions for use.

Checking the alternator charging rate is covered in the section on alternators.

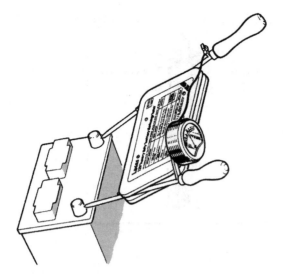

Fig. 10.8 High rate discharge tester

Learning tasks

1. Ask your supervisor or tutor about the workshop rules that apply to the handling of battery charging in your workshop.
   Write a summary of the these rules.
2. On which types of vehicles are you likely to find two 12 volt batteries connected in series, cars and light commercial vehicles or heavy trucks and buses?
3. On some electronically controlled systems, settings such as idle speed are held in the volatile memory of the ECU. That is to say, that the settings are lost when the battery is discharged or disconnected. Write out a brief description of the procedure that is followed to restore such settings.
4. What is a 'rapid' battery charger? What precautions must be observed when using one?
5. Examine a vehicle battery installation and make a simple sketch to show how the battery is secured to the vehicle.
6. Assuming that you are asked to remove the battery that you have examined for question 5, describe the procedure to be followed. State which battery lead to remove first, and include any steps taken to observe safety rules and protect the vehicle.

10.2 The alternator

The alternator supplies the electrical energy to recharge the battery and to operate the electrical/electronic systems on the vehicle, while the engine is running.

The basic vehicle alternator (generator of electricity)

In a vehicle alternator, the magnet is the rotating member and the coil in which current is generated is the stationary part. Figure 10.9 illustrates the general principle. The rotor shaft is driven round by the pulley and drive belt and the stator is fixed to the body of the alternator.

(a)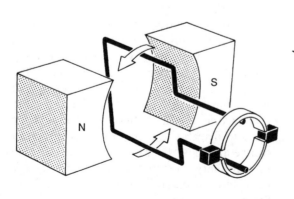

(b)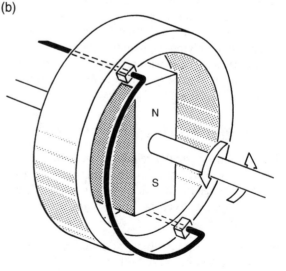

Fig. 10.9 The principle of the generator (a) dynamo stationary field, (b) alternator rotating field

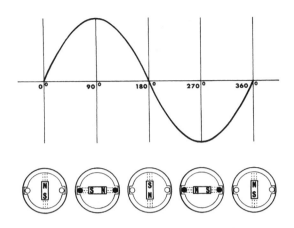

Fig. 10.10 Diagram showing EMF produced in one complete cycle relative to magnet position

Figure 10.10 shows how the electro motive force [e.m.f] (voltage) produced by this simple rotating magnet and stator coil varies during one complete revolution (360°) of the magnetic rotor.

Alternating current (a.c.)

Alternating current, as produced by this simple alternator, is not suitable for charging batteries. The alternating current must be changed into direct current (d.c.). The device that changes a.c. to d.c. is the rectifier. Vehicle rectifiers make use of electronic devices, such as the p–n junction diode. The property of the diode that makes it suitable for use in a rectifier is that it will pass current in one direction but not the other. Figure 10.11 demonstrates the principle of the semiconductor diode.

Using a single diode as a simple half wave rectifier

As a guide to the principle of operation of the actual rectifier as used in a vehicle alternator, we think that it is useful to briefly consider the operation of the simple half-wave rectifier shown in Fig. 10.12.

In this simple rectifier the negative half of the wave is 'blocked' by the diode. The resulting output is a series of positive half waves separated by 190°. In order to get a smooth d.c. supply it is necessary to use a circuit that contains several diodes. The bridge circuit rectifier is the basic approach that is used.

A full wave rectifier. The bridge circuit

Figure 10.13(a) shows a full wave rectifier circuit. There are 4 diodes, A, B, C and D which are connected in bridge form. There is an a.c. input and a voltmeter is connected across the output resistor.

Reference to Fig. 10.13(b) shows how the current flows for one half of the a.c. cycle, and Fig. 10.13(c) shows the current flow in the other half of the a.c. cycle. In both cases, the current flows through the load resistor R, in the same direction. Thus, the output is direct current and this is achieved by the 'one way' action of the diodes.

A typical vehicle alternator

The following description of the parts shown in Fig. 10.14 should help to develop an understanding that will aid the additional study and

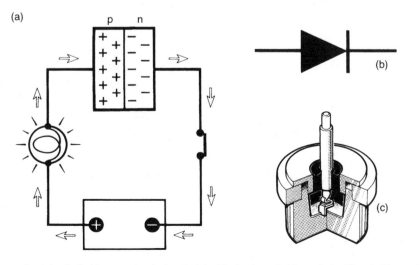

Fig. 10.11 The principle of the diode (a) circuit for diode principle, (b) circuit symbol for a diode, (c) typical heavy duty diode

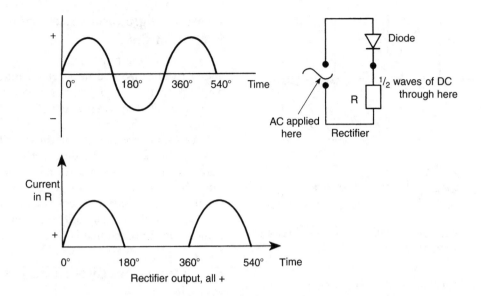

Fig. 10.12 The half-wave rectifier

training that will build the skills necessary for successful work on electrical/electronic systems.

The ROTOR

This is where the magnetic field is generated by the current that flows into the rotor windings through the slip rings and brushes. Figure 10.15 shows 3 types of rotors that are used in Lucas alternators.

The rotor is made from an iron core around which is wound a coil of wire. The

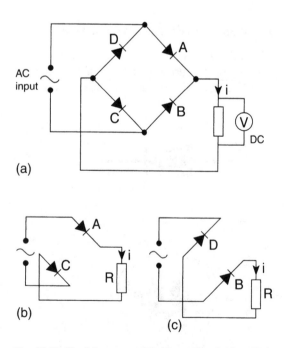

(a)

(b)

(c)

Fig. 10.13 The full-wave rectifier (a) complete bridge circuit, (b) current flow for one half of the voltage waveform, (c) current flow for the other half of the voltage waveform

current in this coil of wire creates a magnetic north pole at one end and a magnetic south pole at the other of the stator coil. The iron claws of the rotor are placed on opposite ends of the rotor (field) winding. The 'fingers' of the claws are north or south magnetic poles, according to the end of the field coil that they are attached to. This makes pairs of north and south poles around the circumference of the rotor. The whole assembly is mounted on a shaft, which also has the drive pulley fixed to it so that the rotor can be driven by the engine.

The STATOR

These stators have a laminated iron core. Iron is used because it magnetises and de-magnetises easily, with a minimum of energy loss. Three separate coils are wound on to the stator core. A separate a.c. wave is generated in each winding, as the rotor revolves inside the stator. The stator windings are either star, or delta connected, according to alternator type.

The RECTIFIER

A 6 diode rectifier

A fairly commonly used bridge circuit of an alternator rectifier uses 6 diodes, as shown in Fig. 10.18. The stator windings are connected to the diodes as shown in this figure.

In operation, a power loss occurs across the diodes and in order to dissipate the heat generated the diodes are mounted on a heat sink which conducts the heat away into the

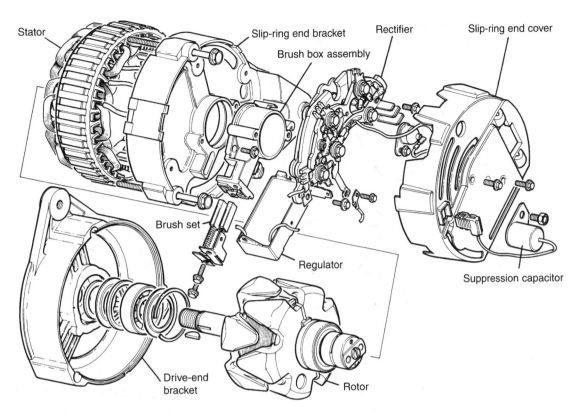

Fig. 10.14 The main component parts of an alternator

atmosphere. Typical diode packs are shown in Fig. 10.19.

The voltage REGULATOR

The voltage regulator is an electronic circuit that holds the alternator voltage to approximately 14 volts on a 12 volt system, and approximately 28 volts on a 24 volt system. It operates by sensing either the voltage at the alternator (machine sensing), or at the battery (battery sensing). The regulator controls the amount of current that is supplied to the rotor field winding and this alters the strength of the magnetic field and hence the alternator output voltage.

The BRUSHES and BRUSHBOX

The brushes carry the current to the slip rings, and from there to the rotor winding. Alternator brushes are normally made from soft carbon because it has good electrical

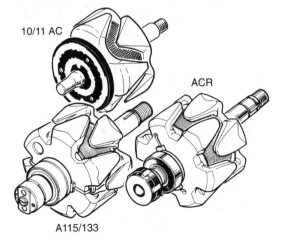

Fig. 10.15 Alternator rotors (electromagnets)

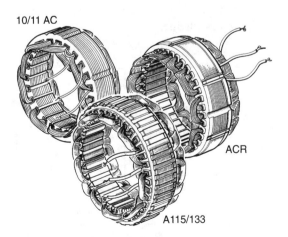

Fig. 10.16 A selection of alternator stators

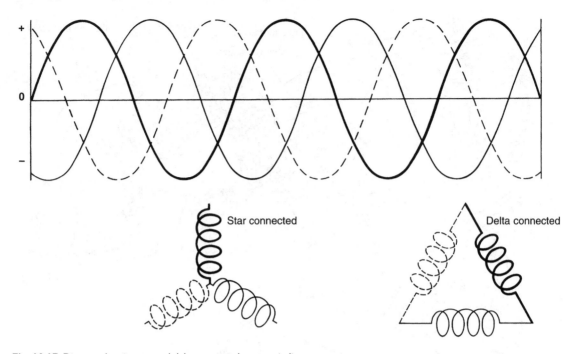

Fig. 10.17 Diagram showing star and delta connected stator windings

conducting properties and does not cause too much wear on the brass slip rings. In time, the brushes may wear and the brush box is normally placed so that brush checking and replacement can be performed without major dismantling work.

BEARINGS

The rotor shaft is mounted on ball races, one at the drive-end, the other at the sliprings-end. These bearings are pre-lubricated and are designed to last for long periods without attention. The drive-end bearing is mounted in the drive-end bracket (Fig. 10.14) and the slipring-end bearing is mounted in the slipring-end bracket. These two brackets also provide the supports that allow the alternator to be bolted to the vehicle engine.

DRIVE PULLEY and COOLING FAN

The drive pulley is usually belt driven by another pulley on the engine. The alternator cooling fan is mounted on the rotor shaft and is driven round

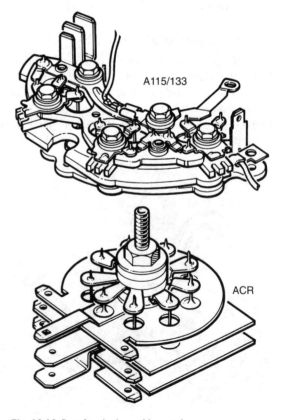

Fig. 10.19 Rectifier diodes and heat sinks

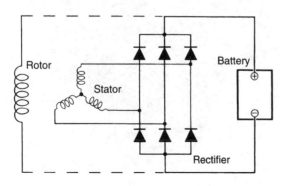

Fig. 10.18 A six-diode rectifier in the alternator circuit

Fig. 10.20 The alternator drive pulley and cooling fan

with it. This fan passes air through the alternator to prevent it from overheating.

Externally, alternators all look very much alike, but they vary greatly in performance and internal construction. The description given here is intended to provide an introduction to the topic. When working on a vehicle, it is essential to have access to the information relating to the actual machine being worked on.

Alternator testing

Before starting on the alternator test procedure, there are a few general points about care of alternators that need to be made:

- Do not run the engine with the alternator leads disconnected. (Note that in the tests that follow the engine is switched off and the ammeter is connected before re-starting the engine).
- Always disconnect the alternator and battery when using an electric welder on the vehicle. Failure to do this may cause stray currents to harm the alternator electronics.
- Do not disconnect the alternator when the engine is running.
- Take care not to reverse the battery connections.

With these points made we can turn our attention to the alternator tests. As with any test a thorough visual check should first be made. In the case of the alternator this will include:

- Check the drive belt for tightness and condition;
- Check leads and connectors for tightness and condition;
- Ensure that the battery is properly charged;
- Check any fuses in the circuit.

Alternator output tests

The question of whether to repair or replace an alternator is a subject of debate. However, one thing that is certain is the need to be able to accurately determine whether or not the alternator is charging properly. The test procedure shown in Fig. 10.21 shows the procedure for checking alternator current output under working conditions.

The test link will be a cable with suitable crocodile clips. This link must be securely attached to ensure proper connections. The ammeter connections must also be secure, with the cables and the meter safely positioned to ensure accurate readings and to avoid mishaps.

The drive belt tension must also be checked, as shown in Fig. 10.22.

The belt tension may also be checked by means of a torque wrench, or a Burroughs type tension gauge, as shown in Fig. 10.23. If the belt is too tight (excessive tension) alternator bearings may be damaged; and if the tension is insufficient the drive belt may slip. This will cause a loss of

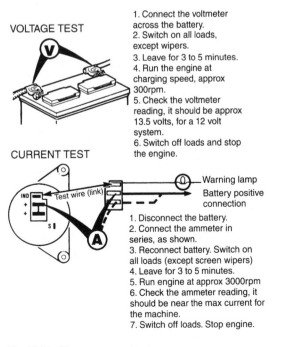

VOLTAGE TEST

1. Connect the voltmeter across the battery.
2. Switch on all loads, except wipers.
3. Leave for 3 to 5 minutes.
4. Run the engine at charging speed, approx 300rpm.
5. Check the voltmeter reading, it should be approx 13.5 volts, for a 12 volt system.
6. Switch off loads and stop the engine.

CURRENT TEST

Test wire (link)

Warning lamp
Battery positive connection

1. Disconnect the battery.
2. Connect the ammeter in series, as shown.
3. Reconnect battery. Switch on all loads (except screen wipers)
4. Leave for 3 to 5 minutes.
5. Run engine at approx 3000rpm
6. Check the ammeter reading, it should be near the max current for the machine.
7. Switch off loads. Stop engine.

Fig. 10.21 Alternator output tests

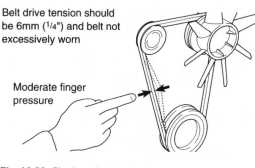

Belt drive tension should be 6mm (¼") and belt not excessively worn

Moderate finger pressure

Fig. 10.22 Checking the drive belt tension

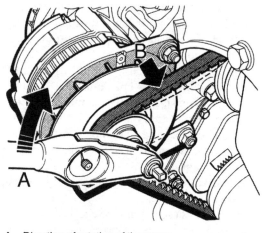

A – Direction of rotation of the gauge
B – Degree of flexure in the drive belt

Fig. 10.23 Using the belt tension gauge

alternator power and, in time, can lead to worn pulleys, hence the reason for knowing what the tension should be and setting it accurately.

10.3 The starter motor

The starter motor shown in Fig. 10.24 is fairly typical of starter motors in modern use.

Most modern starter motors are of the pre-engaged type. This means that the drive pinion on the starter motor is pushed into engagement with the ring gear on the engine flywheel before the starter motor begins to rotate and start the engine.

The amount of torque (turning effect) that the starter motor can generate is quite small and it has to be multiplied to make it strong enough to rotate the engine crankshaft at about 100 revolutions per minute (RPM). This multiplication of torque is achieved by the ratio of the number of teeth on the flywheel divided by the number of teeth on the starter pinion. In the example shown in Fig. 10.25, the ratio is 10 to 1.

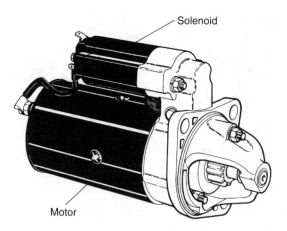

Fig. 10.24 A starter motor

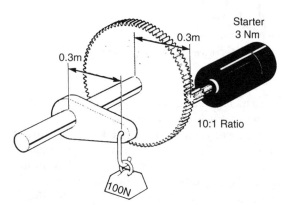

Fig. 10.25 Gear ratio of a starter pinion and flywheel gear

In the starter motor drive shown in Fig. 10.26, the pinion is moved into engagement by the action of the solenoid on the operating lever.

The circuit in Fig. 10.27 shows that there are two solenoid coils that are energised when the starter switch is closed. One coil is the holding coil and the other is the closing coil.

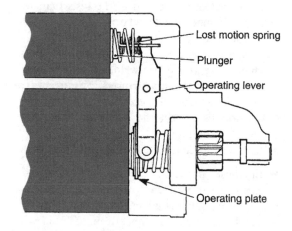

Fig. 10.26 The operating lever

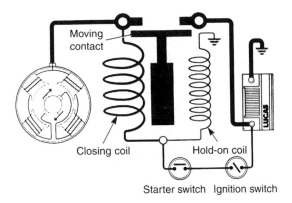

Fig. 10.27 The starter solenoid

The armature (plunger) of the solenoid is attached to the operating lever. When the pinion is properly engaged, the heavy current contacts of the solenoid will close and the starter motor will rotate; this action by-passes the closing coil and the holding coil remains energised until the starter switch is released. The plunger then returns to the normal position, taking the starter pinion out of engagement with the flywheel.

The starter motor pinion must be taken out of engagement with the flywheel immediately the engine starts. An engine will normally start at approximately 100 rpm. If the starter motor remains in engagement, the engine will drive the starter motor at high speed (because of the gear ratio) and this will destroy the starter motor. To guard against such damage, pre-engaged starter motors are equipped with a free wheeling device. A common example of such an over-run protection device is the roller clutch drive assembly shown in Fig. 10.28. The inner member is fixed to the starter motor shaft and the outer member to the starter motor pinion.

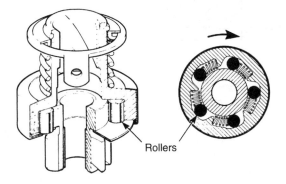

Fig. 10.28 Starter pinion over-run protection

This is a convenient point at which to have a look at some tests that can be used to trace faults in starter motor circuits.

Testing starter motor circuits

As with all diagnostic work, it is advisable to adopt a strategy. One strategy that works well is the six step approach. The six steps are:

1. collect evidence
2. analyse evidence
3. locate the fault
4. find the cause of the fault and remedy it
5. rectify the fault (if different from 4)
6. test the system to verify that repair is correct.

In the case of a starter motor system and assuming that the starter motor does not operate when the switch is on, step 1 of the six steps will include a visual check to see whether all readily visible parts such as cables, etc., are in place and secure. The next step would be to check the battery. An indication of battery condition can be obtained by switching on the headlights. If the starter switch is operated with the headlights on and the lights go dim, there is a case for a battery test, as described in the section on batteries.

If the battery condition is correct, there are some voltmeter tests that can be performed to check various parts of the starter motor circuit. For these tests, it is best to use an assistant to operate the starter switch because it is unwise to attempt to make the meter connections and observe the meter readings in the vicinity of the starter motor and attempt to operate the starter switch simultaneously. The meter must be set to the correct d.c. range and clips should be used so that secure connections can be made to the various points indicated.

To avoid unwanted start up, the ignition system should be disabled. On a coil ignition system, the low tension lead between the coil and distributor should be disconnected. On a compression ignition (diesel) engine, the fuel supply should be cut off, probably by operating the stop control. The instructions that relate to the vehicle being worked on must be referred to; this warning is particularly important when working on Engine Management Systems, because fault codes may be generated.

Figure 10.29 shows the test to check voltage at the battery terminals under load.

Here the battery voltage is being checked while the starter switch is being operated. For

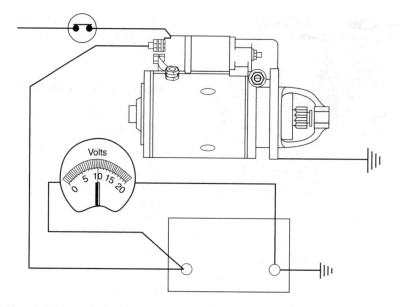

Fig. 10.29 Checking battery voltage under load

a 12 volt system on a petrol engined vehicle, the voltmeter should show a reading of approximately 10 volts. On diesel engined vehicle, the load on the battery is higher and the voltage reading will be lower, approximately 9 volts should be obtained.

The starter terminal voltage

The purpose of the test shown in Fig. 10.30, is to check the heavy current circuit of the starter motor and the condition of the solenoid contacts. When the starter motor switch is in the start position, the voltage recorded between the input terminal and earth should be within 0.5 volt of the battery voltage on load. If that voltage was 10.5 volts, then the voltage recorded here should be at least 10 volts.

Checking the solenoid contacts

Should the starter motor voltage under load be low, it is possible that the heavy current solenoid contacts are defective. The test shown in Fig. 10.31 checks for voltage drop

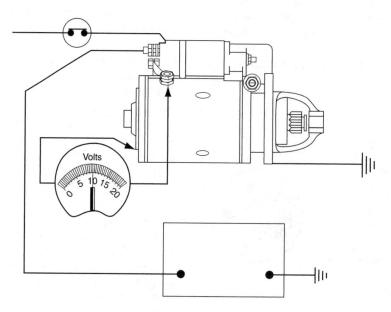

Fig. 10.30 Starter motor voltage test (under load)

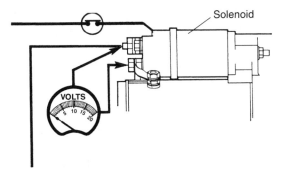

Fig. 10.31 Checking the soleniod contacts

across the solenoid contacts. Good contacts will produce little or no voltage drop. The voltmeter must be connected, as shown, to the two heavy current terminals of the solenoid. When the starter switch is in the off position, the voltmeter should read battery voltage. When the starter switch is in the start position, the voltmeter reading should be zero.

Checking the earth circuit

Because vehicle electrical systems rely on a good earthing circuit, such as metallic contact between the starter and the engine, and other factors such as earthing straps, this part of the circuit needs to be maintained in good order. The test shown in Fig. 10.32 checks this part of the starter circuit. When the starter motor switch is turned to the on position, the voltmeter should read zero. If the reading is 0.5 volt or

more, there is a defect in the earth circuit, a good visual examination should help to locate the problem.

Note

These tests are based on Lucas recommendations. Lucas sell a range of inexpensive, pocket-size test cards. These are recommended for use by those who are learning about vehicle maintenance. The one that relates to 12 volt starting systems has the reference XRB201. We believe that other equipment manufacturers produce similar aids to fault diagnosis.

Learning tasks

Note

In most cases, these the procedure for performing the practical tasks is detailed in the text. However, you should always seek the advice of your supervisor before you start.

Working on vehicles can be dangerous so you must not attempt work that you have not been trained for. When you are in training you must always seek the permission of your supervisor before attempting any of the practical work.

In the case of the first task you will need to have been trained before you can read out the fault codes.

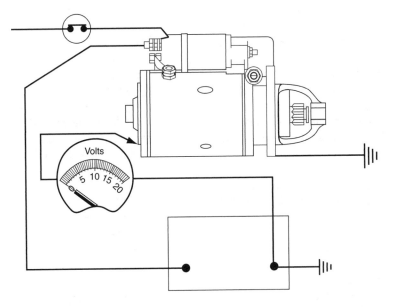

Fig. 10.32 Testing voltage drop on the earth circuit

Part of the purpose of these practical tasks is to assist you in building up a portfolio of evidence for your NVQ. You should, therefore, keep the notes and diagrams that you make neatly in a folder.

1. With the aid of diagrams, where necessary, describe the procedure for obtaining fault codes from an ECU on a vehicle system that you are familiar with.
2. Write down the procedure for removing a battery from a vehicle. Make careful notes of the safety precautions and specify which battery terminal should be disconnected first.
3. Describe the tools and procedures for checking the state of charge of a battery.
4. Carry out a check of the charging rate of an alternator. Make a list of the tools and equipment used and give details of the way in which you connected the test meter. Make a note of the voltage values recorded.
5. Make a sketch of the provision made for tensioning the alternator drive belt on any vehicle that you have worked on. Explain why drive belt tension is important. State the types of problems that may arise if an alternator belt is not correctly adjusted.
6. With the aid of diagrams, explain how to check for voltage at the starter motor terminal when the engine is being cranked.
7. Examine the leads that are connected to the headlights on a vehicle. Make a note of the colours of the cable insulation. Obtain a wiring diagram for the same vehicle and, by reference to the wiring diagram, and the colours that you have observed the cables to be, locate the headlamp cables on the wiring diagram.
8. Make a note of the headlamp alignment equipment that is used in your workshop. Describe the procedure for carrying out a headlamp alignment check. Make a sketch of the method used to adjust headlamp alignment on any vehicle that you have worked on.
9. Give details of the fuseboxes on any vehicle that you are familiar with. State the type and current rating of the side and tail lamp fuse.
10. Write down the precautions that should be taken to protect vehicle electrical equipment before attempting any electric arc welding on the vehicle.

Self assessment questions

1. The function of the heat sink part of the rectifier diode pack in an alternator is to:
   (a) conduct heat away from the diodes in order to protect them
   (b) provide a solid base for the transistors
   (c) act as a voltage regulator
   (d) prevent the stator from overheating
2. Machine sensing, where the alternator output voltage is used to vary the current in the rotor windings, is a method of controlling:
   (a) the phase angle of the a.c. output
   (b) the voltage output of the alternator
   (c) the self exciting voltage of the alternator
   (d) the brush pressure on the slip rings
3. A slack alternator drive belt may cause the:
   (a) engine to overheat
   (b) engine to lose speed
   (c) battery to become discharged
   (d) charging rate to be too high
4. Alternator generate a.c. is converted to d.c. by:
   (a) Zener diode
   (b) the regulator
   (c) the battery
   (d) the slip rings
5. The regulated voltage output of the alternator on a 12 volt system should be:
   (a) exactly 12 volts
   (b) 10.9 volts
   (c) approximately 13 to 15 volts
   (d) 18 volts
6. The maximum permissible voltage drop on a 12 volt starter motor system as measured across the battery terminals when the starter motor is operating is:
   (a) 0.2 volts
   (b) 2 volts
   (c) 5 volts
   (d) 0.5 volts
7. If the gear ratio between a starter motor and the flywheel ring gear is 10:1, and the starter motor generates a torque of 5 Nm, the torque at the flywheel will be:
   (a) 50 Nm
   (b) 105 Nm
   (c) 0.5 Nm
   (d) 50 amperes
8. On a pre-engaged type of starter motor, overspeeding of the armature is prevented by:
   (a) the back e.m.f. in the windings
   (b) Fleming's right-hand rule
   (c) a free-wheeling clutch
   (d) manual withdrawal of the pinion

9. At normal room temperature, the relative density of the electrolyte in a lead acid battery should be:
   (a) 13.6
   (b) 12.6
   (c) 1.28
   (d) 0.128

10. A high rate discharge tester:
    (a) should always be used to test a flat battery
    (b) should only be used if the battery is at least 70% charged
    (c) can only be used on nickel alkaline cells
    (d) determines the Watt-hour capacity of the battery

11
Ignition systems

Topics covered in this chapter

Ignition systems – coil, capacitor (condenser)
Electronic ignition systems,
non-electronic ignition systems
Fault tracing and service and maintenance procedures
Sensors and pulse generators
Ignition timing
Diagnostic trouble codes

The purpose of the ignition system is to provide a spark to ignite the compressed air–fuel mixture in the combustion chamber. The spark must be of sufficient strength to cause ignition and it must occur at the correct time in the cycle of operations. The ignition system is the means by which the 12 volt supply from the vehicle battery is converted into many thousands of volts that are required to produce the spark at the sparking plug, as illustrated by Figs 11.1 and 11.2.

11.1 Producing the high voltage required to cause ignition

Electromagnetism was discussed in the electrical principles section. It is these electro-magnetic principles that are employed in the induction coil type of ignition system.

In this coil there are two windings, a primary winding and a secondary winding. The primary winding consists of a few hundred turns of lacquered copper wire. The secondary winding consists of several thousand turns of thin, lacquered copper wire. The primary winding is wound around the outside of the secondary

winding because the heavier current that it carries generates heat which needs to be dissipated. The layers of windings are normally electrically insulated from each other by a layer of insulating material.

In the centre of the secondary winding there is a laminated soft iron core. The purpose of this

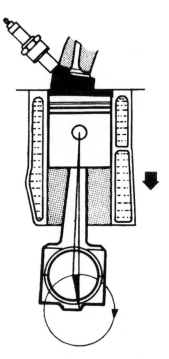

Fig. 11.1 Position of the spark in a reciprocating (piston)

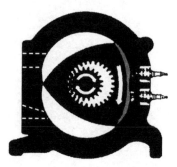

Fig. 11.2 Position of the spark in a rotary engine

laminated core is to concentrate the magnetic field that is produced by the current in the coil windings. The laminations between the coil's outer casing and the windings serves a similar purpose.

The interior parts of the coil are supported on a porcelain base. In some cases the coil may be filled with a transformer oil to improve electrical insulation and to aid heat dissipation.

Principle of operation

The actual size of the secondary voltage (high tension (HT) voltage) is related to the turns ratio of the coil, for example several thousand turns in the secondary winding and a few hundred turns in the primary winding.

In order for the ignition system to operate, the primary winding current must be switched on and off, either by electronic means, or in the case of older systems the contact breaker.

Figure 11.4 shows a simple ignition coil primary circuit. Current from the vehicle battery flows through the primary winding via the ignition switch and the closed contact breaker points, to earth. The current of about 3 amperes in the primary winding causes a strong magnetic field, which is concentrated

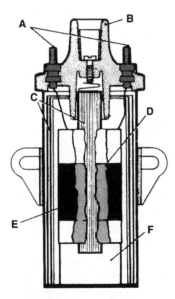

A-Primary terminals
B-High tension terminals
C-Laminations
D-Secondary winding
E-Primary winding
F-Porcelain insulator

Fig. 11.3 The type of ignition coil that has been commonly used on vehicles

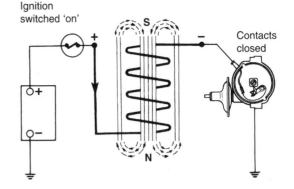

Fig. 11.4 A basic primary circuit of an ignition coil with contacts closed

by the soft iron laminations at the coil centre and the outer sheath of iron. It should be remembered that the secondary winding is inside the primary winding, so the magnetic field from the primary winding also affects the secondary winding. A primary purpose of the primary winding is to create the strong magnetic field that will induce the high voltage in the secondary circuit.

Methods of inducing the high tension (HT) secondary voltage

Switching off the primary current by opening the contact breaker points has two major effects which are caused by the collapsing inwards of the magnetic field. These two effects are:

1. Self induction – a voltage of approximately 250 volts is generated across the ends of the primary winding. The direction of this induced voltage in the primary winding is such that it opposes the change which produced it, that is to say that it attempts to maintain the same direction of current flow.
2. Induction – a high voltage (several thousand volts kV) is generated across the ends of the secondary winding.

Secondary circuit

Figure 11.5 shows the secondary circuit as well as the primary circuit of the simplified system. One end of the secondary winding (thousands of turns of fine copper wire) is connected to the HT terminal in the coil tower, the other connected to the primary winding at one of the low tension (LT) connections. When the primary winding is connected to the secondary

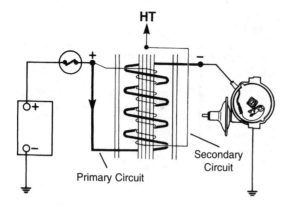

Fig. 11.5 Typical primary and secondary circuits of a coil

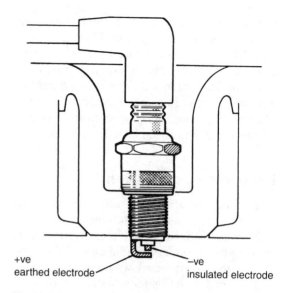

Fig. 11.7 Polarity of spark plug electrodes

winding it is known as 'auto transformer'. The induced primary voltage is added to the secondary voltage.

The capacitor (condenser)

The simplified ignition system described above would not work well because the induced voltage in the primary winding would be strong enough to bridge the contact breaker gap and carry current to earth. This would reduce the secondary voltage and also lead to excessive sparking at the contact breaker points. The points would 'burn' and the efficiency of the system would be seriously impaired. This problem is overcome by fitting a suitable capacitor (approximately 0.2 uF) across (in parallel with) the points.

Figure 11.6 shows the capacitor in circuit. One terminal of the capacitor is connected to the low tension connection on the distributor and the other terminal is earthed.

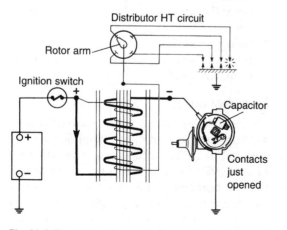

Fig. 11.6 The ignition circuit with the capacitor in place

Capacitors store electricity so that when, in the case of the coil ignition system, the contact points begin to open, the self-induced current from the primary winding will flow into the capacitor instead of 'jumping' across the points gap. This flow of current will continue until the capacitor is fully charged. When the capacitor is fully charged, it will automatically discharge itself back into the primary winding. This capacitor discharge current is in the opposite direction to the original flow and it helps to cause a rapid collapse of the magnetic field. This, in turn, leads to a much higher HT voltage from the secondary winding.

Coil polarity

Electrons flow from negative to positive and thermal activity (heat) also aids electron generation. For this reason, the insulated (central) hotter electrode of the spark plug is made negative in relation to the HT winding.

The result of this polarity direction is that a lower voltage is required to generate a spark across the plug gap. To ensure that the correct coil polarity is maintained the L.T. connections on the coil are normally marked as + and −.

The coil polarity is readily checked by means of the normal garage type oscilloscope. Figure 11.8 shows the inverted waveform that results from incorrect connection of the coil, or the fitting of the wrong type of coil for the vehicle under test.

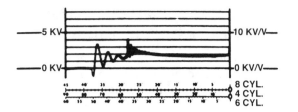

Fig. 11.8 The incorrect (inverted) waveform on the oscilloscope screen

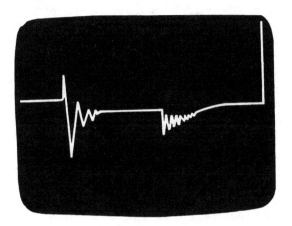

Fig. 11.9 The correct waveform on the oscilloscope screen

Electronic ignition systems

Electronic ignition systems make use of some form of electrical/electronic device to produce the electrical pulse that switches 'on and off' the ignition coil primary current so that a high voltage is induced in the coil secondary winding in order to produce a spark, in the required cylinder, at the correct time.

There are several methods of producing the 'triggering' pulse for the ignition that replaces the 'on and off' action of the contact breaker, and the commonly used types are reviewed in the following sections.

Electromagnetic pulse generators

The constant energy ignition system

Figure 11.10 shows part of a type of electronic ignition distributor that has been in use for many years. The distributor shaft is driven from the engine camshaft and turns the reluctor rotor that is attached to it. The reluctor thus rotates at half engine speed.

Each time that a lobe on the rotor (reluctor) passes the pick-up probe, a pulse of electrical energy is induced in the pick-up winding. The

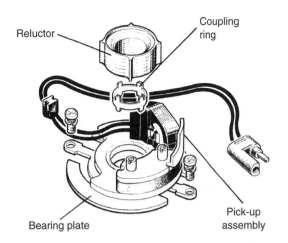

Fig. 11.10 Reluctor and pick-up assembly

pick-up winding is connected to the electronic ignition module and when the pulse generator voltage has reached a certain level (approximately 1 volt) the electronic circuit of the module will switch on the current to the ignition coil primary winding.

As the reluctor (rotor) continues to rotate, the voltage in the pick-up winding begins to drop and this causes the ignition module to 'switch off' the ignition coil primary current; the high voltage for the ignition spark is then induced in the ignition coil secondary winding. The period between switching on and switching off the ignition coil primary current is the dwell period. The effective increase in dwell angle, as the speed increases, means that the coil current can build up to its optimum value at all engine speeds. Figure 11.11 shows how the pulse generator voltage varies due to the passage of one lobe of the reluctor past the pick-up probe.

From these graphs it may be seen that the ignition coil primary current is switched on when the pulse generator voltage is approximately 1 volt and is switched off again when the voltage falls back to the same level. At higher engine speeds the pulse generator produces a higher voltage and the switching on voltage (approximately 1 volt) is reached earlier, in terms of crank position, as shown in the second part of Fig. 11.11. However, the 'switching off' point is not affected by speed and this means that the angle (dwell) between switching the coil primary current on and off increases as the engine speed increases. This means that the build-up time for the current in the coil primary winding, which is the important factor affecting the spark energy, remains virtually constant at

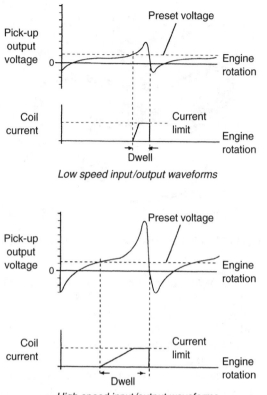

Low speed input/output waveforms

High speed input/output waveforms

Fig. 11.11 Pick-up output voltage at low and high speeds

all speeds. It is for this reason that ignition systems of this type are known as 'constant energy systems'. It should be noted that this 'early' type of electronic ignition still incorporates the centrifugal and vacuum devices for automatic variation of the ignition timing.

Digital (programmed) ignition system – Hall effect sensor

Programmed ignition makes use of computer technology and it permits the mechanical, pneumatic and other elements of the conventional distributor to be dispensed with. Figure 11.12 shows an early form of digital ignition system.

The control unit (ECU or ECM) is a small dedicated computer which has the ability to read input signals from the engine, such as speed, crank position and load. These readings are compared with data stored in the computer memory and the computer then sends outputs to the ignition system. It is traditional to represent the data, which is obtained from engine tests, in the form of a 3-dimensional map, as shown in Fig. 11.13.

Any point on this map may be represented by a number reference. For example, engine speed 1 000 rpm, manifold pressure (engine load) 0.5 bar, ignition advance angle 5 degrees. These numbers can be converted into computer (binary) codewords, made up of 0s and 1s (this is why it is known as digital ignition). The map is then stored in computer memory, where the processor of the control unit can use it to provide the correct ignition setting for all engine operating conditions.

In this early type of digital electronic ignition system, the 'triggering' signal is produced by a Hall effect sensor of the type shown in Fig. 11.14.

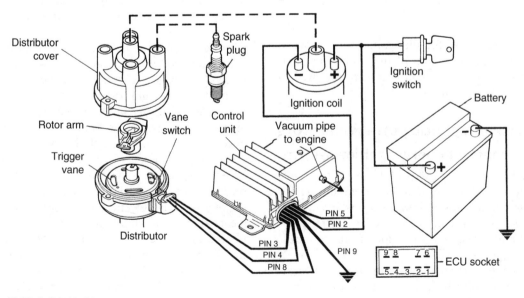

Fig. 11.12 A digital ignition system

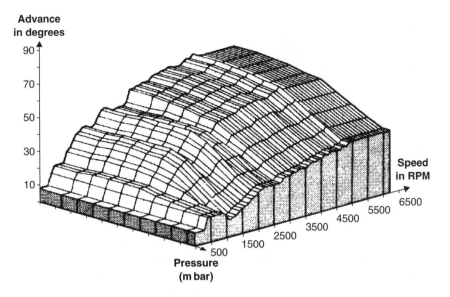

Fig. 11.13 An ignition map that is stored in the ROM of the ECM

When the metal part of the rotating vane is between the magnet and the Hall element the sensor output is zero and when the gaps in the vane expose the Hall element to the magnetic field, a voltage pulse is produced. In this way, a voltage pulse is produced by the Hall sensor each time that a spark is required. Whilst the adapted form of the older type ignition distributor is widely used for electronic ignition systems, it is probable that the trigger pulse generator driven by the crankshaft and flywheel is more commonly used on modern systems.

Opto-electronic sensing for the ignition system

Figure 11.15 shows the electronic ignition photo-electronic distributor sensor used on a Kia. There are two electronic devices involved in the operation of the basic device. One is a light

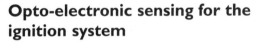

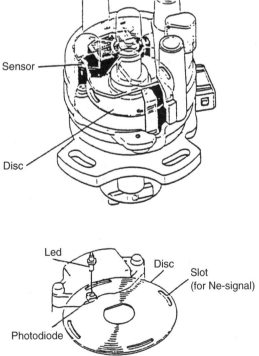

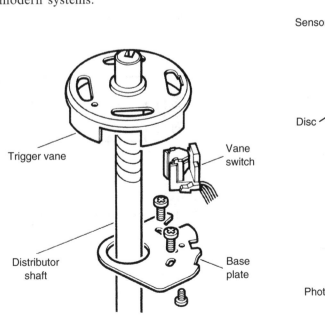

Fig. 11.14 A Hall type sensor

Fig. 11.15 An opto-electronic sensor

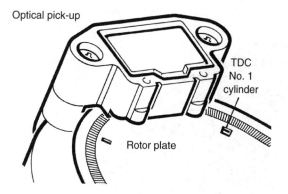

Optical pick-up

TDC
No. 1
cylinder

Rotor plate

Fig. 11.16 An alternative form of opto-electronic sensor

emitting diode (it converts electricity into light), the other is a photo diode that can be 'switched on' when the light from the LED falls on it.

Another version of this type of sensor is shown in Fig. 11.16. Here the rotor plate has 360 slits placed at 1 degree intervals for engine speed sensing, and a series of larger holes for TDC indication that are placed nearer the centre of the rotor plate. One of these larger slits is wider than the others and is used to indicate TDC for number 1 cylinder.

Distributorless ignition system

Figure 11.17 shows an ignition system for a 4-cylinder engine. There are two ignition coils, one for cylinders 1 and 4, and another for cylinders 2 and 3. A spark is produced each time a pair of cylinders reaches the firing point, near top dead centre. This means that a spark occurs on the exhaust stroke as well as on the power stroke.

For this reason, this type of ignition system is sometimes known as the 'lost spark' system.

Figure 11.17 shows that there are two sensors at the flywheel; one of these sensors registers engine speed and the other is the trigger for the ignition. They are shown in greater detail in Figs 11.18 and 11.19 and they both rely on the variable reluctance principle for their operation.

An alternative method of indicating the TDC position is to use a toothed ring attached to the flywheel, which has a tooth missing at the TDC positions, as shown in Fig. 11.19.

With this type of sensor, the TDC position is marked by the absence of an electrical pulse, known as also a variable reluctance sensor. The other teeth on the reluctor ring, which are often spaced at 10 degree intervals, are used to provide pulses for engine speed sensing.

Direct ignition systems

Direct ignition systems are those where each cylinder has a separate ignition coil which is placed directly on the sparking plug, as shown in Fig. 11.20.

The trigger unit is incorporated in the coil housing in order to protect the ECM against high current and temperature.

Ignition system refinements

As the processing power of microprocessors has increased it is natural to expect that system designers will use the increased power to provide further features, such as combustion knock sensing and adaptive ignition control.

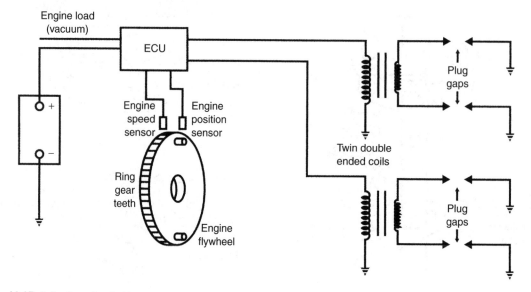

Fig. 11.17 A distributorless ignition system

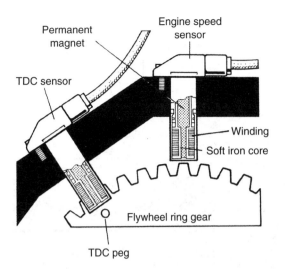

Fig. 11.18 Details of engine speed and crank position sensor

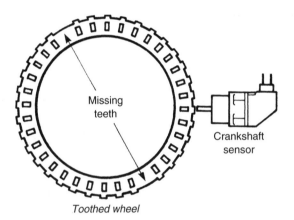

Toothed wheel

Fig. 11.19 Engine speed and position sensor that uses a detachable reluctor ring

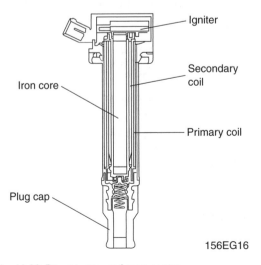

Fig. 11.20 Direct ignition coil cross-section

Knock sensing

Combustion knock is a problem that is associated with engine operation. Early motor vehicles were equipped with a hand control that enabled the driver to retard the ignition when the characteristic 'pinking' sound was heard. After the pinking had ceased the driver could move the control lever back to the advanced position. Electronic controls permit this process to be done automatically and a knock sensor is often included in the make up of an electronic ignition system.

Combustion knock sensors

A knock sensor that is commonly used in engine control systems utilises the piezo-electric generator effect. That is to say that the sensing element produces a small electric charge when it is compressed and then relaxed. Materials such as quartz and some ceramics like PZT (a mixture of platinum, zirconium and titanium) are effective in piezo-electric applications. In the application shown, the knock sensor is located on the engine block adjacent to cylinder number 3. In this position it is best able to detect vibrations arising from combustion knock in any of the 4 cylinders.

Because combustion knock is most likely to occur close to top dead centre, in any cylinder, the control program held in the ECM memory enables the processor to use any knock signal generated to alter the ignition timing by an amount that is sufficient to eliminate the knock. When knock has ceased, the ECM will advance the ignition in steps, back to its normal setting. The mechanism by which vibrations arising from knock are converted to electricity is illustrated in Fig. 11.22.

The sensor is accurately designed and the centre bolt that pre-tensions the piezo crystal is accurately torqued so as to provide the correct setting. The steel washer that makes up the seismic mass has very precise dimensions. When combustion knock occurs, the resulting mechanical vibrations are transmitted by the seismic mass to the piezoelectric crystal. The 'squeezing up' and relaxing of the crystal in response to this action produces a small electrical signal that oscillates at the same frequency as the knock sensor element. The electrical signal is conducted away from the crystal by wires that are secured to suitable points on the crystal.

The tuning of the sensor is critical because it must be able to distinguish between knock from combustion and other knocks that may arise

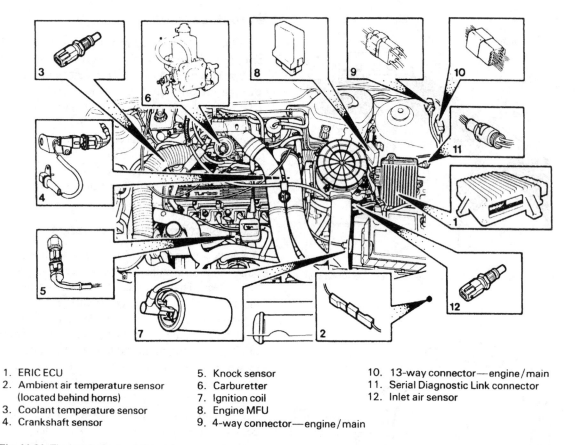

1. ERIC ECU
2. Ambient air temperature sensor (located behind horns)
3. Coolant temperature sensor
4. Crankshaft sensor
5. Knock sensor
6. Carburetter
7. Ignition coil
8. Engine MFU
9. 4-way connector — engine/main
10. 13-way connector — engine/main
11. Serial Diagnostic Link connector
12. Inlet air sensor

Fig. 11.21 The knock sensor on the engine

from the engine mechanism. This is achieved because combustion knock produces vibrations that fall within a known range of frequencies.

Adaptive ignition

The computing power of modern ECMs permits ignition systems to be designed so that the ECM can alter settings to take account of changes in the condition of components, such as petrol injectors, as the engine wears. The general principle is that the best engine torque is achieved when combustion produces maximum cylinder pressure just after TDC. The ECM monitors engine acceleration, by means of the crank sensor, to see if changes to the ignition setting produce a better result, as indicated by increased engine speed as a particular cylinder fires. If a better result is

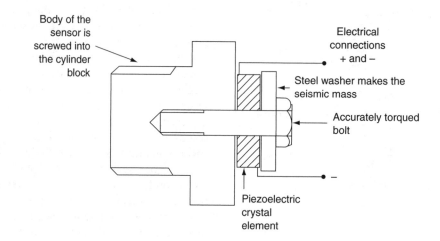

Fig. 11.22 The principle of the piezoelectric combustion knock sensor

achieved, then the ignition memory map can be reset so that the revised setting becomes the one that the ECM uses. This 'adaptive learning strategy' is now used quite extensively on computer controlled systems and it requires technicians to run vehicles under normal driving conditions for several minutes after replacement parts and adjustments have been made to a vehicle.

11.2 Fault tracing, maintenance and repair of ignition systems

The majority of the following descriptions are based on the use of a portable oscilloscope. However, for tests on the primary circuit of an ignition system, a good-quality digital multimeter will also serve most purposes.

Performance checks on pulse generators

If preliminary tests on the ignition circuit indicate that the pulse generator may be defective, checks of the following type can provide useful information.

Magnetic (inductive) type ignition pulse generator

The voltage output of a typical ignition pulse generator of this type varies with engine speed and the actual voltage varies considerably according to manufacturer. However, the voltage waveform is of similar form for most makes and useful test information can be gained from a voltage test. There are two possible states for a voltage test: one is with the pulse generator disconnected from the electronic module when a waveform of the type shown in Fig. 11.23 Bosch should be produced and the other is when the pulse generator is connected to the electronic module when the waveform shown in Fig. 11.24 should be produced. The difference is accounted for by the fact that lower part of the pattern is affected by the loading from the electronic module.

A digital voltmeter set to an appropriate a.c. range may also be used to check the output of this type of pulse generator. The voltages for each pulse should be identical and the maximum value should be at the level given in the vehicle repair data.

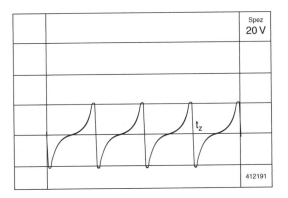

Fig. 11.23 Waveform, pulse generator - unloaded

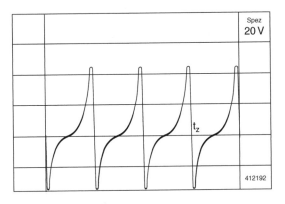

Fig. 11.24 Waveform, pulse generator - electronic module connected

Other checks that can be performed on the inductive type pulse generator
Resistance check

If the sensor voltage is not correct it is possible to conduct a resistance test that gives an indication of the condition of the winding part of the sensor. Figure 11.25 gives an indication of the connections used for this process.

Hall type sensor tests

A Hall type sensor is known as a passive device because it operates by processing electricity that is supplied to it and it can be expected to have at least three cables at its connection. Figure 11.26 shows the approximate position of the oscilloscope test probes on a Hall type ignition distributor.

If the test is being performed because the engine will not start, the purpose of the test will be to check that there is a signal from the ignition trigger sensor. The oscilloscope, or meter, would be connected as shown and the engine cranked over. The voltage pattern would be observed and,

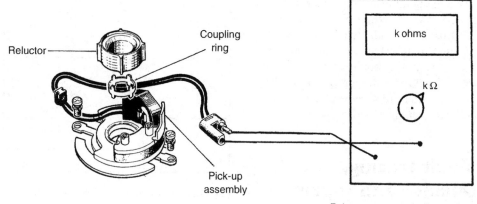

Reluctor

Coupling ring

Pick-up assembly

Reluctor and pick-up assembly

k ohms

k Ω

Pulse generator winding resistance

Fig. 11.25 Testing the pulse generator coil

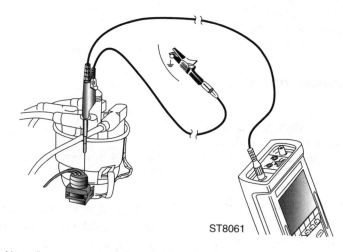

ST8061

Fig. 11.26 Using a portable oscilloscope to test a Hall type ignition pulse sensor

provided the signal is adequate, the fault would be sought elsewhere in the system.

If the engine starts up and the test is being conducted to check for other defects, the test connections remain as for the 'no start' condition but the sensor signal may be observed at number of engine speeds. An approximate shape of a voltage trace for a Hall type sensor is shown in Fig. 11.27. The frequency of the voltage pulse will vary with engine speed.

The reference voltage that is referred to is the regulated voltage that is supplied to the sensor.

Opto-electronic trigger sensor

This is also a passive device and will have a connection for a supply of electricity. The test procedure is similar to that described for the Hall type sensor and the type of voltage output to be expected in the test is shown in Fig. 11.28.

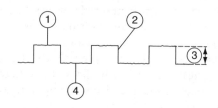

1. The upper horizontal lines should reach reference voltage.
2. Voltage transitions should be straight and vertical.
3. **Peak-Peak** voltages should equal reference voltage.
4. The lower horizontal lines should almost reach ground.

Fig. 11.27 Oscilloscope trace for Hall type ignition trigger sensor

Servicing an electronic type distributor

The rotor (reluctor) to magnetic pick-up air gap on an electronic ignition distributor is an important setting that requires occasional

- Primary distributor triggering (optical)

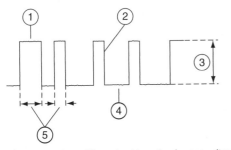

1. The upper horizontal lines should reach reference voltage.
2. Voltage transitions should be straight and vertical.
3. **Peak-Peak** voltage should equal reference voltage.
4. The lower horizontal lines should almost reach ground.
5. Signal pulse width may vary due to size variations in the trigger wheel window.

! Specifications may vary. Consult manufacturer's specifications.

Fig. 11.28 Oscilloscope voltage trace for opto-electronic type ignition trigger sensor

checking. Figure 11.29(a) shows the rotor and magnetic pick up and Fig. 11.29(b) shows non-magnetic, i.e. brass or plastic feeler gauges, being used to check the gap. The use of non-magnetic feeler gauges is important because, if steel is placed in this air gap, the electric charge generated may damage to electronic circuit.

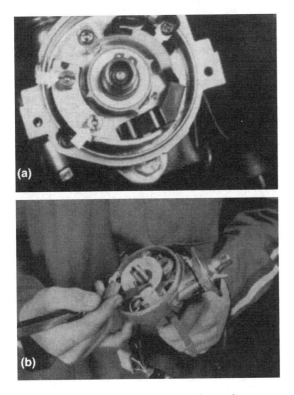

Fig. 11.29 Checking the air gap between pick up and rotor

Setting the static timing of an electronic type distributor

The procedures described here applies to two types of distributors. One is the variable reluctance pulse generator type and the other is a Hall effect pulse generator. It should be noted that the procedure will vary according to engine type and also to the type and make of distributor. It is essential to have access to the details relating to a specific application. Of course, if you are familiar with a particular make of vehicle it is quite possible that you will be able to memorise the procedures.

Static timing – variable reluctance pulse generator

The cylinder to which the distributor is being timed must be on the firing stroke and the timing marks, on the engine, must be accurately aligned. The distributor rotor must be pointing towards the correct electrode in the distributor cap and the corresponding lobe of the reluctor rotor must be aligned with the pick up, as in the case shown in Fig. 11.29. When the distributor is securely placed and the electrical connections made, the final setting of the timing is achieved by checking the timing with the strobe light. Final adjustment can then be made by rotating the distributor body, in the required direction, prior to final tightening of the clamp.

Static timing Hall effect pulse generator

Figure 11.30 shows a Hall effect distributor, for digital electronic ignition, where the distributor itself is provided with timing marks. With the engine timing marks correctly set, the cut-out in the trigger vane must be aligned with the vane switch, as shown.

Should the setting require adjustment, the distributor clamp is slackened, just sufficient to

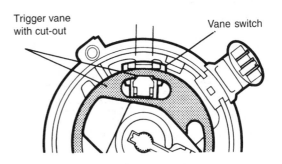

Trigger vane with cut-out

Vane switch

Fig. 11.30 Static timing Hall effect pulse generator

permit rotation of the distributor body. The distributor is then rotated until the marks are aligned. After tightening the clamp and replacing all leads, etc., the engine may be started and the timing checked be means of the stroboscopic light.

Diagnosis – secondary (high voltage) side of ignition system

Firstly, it is important to **Work in a Safe Manner**. Electric shock from ignition systems must be avoided; not only is the shock dangerous in itself but it can also cause involuntary muscular actions which can cause limbs to be thrown into contact with moving or hot parts.

Tests on distributor type ignition systems

The king lead and the HT lead to number 1 cylinder are normally connected to the oscilloscope, as shown in Fig. 11.31.

A typical oscilloscope display for a distributor type ignition is shown in Fig. 11.32.

These displays show the firing voltage for number 1 cylinder. The parade pattern shows the voltage of the remaining cylinders as displayed across the screen, in the firing order, from left to right.

Tests on distributorless ignition systems

In this case there is no king lead so the inductive pick up is clamped to the HT leads individually, as close as possible to the spark plug, as shown in Fig. 11.33.

The screen display obtained from such a test shows data about engine speed and burn time and voltage. These and other details may more easily be seen by taking an enlargement of the trace for a single cylinder as shown in Fig. 11.34.

The details are:

1. Firing line – this represents the high voltage needed to cause the spark to bridge the plug gap;
2. The spark line;
3. Spark ceases;
4. Coil oscillations;
5. Intermediate section (any remaining energy is dissipated prior to the next spark);

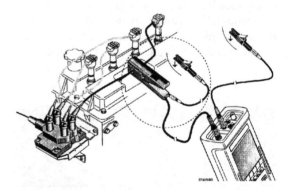

Fig. 11.33 The oscilloscope set-up for obtaining a secondary voltage trace from a DIS

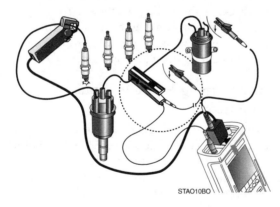

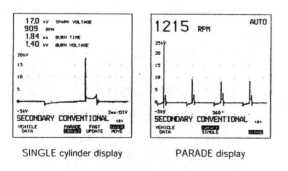

SINGLE cylinder display   PARADE display

Fig. 11.32 Oscilloscope displays of secondary ignition voltage – single pattern and parade

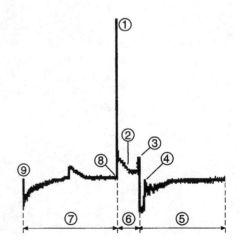

Fig. 11.34 Details of the HT voltage trace for a single cylinder

Fig. 11.31 Oscilloscope connections for obtaining ignition trace

6. Firing section (represents burn time);
7. Dwell section;
8. Primary winding current is interrupted by transistor controlled by the ECM;
9. Primary winding current is switched on to energise the primary. The dwell period is important because of the time required for the current to reach its maximum value.

A comparison of the secondary voltage traces for each of the cylinders should show them to be broadly similar. If there are major differences between the patterns, then it is an indication of a defect.

For example, a low firing voltage (1) indicates low resistance in the HT cable or at the spark plug. The low resistance could be attributed to several factors including oil or carbon fouled spark plug, incorrect plug gap, low cylinder pressure or defective HT cable. A high voltage at (1) indicates high resistance in the HT cable or at the spark plug. Factors to consider here include a loose HT lead, wide plug gap or an excessive amount of resistance has developed in the HT cable. Table 11.1 below summarises the major points:

Table 11.1 Factors affecting firing voltage

Factor	High firing voltage	Low firing voltage
Spark plug gap	Wide	Small
Compression pressure	Good	Low
Air–fuel ratio	Weak	Correct
Ignition point	Late	Early

Sparking plugs can be removed and examined; HT leads can also be examined for tightness in their fittings and their resistance can be checked with an ohmmeter. Resistive HT leads are used for electrical interference purposes and they should have a resistance of approximately **15 000 to 25 000 ohms per metre length**.

As in all cases, it is important to have to hand the information and data that relates to the system being worked on.

Direct on sparking plug systems

Direct ignition systems of the type shown in Fig. 11.19, are normally controlled by the engine management computer (ECM) which incorporates self diagnostics that give detailed information about ignition faults. Should a misfire be attributed to one of the ignition coils, the condition of the coil may be checked by conducting a resistance test on the primary and secondary windings. The primary winding of the coil will have a resistance of approximately **1 ohm**, whilst the secondary winding resistance is normally **5 000 ohms to 10 000 ohms**.

Diagnostic trouble codes (DTCs)

Modern engine management use European or US on-board diagnostic standards. European on-board diagnostic systems (EOBD) require that emissions related trouble codes (DTCs) are available through a standard connection that permits an approved repairer to gain access to the DTCs by means of an appropriate scan tool. The form of the EOBD diagnostic plug and the recommended position of it on the vehicle is shown in Fig. 11.35.

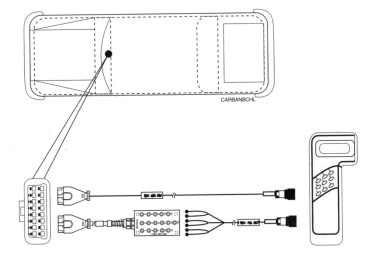

Fig. 11.35 The leads for the Bosch KTS 300 pocket system tester

Because the vehicle emissions are affected by a number of vehicle systems, such as fuelling and ignition, the ECM is programmed to record and report on a range of faults in these systems that may contribute to a vehicle's failure to meet emissions regulation limits.

An example of a DTC is:

DTC   **Function**
PO300   **Random/multiple cylinder misfire detected**

Analysis of this diagnostic trouble code

This DTC tells us that there is a misfire. To determine the cause will require further testing of the ignition related systems in order to trace the cause and effect a repair. Some on-board diagnostics may go deeper into the causes of the misfire, in which case the task of fault tracing should be made simpler.

11.3 Contact breaker type ignition systems

Each time a spark is required the contact breaker points must open and close. For a four stroke engine this means that a spark is required once for each two revolutions of the engine crank. To achieve this the cam that operates the contact breaker is normally driven from the engine camshaft. For multi-cylinder engines the cam normally has the same number of lobes as there are cylinders. A typical example of a cam and contact breaker mechanism for a four cylinder engine is shown in Fig. 11.36.

In order to get an idea of the speed of operation of the contact breaker points, consider what happens at an engine crank speed of 3 600 revolutions per second (rpm). The cam of the distributor is rotating at half this speed, i.e. 1 800 rpm. That is equivalent to 30 rpm. In our 4 cylinder engine the contact points will open and close 4 times for each revolution of the cam and this means that they will open and close $30 \times 4 = 120$ times a second. This speed obviously varies with engine speed but it is evident that the contact breaker and its operating mechanism is subject to a lot of hard work. To give reliability and durability the cam is made from high-quality steel with a hard wearing surface. The moving contact is operated by a 'heel' made from hard wearing, electrically insulating material, usually high-quality plastic. The contact points are made from a steel alloy containing tungsten, enabling the points to cope with the burning action that arises from electrical action as the points open. This electrical action at the points will, over time, lead to pitting of one contact and the building up of a corresponding 'pip' on the other point; this is known as 'pitting and piling'. The electrical action that leads to this effect also causes a dark layer of oxide to form on the contact point faces. These factors lead to a deterioration of ignition system performance and requires the 'points' to be checked and, if they are pitted and blackened, the usual procedure is to replace them.

Sliding contact (self cleaning) type of contact breaker

A method that is used to overcome the problem of pitting is to use sliding contacts. In sliding contacts, the fixed contact has a larger face area than the moving one. In addition to the normal opening and closing action, the moving contact is made to slide across the face of the fixed contact.

Figure 11.37, shows a sliding contact assembly. The contact breaker heel has two

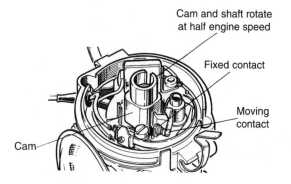

Fig. 11.36 The distributor cam and contact breaker

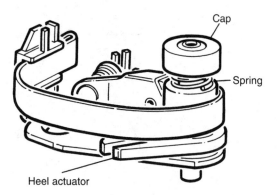

Fig. 11.37 A sliding contact assembly

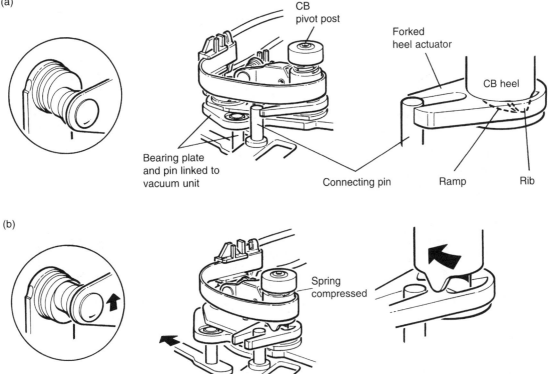

Fig. 11.38 The sliding contact mechanism (a) static position, (b) advanced position

ribs at its base and these ribs rest in ramps formed in the forked heel actuator. This forked heel actuator pivots on the contact breaker pivot post. The fork in the heel actuator engages with a pin which is fixed to the distributor base plate. When the vacuum advance and retard mechanism rotates the bearing plate, the heel ribs ride up the ramps in the actuator plate and cause the moving contact to slide across the fixed contact, as shown in Fig. 11.38. This action is controlled by a small coil spring which is secured to the end of the pivot post.

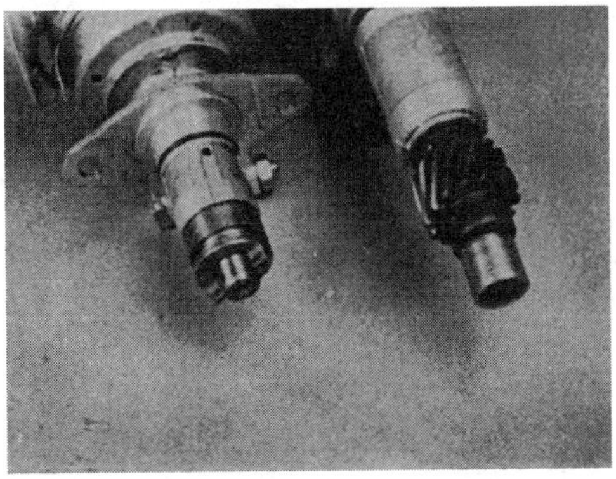

Fig. 11.39 Two types of distributor drive

Distributor drives

Two types of drive are commonly used to rotate the distributor shaft that rotates the cam. One of these is the skew gear and the other is an Oldham type coupling. The slot in the Oldham coupling is normally offset a little, to one side of the distributor shaft centre. This aids re-timing should the distributor be removed from the engine. Both types of distributor drive are shown in Fig. 11.39.

Dwell angle

The dwell angle is the period of time, as represented by angular rotation of the cam, for which the contact breaker points are closed. Figure 11.40 shows the dwell angles for the cams of 4, 6 and 8 cylinder engines.

The dwell angle is important because it controls the period of time during which the primary current is energising the primary winding of the coil. In order for the magnetic field to reach its maximum strength, the primary

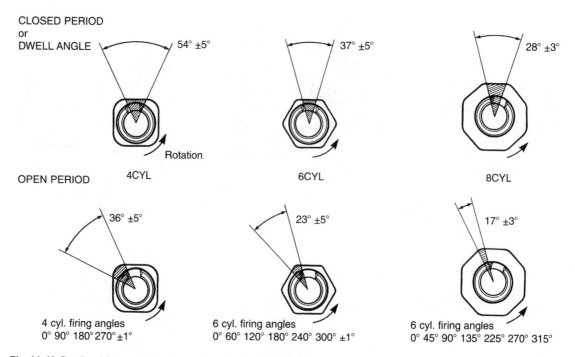

Fig. 11.40 Dwell and firing angles for four-, six- and eight-cylinder engines

current must reach its maximum value, and this takes time. In an inductance, which is what the primary coil is, it takes several milliseconds for the current to rise to its maximum value, as shown in Fig. 11.41. The actual time of current build up is dependent on the inductance (Henry) and the resistance (Ohms) of the primary winding, for a typical ignition coil the time constant is of the order of 15 milli seconds.

The dwell angle, which determines current build-up time, is affected by the position of the heel of the moving contact in relation to the cam, as shown in Fig. 11.42.

Timing of the spark

The process of ensuring that the spark occurs at the right place and at the right time is known as

timing the ignition. In the first place, there is **static timing**. This is the operation of setting up the timing when the engine is not running. Ignition timing settings are normally expressed in degrees of crank rotation. The static timing varies across the range of engines in use today. A reasonable figure for static timing would be approximately 5° to 10°. **Please note: this setting is critical and must be verified for any engine that is being worked on, never guessed.** When the spark is made to occur before this static setting, it said to be **advanced**, and when the spark occurs after the static setting, it is said to be **retarded**.

Because vehicle engines are required to operate under a wide range of varying operating conditions, it is necessary to alter the ignition timing while the vehicle and engine are in operation. The devices used for this purpose are often referred to as 'the automatic advance and retard devices'. There are two basic forms of automatic 'advance and retard' mechanisms; one is **speed sensitive**, the other **load sensitive**.

The speed sensitive (centrifugal) timing device

Ideally ignition should commence so that full combustion is achieved and maximum gas

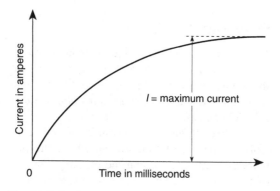

Fig. 11.41 Time and current in a coil

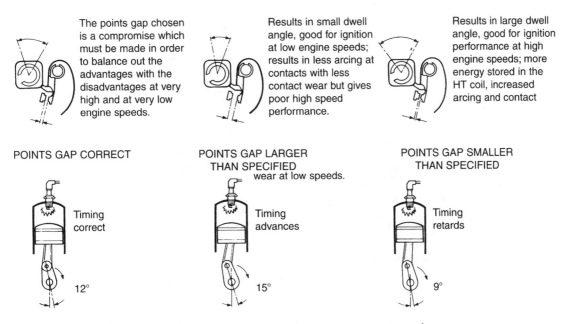

POINTS GAP CORRECT

The points gap chosen is a compromise which must be made in order to balance out the advantages with the disadvantages at very high and at very low engine speeds.

Timing correct

12°

POINTS GAP LARGER THAN SPECIFIED

Results in small dwell angle, good for ignition at low engine speeds; results in less arcing at contacts with less contact wear but gives poor high speed performance.

wear at low speeds.

Timing advances

15°

POINTS GAP SMALLER THAN SPECIFIED

Results in large dwell angle, good for ignition performance at high engine speeds; more energy stored in the HT coil, increased arcing and contact

Timing retards

9°

Fig. 11.42 Effect of contact points gap variations on dwell angle and ignition timing or engine perfomance

pressure reached when the piston is near top dead centre. Because combustion takes some time to happen it is necessary to make the spark occur earlier (advance it) as the engine speed increases. Assume that the spark is set to occur at 15 degrees before top dead centre. At an engine speed of 1 200 rpm, it takes approximately 0.002 second to move the crank (and piston) through 15 degrees. Doubling the engine

speed will halve this time, and so on. In order to increase the time, it is usual to advance the ignition to compensate.

Figure 11.45 shows the construction of a typical automatic advance mechanism. The whole assembly shown here is rotated at half engine speed by the distributor drive shaft. The arrangement allows the cam to rotate relative to the distributor drive shaft and it is thus possible to open the points earlier according to the position that the cam is moved to by the action of weights.

In Fig. 11.46, the weights are pulled in by the springs and the cam is in the static timing

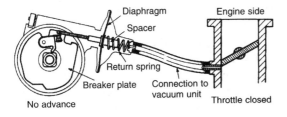

Fig. 11.43 A typical speed sensitive mechanism

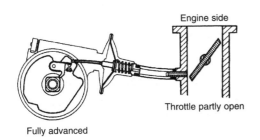

Fig. 11.44 A typical load sensitive mechanism

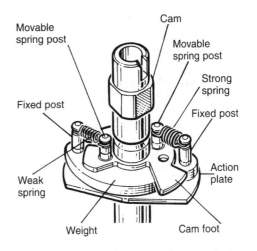

Fig. 11.45 Construction of automatic advance mechanism

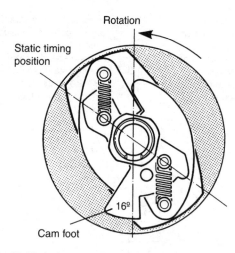

Fig. 11.46 A plan view of the automatic timing mechanism – zero centrifugal advance

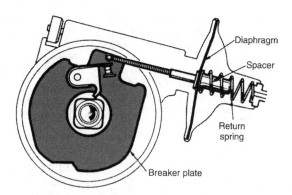

Fig. 11.48 The action of the vacuum operated timing control – no advance

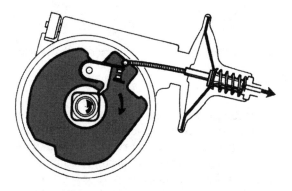

Fig. 11.49 The action of the vacuum operated timing control – fully advanced

position. As the engine speed and the distributor shaft speed increases the weights are forced outwards by centrifugal force against the tension of the springs, as shown in Fig. 11.47. Through the mechanism of the action plate, the weights and the cam foot, the cam is moved round slightly in the direction of rotation on the distributor shaft. This has advanced the ignition timing by 16 degrees in the case shown. The extension on the cam foot eventually contacts one of the fixed posts on a weight and this prevents further advance. Evidently the design of this cam foot extension will determine the maximum amount of advance for a given application.

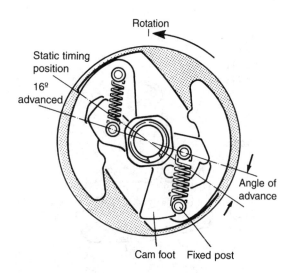

Fig. 11.47 A plan view of the automatic timing mechanism – 16° of centrifugal advance

Load sensitive automatic timing control

Under light load, cruising conditions, it is possible to improve fuel economy by weakening the mixture. Weak mixtures take longer to burn and, if economy devices are to work effectively, it is necessary to provide extra spark advance. In a throttle controlled engine there is a relationship between the level of vacuum (absolute pressure) in the manifold and the load on the engine. At light engine loads, the manifold vacuum is high and at heavy engine loads it is low. The vacuum sensitive control adjusts the ignition timing to suit engine load.

11.4 Distributing the high voltage electrical energy

The timing of the spark is achieved by setting the points opening to the correct position in relation to crank and piston position, or the rotor in a rotary engine such as the Wankel engine. Distributing the resultant high voltage electricity to produce the spark at the required cylinder is achieved through the medium of the rotor arm and the distributor cap, as shown in Fig. 11.50.

In Fig. 11.50, a spring loaded carbon brush is situated beneath the king lead (the one from the coil). This carbon brush makes contact with the metal part of the rotor and the high voltage current is then passed from the rotor via an air gap to the plug lead electrode and, from there through an HT lead to the correct sparking plug.

The rotor arm shown in Fig. 11.51 is typical of the type in common usage. It is sometimes the case that the metal nose of the rotor is extended, as shown in Fig. 11.52. This is intended as an aid to prevent the engine running backwards by making the spark occur in a cylinder where the piston is near bottom dead centre, should the engine start to rotate backwards.

Note the slot above the cam. Inside the rotor arm, in the insulating plastic, is a protrusion

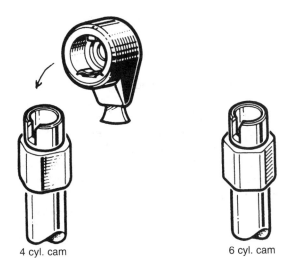

4 cyl. cam   6 cyl. cam

Fig. 11.51 Typical four-/six-cylinder rotor arm and cams

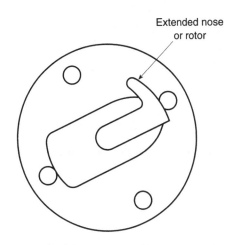

Extended nose or rotor

Fig. 11.52 The extended metal nose

which locates the rotor arm in the slot on the distributor cam. The rotor is thus driven round by the distributor cam. The HT leads are arranged in the cap so that they are encountered in the correct firing sequence, for example, 1 342 for a 4 cylinder and probably 153 624 for a 6 cylinder in-line engine.

11.5 Cold starting

When an engine is cold the lubricating oil in the engine is more viscous than when it is hot. This requires greater torque from the starting motor and this, in turn, places a greater drain on the battery.

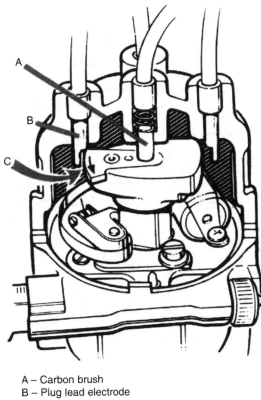

A – Carbon brush
B – Plug lead electrode
C – Rotor gap

Fig. 11.50 The HT circuit inside the distributor cap

Figure 11.53 shows how the voltage required to produce a spark at the spark plug gap varies with temperature. The decimal figures refer to 3 different sparking plug gaps. **A much higher voltage is required for cold temperatures**.

One of the factors that sometimes make an engine more difficult to start is when it is cold rather than hot. To ease this problem, coil ignition systems are often designed to give a more powerful spark for starting up than is used for normal running. This process is achieved by the use of a ballast resistor and a coil that gives normal operating voltage at a voltage value lower that the battery voltage, for example, an 8 volt coil in a 12 volt system. Under running conditions the coil operates on current supplied from the ignition switch through the ballast resistor to the primary winding. The size of the ballast resistor, in Ohms, is that which produces the desired operating voltage for the coil. Under starting conditions the ballast resistor is by-passed and the coil is supplied with full battery voltage. This gives a higher voltage spark for starting purposes.

Figure 11.54 shows the circuit for a ballasted ignition system. In this system the ballast resistor is a separate component but some other systems use a length of resistance cable between the ignition switch and the coil, which achieves the same result. Whichever type of ballast resistor is used, its function will be the same, which is to produce a voltage drop that will give the coil its correct operating voltage. If the coil operates at 8 volts, and the battery voltage is 12 volts, the ballast resistor will provide a voltage drop of 4 volts.

11.6 Sparking plugs

The sparking plug is the means of introducing the spark into the combustion chamber of internal combustion engines.

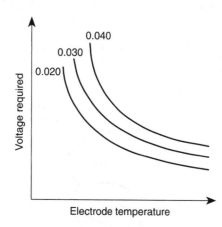

Fig. 11.53 Spark voltage and plug temperature

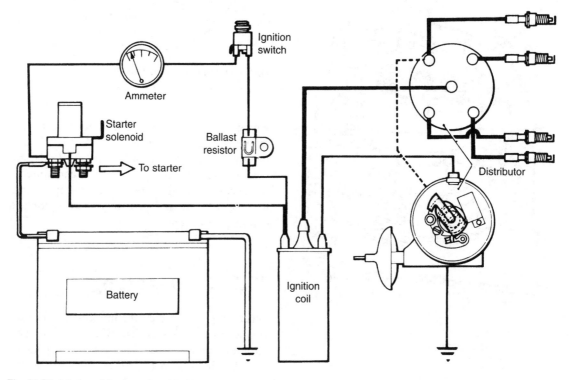

Fig. 11.54 A ballasted (series resistor) ignition circuit

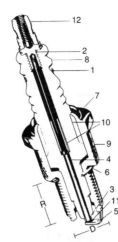

Fig. 11.55 Sparking plug construction

1 – Brand name and product code
2 – Cement
3 – Core nose
4 – Centre electrode
5 – Earth electrode
6 – Attached gasket
7 – Hexagon
8 – Anti-flashover 5-ribbed insulator
9 – Shell
10 – Gas-tight 'sillment seals
11 – Spark gap
12 – Terminal
D – Thread diameter
R – Thread reach

Figure 11.55 shows details of a widely used make of sparking plug. In order to function effectively, sparking plugs must operate over a wide range of varying conditions. These conditions are considered next, together with some features of sparking plug design.

Compression pressure

The voltage required to provide a spark at the plug gap is affected by the pressure in the combustion chamber. It is approximately a linear relationship, as shown in Fig. 11.56. Note that the higher the compression pressure, the higher the voltage required to produce a spark.

The spark gap (electrode gap)

An optimum gap width must be maintained. If the gap is too small the spark energy may be inadequate for combustion. If the gap is too wide it may cause the spark to fail under pressure. Both the centre electrode and the earth electrode are made to withstand spark erosion and chemical corrosion arising from combustion. To achieve these aims, the electrodes are made from a nickel alloy, the composition of which is varied to suit particular plug types. For some special applications, the alloys used for the electrodes may contain rarer metals such as silver, platinum and palladium.

Figure 11.57 shows the spark gap on a new plug. In time the electrodes will wear and this means that spark plugs should be examined at regular intervals. The type of electrode wear that might occur in use is shown in Fig. 11.58. The shape of worn electrodes varies. Small amounts of wear can be remedied by bending the side electrode to give the required gap but, in cases of excessive wear, the only remedy is to replace the plug.

For good performance the spark plug insulator tip must operate in the range from approximately 350 °C to 1050 °C. If the insulator tip is too hot, rapid electrode wear will occur and pre-ignition may also be caused. If the insulator tip is too cool, the plug will 'foul'

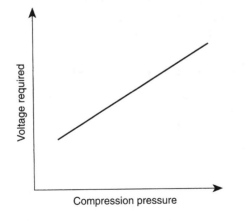

Fig. 11.56 Variation of sparking voltage with compression pressure

(y-axis: Voltage required; x-axis: Compression pressure)

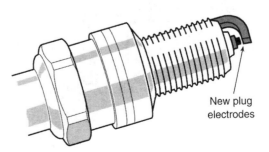

New plug electrodes

Fig. 11.57 The spark plug electrodes gap

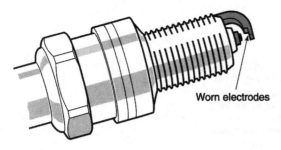

Fig. 11.58 Worn electrodes

and misfiring will occur. The heat range of the spark plug is important in this respect.

Heat range of sparking plugs

Figure 11.59 shows the approximate amounts of heat energy that are dissipated through the various parts of the spark plug, by far the greatest amount passes through the threaded portion into the metal of the engine.

A factor that affects the dissipation of heat away from the centre electrode and the insulator nose, is the amount of insulator that is in contact with the metallic part of the plug. This is known as the heat path. Spark plugs that 'run' hot have a long heat path, and spark plugs that 'run' cold have a short heat path, as shown in Fig. 11.60.

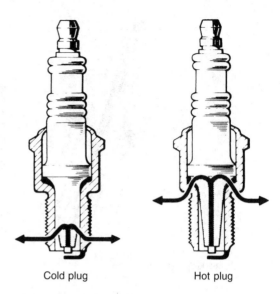

Cold plug   Hot plug

Fig. 11.60 The heat path

In general, hot running plugs are used in cold running engines, and cold running plugs are used in hot running engines. The type of plug to be used in a particular engine is specified by the manufacturer. This recommendation should be observed. In order not to restrict choice, it is possible to obtain charts which give information about other makers plugs that are equivalent to a particular type.

Plug reach

The length of the thread, shown as dimension R in Fig. 11.55, is a critical dimension. If the correct type of plug is fitted to a particular engine, there should be no problem and the plug will seat, as shown in Fig. 11.61.

However, if the reach is too long, the electrodes and threaded part of the plug will protrude into the combustion chamber and will probably damage the valves and piston.

Gasket or tapered seat

Spark plugs will be found with a gasket type seat, as shown in Fig. 11.62, or a tapered seat, as shown in Fig. 11.63.

Gasket seat

Spark plugs with a gasket type seat need to be installed 'finger tight' until the gasket is firmly on to its seat. Having previously ensured that the seat is clean, the final tightening should be to the engine maker's recommended torque.

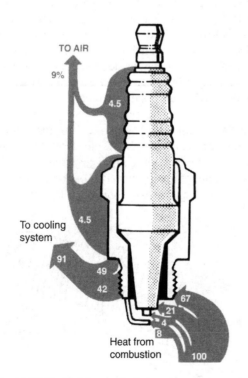

Fig. 11.59 The heat flow in a typical sparking plug

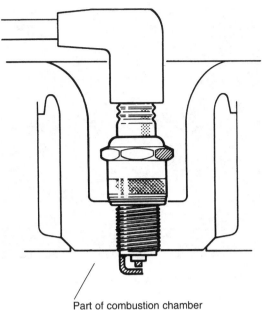

Fig. 11.61 The importance of plug 'reach'

Fig. 11.62 A gasket-type plug seat

Tapered seat

These spark plugs do not require a gasket (commonly known as a plug washer). The tapered faces of both the gasket and the cylinder head need to be clean. The plug must be correctly started in its threaded hole, by hand, and then screwed down, finger tight. The final tightening must be done very carefully because the tapered seat can impose strain on the thread. As with the gasket type seat, the final tightening must be to the engine maker's recommended torque.

Fig. 11.63 A tapered-type plug seat

Caution

Alignment of the thread when screwing a spark plug in is critical; always ensure that the plug is not cross threaded before exerting torque above that which can be applied by the fingers. In aluminium alloy cylinder heads it is very easy to 'tear' the thread out. Repairing such damage is normally very expensive.

Maintenance and servicing of sparking plugs

In common with many other features of vehicle servicing, the service intervals for sparking plugs have also lengthened and it is quite common for the servicing requirement to be to keep the exterior of the ceramic insulator clean and to replace sparking plugs at prescribed intervals, perhaps 20 000 miles or so. However, there are still quite a few older vehicles in use and it is useful to have an insight into the methods of spark plug maintenance that may be applicable to such vehicles.

Cleaning spark plugs

Early spark plugs were made so that they could be taken apart to be cleaned. This meant that all parts could be cleaned satisfactorily by the use of a wire brush. When the non-detachable spark plug was introduced it became necessary to clean by abrasive blasting. This requires the use of a special machine. Figure 11.64 shows such a machine in use. It should be noted that protective eye wear must be worn for this operation because it is fairly easy to make a slip and this could lead to injury.

Fig. 11.64 A spark plug cleaning machine

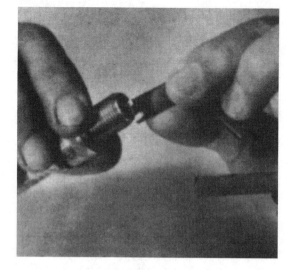

Fig. 11.66 Bending the side electrode to set the gap

In addition to the abrasive (sand) blast facility, the plug cleaning machine contains an 'air blast' facility that will remove unwanted abrasive from the spark plug. When the cleaning operation is completed the plug is removed from the machine. The plug thread can then be cleaned and the electrodes inspected. If the electrodes are spark eroded then the surfaces should be filed flat, as shown in Fig. 11.65.

After this operation has been performed, the spark gap electrodes should be adjusted to the correct gap, and this gap is checked by placing a feeler gauge, of the required thickness, between the electrodes. Any adjustment of the gap is carried out by bending the side contact with the aid of a small 'wringing iron', as shown in Fig. 11.66.

Testing the spark plug

The spark plug servicing machine normally includes a facility for testing the spark plug. This facility is a small pressure chamber into which the spark plug is screwed. A control allows the operator to admit compressed air to pressurise this chamber. The degree of compression is registered on a pressure gauge. The machine is also supplied with an H.T. coil which is connected by a suitable H.T. lead to the terminal of the spark plug. When the desired test pressure is reached, the 'spark' button is pressed. The pressure chamber also includes a transparent window and it is possible to examine the spark through this window.

11.7 Fault tracing in contact breaker ignition systems

Safety

The H.T. voltage is of the order of many thousands of volts, perhaps as high as 40 000 volts. The danger of electric shock is ever present and steps must be taken to avoid it. In addition to electric shock, the involuntary 'jerking' of limbs, when receiving a shock, may cause parts of the human frame to be thrown into contact with moving parts, such as drive belts, pulleys and fans.

Fig. 11.65 Filing the electrodes

Care must always be taken to avoid receiving an electric shock. It is possible to obtain special electrically insulated tongs for handling H.T. cables when it is necessary to handle the cables of running engines.

Most of the details about oscilloscope testing of electronic ignition systems also apply to contact breaker systems. The details in this section refer to a more practical type of work that may be required when repairs are required outside a workshop, as may happen in the case of a roadside breakdown.

Fault finding principles

Before considering ignition system faults and how to deal with them, it is wise to remember that much of the skill required to perform diagnosis and repair of electrical and electronic systems is the same as that which is required for good-quality work of any type. It is important to be methodical and unwise to start testing things randomly, or even to try changing parts in the hope that you might hit on the right thing by chance.

Many people do work methodically and they probably employ a method similar to the 'Six Step Approach', which is a good, commonsense approach to problem solving in general. The six step approach provides a good starting reference, although it requires some refinement when used for diagnosis of vehicle systems.

We will briefly consider the six steps and at a later stage take into account the refinements that are considered necessary for vehicle systems.

The 'six step approach'

This six step approach may be generally recognised as an organised approach to problem solving. As quoted here, it may be seen that certain steps are recursive. Therefore it may be necessary to refer back to previous steps as one proceeds to a solution. Nevertheless, it does provide a proven method of ensuring that vital steps are not omitted in the fault tracing and rectification process. The six steps are:

1. Collect evidence.
2. If it appears to be a flat battery, what checks can be applied, for example, switch on the headlamps.
3. Assume that it is a flat battery.
4. What caused the battery to become discharged?
5. Assume, in this case, that the side and tail lights had been left on. So, in this case,

re-charging the battery would probably cure the fault.
6. Testing the system would, in this case, probably amount to ensuring that the vehicle started promptly with the recharged battery. However, further checks might be applied to ensure that there was not some permanent current drain from the battery.

We hope that you will agree that these are good, commonsense steps to take and we feel sure that most readers will recognise that these steps bear some resemblance to their own method of working.

Visual checks

Figure 11.67 shows a preliminary check of the type that is advised in most approaches to fault tracing. Here the technician is looking for any obvious external signs of problems, such as loose connections or broken wires, cracked or damaged H.T. cables, dirty or cracked distributor cap and coil tower insulator and dirty and damaged spark plug insulators.

Checking for H.T. sparking

A suitable neon test lamp is useful for this purpose. However, Fig. 11.68 shows the king lead removed from the top of the distributor. Here, it is held 6 mm from a suitable earth point. The engine is then switched on and the starter motor operated and there should then be a regular healthy spark each time the C.B. points open. A regular spark in this test shows that the coil and L.T. circuits are working properly and that the fault lies in the H.T system, i.e., the rotor arm, the distributor cap, the H.T. leads or the spark plugs.

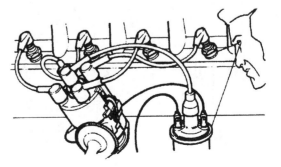

Fig. 11.67 Performing a visual check

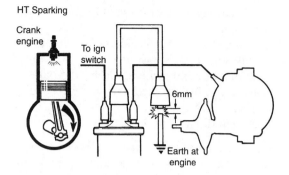

Fig. 11.68 Checking for H.T sparking

Checking the rotor

The rotor arm can be checked by the procedure shown in Fig. 11.69. Here the distributor cap end of the king lead is removed from the distributor cap. The end of the king lead is held about 3 mm above the metal part of the rotor. With the C.B. points closed and the ignition switched on the points can be flicked open with a suitable insulated screwdriver, or an assistant can be used to turn the engine over on the starter motor. There should be no spark. If there is a spark, it means that the rotor insulation has broken down and means that the rotor must be replaced by a new one.

If there is no spark, then this indicates that the rotor is satisfactory. The next step would be to check the H.T. leads. If, as is likely, they are resistive leads, it will be necessary to look up the permitted resistance of the leads before performing an ohm-meter test. The resistance value must fall within the stated limits. If they are not resistive leads, a straightforward continuity test will prove the conductivity of the H.T. leads. If these checks prove satisfactory, the sparking plugs must be examined. I have already described the procedure for checking sparkingplugs.

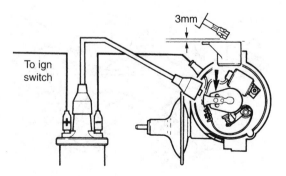

Fig. 11.69 Checking the insulation of the rotor

If these facilities are not available, the only solution may be to inspect the plugs internally, and the most likely step will be to fit a new set of spark plugs. It is unusual to find that a set of spark plugs has failed completely.

Checking the condition of the contact points

Should the first test (Fig. 11.68) have failed, i.e., no spark, it is advisable to check the contact breaker condition. The contact surfaces should be clean and not badly pitted and piled, and the gap should be checked to ensure that it is within the recommended limits. The securing screws should be checked for tightness and there should be no evidence of fouling of cables inside the distributor. If the points are badly worn and burnt, a new set of contacts should be fitted. After this, the first check for H.T. spark should be repeated. If this is successful then there may be no need to proceed further. However, if there is still no spark, further tests will be required. These tests are now described. The distributor cap should be removed and the engine rotated, with the ignition off, until the C.B. points are closed. The voltmeter is then connected between the switch side of the coil and earth. The ignition should then be switched on. The voltmeter should then read battery voltage (very closely) or, if it is a ballasted coil, a lower voltage, probably 6 to 8 volts.

Checking the voltage on the C.B. side of the coil

If these tests are satisfactory, the next step is to proceed to check the voltage at the contact breaker side of the coil. The voltmeter position for this test is shown in Fig. 11.71.

The C.B. contacts must be open. The voltmeter is then connected between the C.B. terminal of the coil and earth. The ignition should now be switched on and the voltmeter reading observed. If the coil primary circuit is satisfactory, a reading that is very close to that obtained in the test shown in Fig. 11.70 should be seen. If the reading is correct, proceed to test the voltage at the coil C.B. terminal with the contact points closed.

If the voltage reading is zero, disconnect the L.T. lead to the distributor. The voltmeter should now read the same voltage as on the switch side of the coil. If it still reads zero, it indicates that there is a broken circuit in the primary side of the coil and this calls for a new coil.

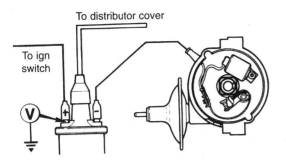

Fig. 11.70 Voltmeter check on coil supply

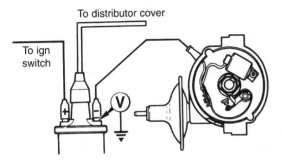

Fig. 11.71 Voltmeter check on CB side of coil

Voltage at the C.B. terminal of the coil – point closed

This test checks the condition of the contact points and also checks whether there are any current leakages to earth. The voltmeter should show a zero reading if all is correct, or less than 0.2 volts if it is a very sensitive meter. If the reading is more than zero volts, the contacts should be checked to see if they are clean and properly closed; the insulation on the contact to condenser lead and on the L.T. lead should also be checked. If these are in order, it may be that the capacitor itself is providing a leakage path to earth.

Testing the capacitor

Now that the H.T. lead has been proved, and if there is still no spark from this test, then the capacitor should be tested.

Figure 11.73 shows how a test capacitor can be temporarily connected to perform this test. The original capacitor is unscrewed and lifted away so that its casing does not make contact with earth. A test capacitor, with suitable crocodile clips and leads, is then connected, as shown. The ignition is switched on again and the engine turned over, on the starter motor. If a spark is obtained, this is convincing evidence that the capacitor should be renewed. If there is still no spark after proving the capacitor, the coil should be replaced.

Checking the electrical condition of the ignition coil

The ignition coil can be checked, as shown in Fig. 11.74. The secondary winding resistance should be high, probably 10 kohms. (Note: In some ignition coils the secondary winding is earthed. In such cases, the check for secondary winding resistance would be made between the HT connection and the coil earth.)

The primary winding resistance should be low (the meter needs to be set to a very low Ohms scale) 0.5 ohms to 1 ohm.

The tests shown here are of a general nature. They do not apply to a specific system and, as with most other work on vehicle systems, it is important to have the figures that relate to the specific system being worked on. However, the tests do show the practical nature of tests that can be applied with the aid of a good-quality multi-meter, sound knowledge of circuits and the relevant information.

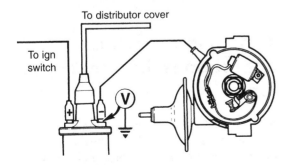

Fig. 11.72 Voltmeter test for voltage drop across the points

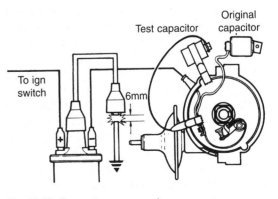

Fig. 11.73 Connecting a test capacitor

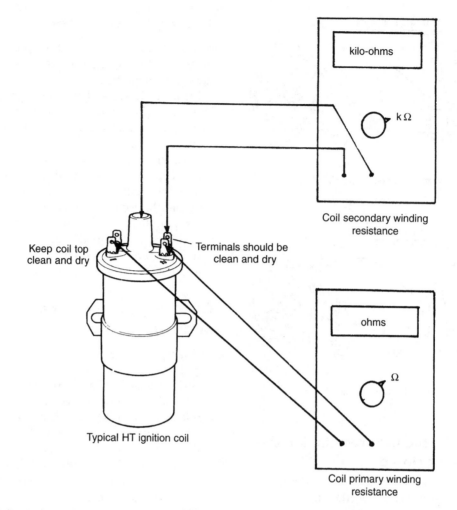

Fig. 11.74 Checking the HT coil

11.8 High tension (HT) leads

The HT leads are the means by which the high voltage electrical energy that makes the spark is conducted to the sparking plug. The high voltage, perhaps as high as 40 kV, will wish to seek the shortest path to earth. In order to prevent HT leakage from the cables, they must be heavily insulated electrically. The insulating material must also be resistant to heat and to oil, water and any corrosive agents that they may encounter. PVC is often used for the insulation, because it possesses most of these properties. To reduce radio interference, the conducting parts of HT leads are made to have electrical resistance. This resistance can change in use and HT leads should be checked to ensure that the leads have the correct resistance. The resistance of HT leads varies considerably, a rough guide being **18 000 ohms per metre length**. This means that an HT cable of **300 mm length**

would have a resistance of 5 400 ohms. It is a value that can be checked by means of an ohm-meter and the resistance values and tolerances will be given in the workshop manual.

As with any electric current in a conductor, the HT leads set up a magnetic field. In order to prevent one HT cable inducing current in a neighbouring one, by mutual induction, cable routing, as arranged by the vehicle manufacturer, should always be adhered to.

11.9 Some practical applications, i.e., doing the job

A substantial amount of background theory has been covered and we have shown how this theory is used to make vehicle systems work. When systems break down or cease to function correctly, it is the job of the vehicle service and repair technician to find out what is wrong and to put it right and this is where practical skill, or competence, comes in to play. In a book of this type, it is not possible to cover every type of vehicle system and we will not attempt to do so. As with all work, proper safe working practices must be observed. Personal safety and the safety of others nearby must be protected, and the vehicle itself must be protected against damage, by ensuring that the correct procedures prescribed by the manufacturer are followed and that wing and upholstery covers, etc. are used.

Because of the global (international) nature of the vehicle industry, vehicle technicians will encounter a range of products and it is unwise to attempt even simple work without having a good knowledge of the product, or access to the information and data relating to the vehicle and system to be worked on.

The practical work described in this section generally relates to a specific product and that is made clear in the text. However, the devices chosen and the work described have been selected because they highlight types of practical procedures and tests that can be applied across a range of vehicles. It is intended that this book should be used in conjunction with courses of practical instruction and learning on the job.

The main source of information and guidance about repair and maintenance procedures is the manufacturers workshop manual which is often available on CDROM.

Ignition timing

Because ignition timing varies with engine speed, it is necessary to check both the static timing and the dynamic timing.

Static timing

The following list shows a the ignition timing details for a 4 cylinder in-line engine.

Firing order 1 3 4 2. Static advance 10°.
Contact breaker gap 0.40 mm
Dwell angle 55°.

For static advance setting, the figures of interest are the static advance angle of 10°, the contact breaker gap of 0.40 mm and the firing order.

When re-setting the static timing after performing work on the engine, such as on removing and refitting the distributor, it is necessary to ensure that the piston in the cylinder used for timing (usually number 1) is on the compression stroke. The firing order helps here, because in this case the valves on number 4 cylinder will be 'rocking' when number 1 piston is at top dead centre (TDC) on the compression stroke. The 10° static advance tells us that the contact breaker points must start to open at 10° before TDC. Engines normally carry timing marks on the crankshaft pulley and these are made to align with a pointer on the engine block, or timing case, as shown in Fig. 11.75.

To position the piston at 10° before TDC, the crank should be rotated by a spanner on the crankshaft pulley nut, in the direction of rotation of the engine. This ensures that any 'free play' in the timing gears is taken up. To make the task more manageable, it is probably wise to remove the sparking plugs. If the alignment is missed the first time round, the crank should continue to be rotated in the direction of rotation until the piston is on the correct stroke and the timing marks are correctly aligned.

When the piston position of number 1 cylinder is accurately set, the distributor should be replaced with the rotor pointing towards the HT segment, in the distributor cap, that is normally connected to number 1 sparking plug. This process is made easier on those distributors that have an off-set coupling.

Setting the timing

The contact breaker gap should be set to the correct value and the distributor clamp loosened. When the low tension terminal of the distributor has been re-connected to the coil, a timing light may then be connected between

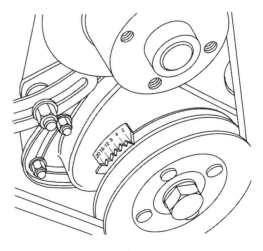

Fig. 11.75 Engine timing marks

the contact breaker terminal of the coil and earth. With the ignition switched on and the contact breaker points closed, the timing lamp will be out. When the points open, the primary current will flow through the timing light which will light up. An alternative is to disconnect both coil LT connections and then connect the timing light in series, as shown in Fig. 11.76. In this case the timing light will be on when the points are closed, and off when they are open.

When the correct setting has been achieved, the distributor clamp should be tightened, the timing light removed and all leads re-connected. The timing light (lamp) can be a light bulb of suitable voltage (12 V) mounted in a holder and fitted with two leads to which small crocodile clips have been securely attached.

Dynamic timing

As the name suggests, dynamic timing means checking the timing with the engine running. At this present level of study we will restrict ourselves to the use of the stroboscopic timing lamp, often referred to as a strobe lamp. The strobe lamp 'flash' is triggered by the H.T. pulse from a sparking plug, usually number one cylinder. When the strobe light is directed on to the timing marks, the impression is given that the mark on the rotating pulley is stationary in relation to the fixed timing mark.

Safety note

This useful stroboscopic effect carries with it potential danger because the impression is given that other rotating parts, such as fan blades, drive belts, etc., are also stationary. It is therefore important not to allow hair and clothing or any part of the body to come into contact with moving parts. It is also important to avoid electric shock. As with other work that involves running engines, exhaust extraction equipment should be used.

Checking the static timing with the engine running

The static timing setting can be checked by means of the strobe lamp because, at idling speed, the centrifugal advance and the vacuum advance mechanisms should not be adding to the static advance angle. The static timing mark on the engine should be highlighted with white chalk or some other suitable marking substance. This will enable you to locate the correct marks when conducting the test. Figure 11.77 shows the strobe light, this one incorporating an advance meter, being used to check the static timing. For this test, the engine will be run at the manufacturer's recommended speed, probably idling speed. If the marks align correctly, no further action is required. However, if the marks do not align, then the distributor clamp must be slackened and the distributor body rotated until the marks are

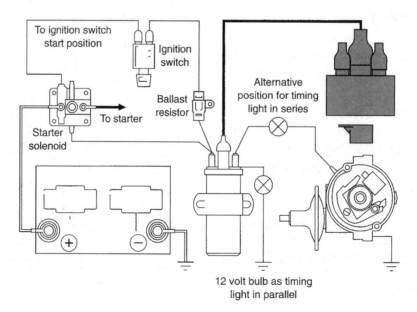

Fig. 11.76 Using a timing light

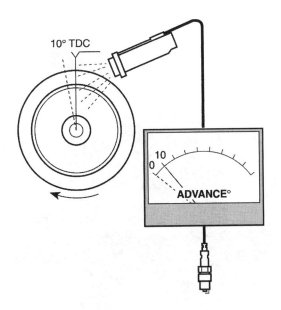

Fig. 11.77 The stroboscopic timing lamp

aligned correctly. When the correct alignment is achieved, the clamp must be re-tightened.

Checking the centrifugal advance mechanism

The timing details that we are using as an example gave us a static advance figure of 10° and the following for vacuum and centrifugal.

> **Automatic centrifugal advance 25°**
> **Vacuum advance 12°**

If the vacuum advance is disabled by disconnecting the pipe at the engine end, and blanking off the hole with a suitable device, the maximum amount of advance obtainable will be 35°, that is 25° from the centrifugal device and 10° static. In this example this maximum advance occurs at 4 300 rev/min of the engine and a tachometer (rev counter) will be needed to record the engine speed. With the vacuum pipe disconnected, the tachometer and strobe light connected, and all leads checked to ensure that nothing is touching moving parts, or hot exhausts, the check may proceed. It will probably require another person to operate the accelerator and observe the tachometer. The strobe light should be aimed at the timing marks and when the engine is brought up to the correct speed, the control on the advance meter is adjusted until the timing marks align. The number of degrees shown on the advance meter should then show the maximum advance angle. If this is correct, the vacuum advance mechanism can be tested.

Checking the vacuum advance mechanism

After removing the blanking device, the vacuum pipe should be re-connected and the engine speed will increase slightly if the vacuum device is working.

Dwell angle

In contact breaker type ignition systems, the dwell angle is the period during which the contact breaker points are closed. The dwell angle is the period during which the electrical energy builds up in the coil primary winding. If the dwell period is too short, the primary current will not reach its maximum value and the H.T. spark will be accordingly weaker. Figure 11.78 shows dwell angle, which has already been discussed in the context of the operating principles of the ignition system.

On a four lobed cam, as shown here, the dwell angle is 54° + or − 5°. However, the dwell angle varies according to distributor type and the manufacturer's data should always be checked to ensure that the correct figure is being used. The dwell angle is affected by the points gap. If the gap is too large, the heel of the moving contact will be closer to the cam and this will cause the dwell angle to be smaller than it should be (Fig. 11.79).

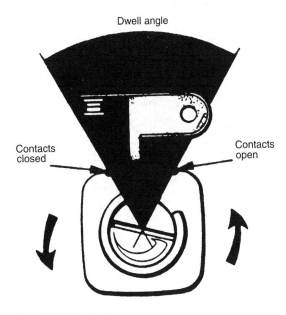

Fig. 11.78 Dwell angle

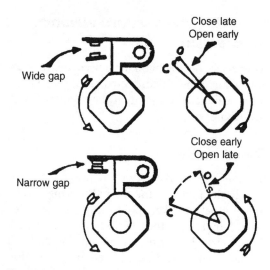

Fig. 11.79 Points gap on dwell angle

Adjusting the contact breaker points gap

Not only does the points gap affect the dwell angle but it also affects the ignition timing. A wide points gap advances the ignition and too small a gap retards the ignition. Setting the contact breaker points to the correct gap is, therefore, a critical engineering measurement in a technicians work. Figure 11.80 shows a set of feeler gauges being used to check the gap, and the small inset shows a method of achieving the fine adjustment.

In performing this task, it is essential to use clean feeler gauges and it should be noted that feeler gauges do wear out. For example, if feeler

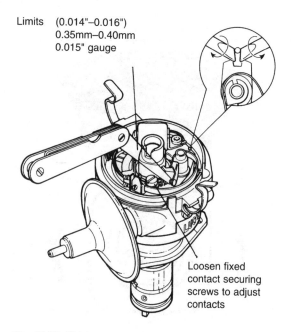

Limits   (0.014"–0.016")
0.35mm–0.40mm
0.015" gauge

Loosen fixed contact securing screws to adjust contacts

Fig. 11.80 Checking points gap

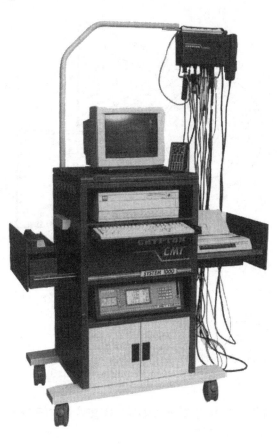

Fig. 11.81 A full size diagnostic machine

gauges have been used to check valve clearances with the engine running, it is quite possible for the blades to be 'hammered' thin. An occasional check with a micrometer will verify the accuracy of the feeler gauges. Back to the points gap setting. The points gap is checked with the ignition switched off. The feeler gauge is inserted between the contacts, as shown in Fig. 11.80, and very light force should be used to 'feel' the setting. Also, care must be taken to keep the feeler gauge blade in line with the contact face. If the points gap is set accurately, the dwell angle should be correct. However, it is common practice to check the dwell angle by the use of a dwell meter and tachometer. These instruments are normally part of an engine analyser, such as the Crypton CMT 1000 shown in Fig. 11.81.

11.10 Servicing a contact breaker type distributor

Figure 11.82 shows the main points that require periodic attention. Oil should be applied sparingly and care taken to prevent contamination

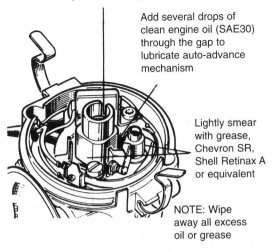

Add 2 or 3 drops of clean engine oil (SAE30)

Add several drops of clean engine oil (SAE30) through the gap to lubricate auto-advance mechanism

Lightly smear with grease, Chevron SR, Shell Retinax A or equivalent

NOTE: Wipe away all excess oil or grease

Fig. 11.82 Service details for a distributor

of the contact points. While the distributor cap is removed it should be wiped clean, inside and out, and checked for signs of damage and tracking (H.T. leakage). The rotor arm should also be inspected.

11.11 Explanation of terms associated with ignition systems

Misfiring

Misfiring is a term that is normally applied to the type of defect that shows up as an occasional loss of spark on one or more cylinders. It can be caused by almost any part of the ignition system and the cause is often difficult to locate. Much depends on the way in which the fault occurs. If it is a regular and constant misfire one could start by checking for H.T. at each sparking plug. If each plug is receiving H.T., then the most likely cause is a spark plug. Plugs should then be removed and tested. In some cases, where the engine has not been running for very long, the insulator of a non-firing spark plug will be significantly cooler to the touch than the other plug insulators. This may help in locating the misfire. Here again, an electronic engine analyser is much more satisfactory, because a power drop test will show which cylinder is not firing.

Cutting out

Cutting out is a term that is used to describe the type of fault where the engine stops completely,

perhaps only for a split second. The most likely cause here is that there is a broken or loose connection that breaks the circuit momentarily. Careful examination of all connections and 'wiggling' of cables whilst conducting a voltage drop test across the connectors, should help to locate such faults.

Hesitation

This normally happens under acceleration and the symptoms are that the engine does not respond to throttle operation and can cause problems when overtaking. The voltage required to produce a spark rises with the load placed on the engine as happens under acceleration, and probable causes are wide spark plug gaps or spark plugs breaking down under load.

Excessive fuel consumption

Incorrect ignition timing is usually associated with high fuel consumption.

Low power

Incorrect ignition timing, weak spark and dirty or badly adjusted plugs are defects associated with low power.

Overheating

The ignition defect most often associated with overheating is the ignition timing, but other defects that cause pre-ignition and detonation can also lead to overheating of the engine.

Running on

This happens when the engine continues to run, even after the ignition is switched off. An obvious cause is that parts of the engine interior remain sufficiently hot to cause combustion. This may be caused by a dirty engine, with heavy carbon deposits in the combustion chamber and dirty, worn or incorrect type of sparking plugs. It is overcome to some extent by fitting a cut-off valve in the engine idling system. This prevents mixture from entering the combustion chamber through the engine idling system. This is the probable source of fuelling as the throttle is virtually closed.

Detonation and pre-ignition

Both detonation and pre-ignition give rise to a knocking sound which, at times, can be quite violent. On other occasions, a high pitched 'pinking' sound may be heard and, as the term implies, this sound is known as pinking. Pre-ignition arises when combustion happens before the spark occurs. Detonation happens after the spark has occurred.

Pre-ignition

Pre-ignition may be caused by 'hot' spots in the combustion space. These hot spots may be caused by sharp edges and rough metallic surfaces, glowing deposits (carbon), overheated spark plugs or badly seated valves. As a result of ignition starting before the spark occurs, the pressure and temperature in the cylinder rise to high levels at the wrong time. This gives rise to poor performance, engine knock, and eventual engine damage. Some of the probable causes of pre-ignition are listed below.

Probable causes of pre-ignition

Sharp edges and rough metallic surfaces

These are likely to be caused by manufacturing defects and they are normally rectified at that stage. However, if an engine has been opened up for repair work it is possible that careless use of tools may damage a surface. Should this be the case, the remedy would be to remove the roughness by means of a scraper, a file or a burr.

Glowing deposits (carbon)

This is most likely to occur in an engine that has covered a high mileage. Decarbonising the engine (a decoke) would probably be the answer here.

Overheated spark plugs

In this case there are various factors to consider, such as:

- Are the spark plugs of the correct type and heat range?
- Are the spark plugs clean?
- Are the electrodes worn thin?

Badly seated valves

If a poppet type valve does not seat properly, it is possible for part of the valve head to form a hot spot. Failure to seat squarely could be caused by a bent valve, a misaligned valve guide and, in the case of older engines, a badly reconditioned valve. In each case the remedy would be to rectify the fault indicated.

In addition to the above defects combustion knock can arise from use of the incorrect fuel. The Octane rating of a fuel is the critical factor here. Tetra-ethyl lead (leaded fuel) used to be the method of altering the octane rating (anti-knock value) of petroleum spirit. But now that most spark ignition engines are fitted with catalytic convertors for emissions control, leaded fuels cannot be used. There are different octane grades of unleaded fuel and many electronically controlled engines are provided with a simple means of altering settings if a driver wishes to use a different octane rating fuel from the one that she/he has been using.

We hope that you will agree that these are good, commonsense steps to take and we feel sure that most readers will recognise that these steps bear some resemblance to their own method of working.

In the following cases it is assumed that the above steps have been followed and that the fault has been found to exist in the area of the ignition system.

Poor starting or failure to start

For current purposes we will assume that the fuel and electrical systems are in good order. In particular, that the battery is fully charged and that the starter motor is capable of rotating the engine correctly. We shall also assume that the mechanical condition of the engine is good and that the fault has been identified as being in the area of the ignition system.

Poor starting

If the controls are operated correctly, the engine should start promptly. The time taken to start up may vary with air temperature. For example, when it is very cold the cranking speed of the engine will be affected by the viscosity of the lubricating oil, and the mixture entering the engine is affected by contact with cold surfaces. Generally, an engine will start more promptly when it has warmed up than it will when cold.

For 'poor' starting we are, therefore, looking at cases where, taking account of the above factors, the engine either takes a long time to start or is reluctant to start at all.

The ignition system faults that have a bearing on the problem of poor starting are:

(a) Weak spark

This could be caused by loss of electrical energy in the ignition circuit or failure to generate sufficient energy for a spark, in the first instance.

Loss of electrical energy could be due to defective connection and/or defective insulation in either the low tension circuit or the high tension circuit. A range of tests and checks that can be carried out, is described in the section on servicing and repair.

(b) Spark occurring sometimes but not at other times

This is known as an intermittent fault. Here again, the problem could lie in either or both the L.T. and the H.T. circuits.

(c) Spark occurring at the wrong time

In most cases this means that the ignition timing is incorrectly set.

Failure to start

As it is ignition faults that we are considering, the obvious pace to start is the sparking plug gap. Failure to start is probably due to there being no spark, or a spark that is too weak to cause combustion. However, if the ignition timing is not correct, or the H.T. leads have been wrongly connected (not in accordance with the firing order), the engine may also fail to start.

Learning tasks

Do not attempt any practical work until you have been properly trained. You must always seek the permission and advice of your supervisor before you attempt any of the practical tasks.

Much of NVQ assessment is based on work that you have actually performed. Each time you complete the tasks that you will keeping a record or you should note down the tools and equipment that you have used.

1. Make a list of the precautions to be taken when working on ignition systems.
2. Describe the procedure for setting the static ignition timing on an engine that has a contact breaker type ignition system.
3. Give details of the procedure and type of feeler gauges used to set the pick-up to rotor air gap in an electronic, contact breakerless, ignition system.
4. Take a set of ignition H.T. leads and measure the length in centimetres and the resistance in kOhms of each lead. Compare the figures with those given in this book or in the workshop manual for the vehicle. Write down the result and state whether or not the leads are in good condition.

---

5. Examine a set of sparking plugs from an engine. Measure and record:
   (a) the spark gaps
   (b) the reach of each plug
   (c) the thread diameter
   (d) the recommended tightening torque
   (e) note the type of spark plug (not the make) and state whether they are intended for use in a hot running engine or a cold running engine.
6. Examine a distributorless ignition system. Write down the number of ignition coils and the position of the sensor that triggers the ignition pulses.
7. Describe, with the aid of diagrams, a simple practical test that can be used to check the electrical condition of a rotor arm.
8. Perform a test to check the electrical output of a magnetic pulse type generator that is used in an electronic ignition system.
9. Obtain an ignition coil and measure the resistance of the primary winding and the secondary winding. Record the figures and compare with the figures given in this book, then write down your opinion about the condition of the coil. State whether or not the coil is for use in a ballasted ignition system.
10. Describe the procedure for setting up a garage type oscilloscope to check ignition coil polarity.

Self assessment questions

1. A Hall effect ignition system sensor:
   (a) requires a supply of electricity
   (b) generates electricity by electro magnetism
   (c) can only be used on ignition systems that have a distributor
   (d) can only be used on 4 cylinder engines
2. In a test on a contact breaker type ignition system with the points closed and the ignition switched on, the voltage drop across the points should be:
   (a) 14 volts
   (b) not more than 0.2 volts
   (c) 2.5 volts
   (d) 2 volts

3. In a 'lost spark' ignition system each cylinder has a spark:
   (a) on the induction stroke as well as at the end of the compression stroke
   (b) on the exhaust stroke as well as at the end of the power stroke
   (c) twice on the power stroke
   (d) on the exhaust stroke as well as at the end of the compression stroke

4. Ignition related faults that affect the emissions controls on the vehicle:
   (a) cause the MIL lamp to illuminate on EOBD systems
   (b) cause the vehicle to be disabled
   (c) require the vehicle to be taken off the road immediately
   (d) should be left to rectify themselves

5. A set of ignition HT leads are required to have a resistance of 18 k ohms per metre of length. One of these leads that is 430 mm long will have a resistance of:
   (a) 41.86 ohms
   (b) 7740 ohms
   (c) 4186 ohms
   (d) 0.7 k ohms

6. Sparking plugs that 'run' hot have a long heat path and sparking plugs that 'run' cold have a short heat path. The sparking plugs fitted to an engine that runs cool would have:
   (a) a short heat path
   (b) tapered seats
   (c) a long heat path
   (d) short reach

7. The 'firing line' on an oscilloscope trace of ignition HT voltage represents:
   (a) the high voltage that is needed to cause the spark to bridge the sparking plug gap
   (b) the current in the primary winding of the ignition coil
   (c) the condition of the condenser
   (d) the air–fuel ratio of the mixture in the cylinder

8. A knock sensor is used:
   (a) to allow the ECM to alter the ignition timing
   (b) to alert the driver to mechanical noise from wide valve clearances
   (c) to allow catalytic converters to operate on leaded fuel
   (d) on contact breaker type ignition systems only

9. Pre-ignition happens when:
   (a) the static ignition timing is too far advanced
   (b) combustion occurs before the spark
   (c) pressure waves cause a region of high pressure, in the combustion chamber, after the spark has occurred
   (d) the ignition is retarded

10. In engine management, the ignition timing can be altered to suit engine wear. When new parts are fitted to replace worn or damaged ones on this type of ignition system, the vehicle should:
    (a) have a new ECM fitted
    (b) be operated under normal running conditions for a length of time that allows the ECM to adjust to the new parts
    (c) be returned to the customer without advising them that it may take time for the vehicle to return to its best working condition
    (d) have the entire ignition system renewed

11. The vacuum operated timing control on distributor type ignition systems:
    (a) changes the ignition timing to suit engine speed
    (b) changes the ignition timing to suit engine temperature
    (c) changes the ignition timing to suit load on the engine
    (d) changes the air–fuel ratio

12
Electrical systems and circuits

Topics covered in this chapter

Central locking
Screen wipers
Vehicle security alarm
Electric windows
Screen washers
Electric mirrors
Vehicle lighting systems
Supplementary restraint systems – airbags and seatbelt pre-tensioners

12.1 Central door locking systems

Manual operation

A central door locking system permits all doors, often including the boot lid or tailgate, to be locked centrally from the driver's door and, normally includes the front passenger's door also.

The locking and unlocking action is produced by an electro-mechanical device, such as a solenoid or an electric motor. The door key operates a switch which controls the electrical supply that actuates the door lock mechanism and the remote control activates a switching circuit in the electronic control module (ECM) that also actuates the locking mechanism. For safety and convenience, the interior door handles also operate the unlocking mechanism.

Remote operation

The remote control operates in much the same way as the remote control for a domestic television. A unique code is transmitted from the remote control which is normally contained in the ignition key. Two types of transmitter are used, infra-red and radio frequency. The electronic control module (ECM) that controls the supply of current to the door lock actuators contains a sensor that detects infra-red signals or, in the case of radio frequency systems, an aerial to detect the radio frequency signal. The remote signal is normally effective over a range of several metres. Figure 12.1 shows the general principle of central locking systems.

The remote controller

The encoded signals that are transmitted by the remote controller are produced by electronic circuits that are energised by a small battery that is housed in the controller. A pad on the controller is pressed for unlocking and another pad is pressed for locking purposes. The device may also be provided with an LED that indicates battery level.

A simple central locking circuit

Both the manual central lock system and the remote central lock systems employ similar circuits and actuators inside the doors. A general description of central locking system circuits, of the type shown in Fig. 12.2, should serve to bring about the necessary level of understanding.

Solenoid

The type of solenoid used in door locking system is shown in Fig. 12.2(a). The flow of current in one direction causes the doors to lock and when current flows in the opposite direction the doors are unlocked. The reversal of current is performed by the ECM.

Motor

The motors (Fig. 12.2(b)) used for door locking systems are normally of the permanent magnet type because they are capable of being reversed by changing the direction of current flow through the motor. The motorised operation is considered to be slightly smoother in operation

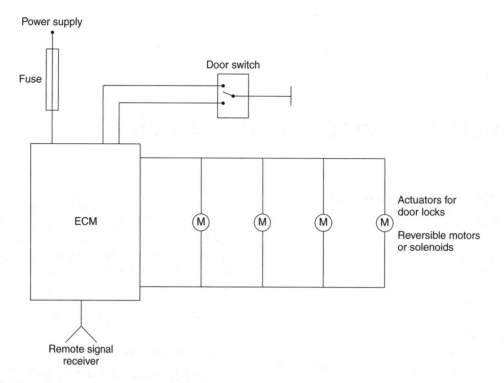

Fig. 12.1 Block circuit diagram for remote central locking

than in the solenoid system. The reversal of current flow is achieved through the switching action provided by the ECM.

Circuit protection

In addition to a fuse, door locking circuits are often equipped with a thermal circuit breaker.

Maintenance and fault tracing

The wiring into the doors is flexible and is protected by a weatherproof sheath. The control rod from the solenoid, or motor, to the mechanical door lock is normally enclosed in a sheath to guard against ingress of water or dust. Periodic visual inspections of the exterior should suffice for these parts. The external mechanical parts of the locks should be lubricated with an approved lubricant whilst ensuring that such lubricant is applied sparingly and carefully so as not to harm the upholstery.

In the event of a locking system failure, every attempt to determine the source of the problem should be made before door trim is removed and work started on the circuit and mechanism inside the door trim. It is recommended that manufactures' guides are used because of the considerable variations that are to be found throughout the vehicles that are in use. Some of the more obvious general points are covered here:

Defect	Action to be taken
Central locking fails to operate.	Check remote control battery Make sure that correct key is being used. If key is lost, follow instructions for coding etc that are normally found in the driver's handbook Check vehicle battery condition Check fuse, or circuit breaker. If fuse is blown, check for electrical Problems. Replace fuse – if fuse blows again shortly after replacing, start search For the cause of the problem.
A single solenoid or motor does not operate.	**Electrical** Check wiring into the door. Check operation of solenoid or motor **Mechanical** Check for broken or seized parts

12.2 Screen wipers

Screen wipers are required so that the driver's ability to see out of the vehicle is maintained in all types of conditions. It is a legal requirement that screen wipers must be maintained in good working condition.

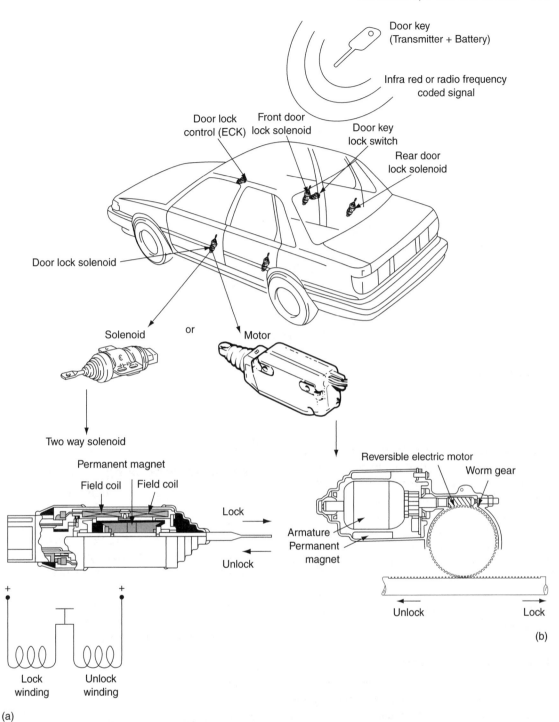

Fig. 12.2 A typical remote central locking system

Modern wipers are electrically operated and a single electric motor normally drives both front screen wipers through a mechanical linkage. The principle is shown in Fig. 12.3.

Modern styling of vehicles requires the rear screen to be fitted with a separate screen wiper and this is equipped with a separate electric motor. Permanent magnet motors are normally used for screen wiper applications because they are compact and powerful and they are readily equipped with two speed operation.

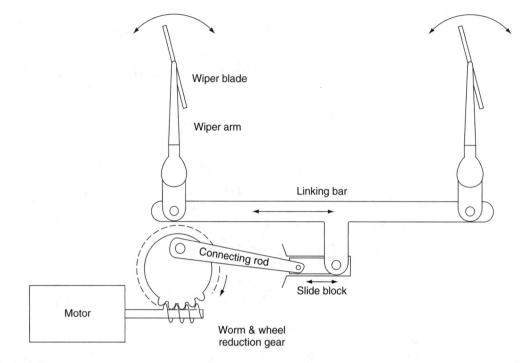

Fig. 12.3 Screen wiper system

Operation

The armature shaft of the motor has a worm gear machined into it and this mates with a worm wheel that provides a gear reduction. The worm wheel carries a crankpin which is connected by a rigid link to the screen wiper linkage and, by this means, the rotary movement of the worm wheel gear is converted into the reciprocating (to and fro) movement of the wiper linkage. The wiper arms are pivoted near the screen and the shaft on which the blade is mounted is connected by a lever to the reciprocating wiper linkage. This results in angular movement of the wiper arm and blade.

Control of the wipers

The control switch for the screen wipers is mounted in a convenient position adjacent to the steering wheel. The switch is multi-functional because it is used to select the options that are available, in addition to the basic on off operations.

Parking the wiper blades

When the wipers are not in use, the arms and blades are placed (parked) in a position that does not obstruct the view from inside the vehicle. This operation is automatically achieved by the use of a cam operated limit switch that is located inside the wiper motor housing. A simplified circuit for this operation is shown in Fig. 12.4.

Regenerative braking

When the wipers are in action, the mechanism possesses considerable momentum and a form of electric braking is utilised the helps to bring the wiper mechanism to rest in the desired position without placing undue strain on it. The circuit in Fig. 12.5 shows how the limit switch cam activates an additional set of electrical contacts.

When these contacts are closed, an electrical load is placed across the screen wiper motor armature and the generator effect of the moving armature places an effective braking effect on the wiper motor and causes it to stop in the required position.

Two speed operation

Permanent magnet screen wiper motors are capable of operating at two different speeds to suit different weather conditions. The speed variation is achieved by means of a third brush. This third brush is connected to a smaller number of armature windings than the other main brushes and this causes a higher current to flow in the armature coils and this, in turn, leads to a higher speed of rotation. Figure 12.6 shows the arrangement.

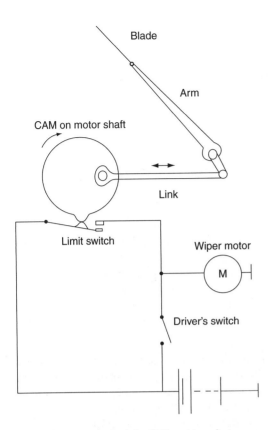

Fig. 12.4 Limit switch circuit for 'Off' position of wiper arm

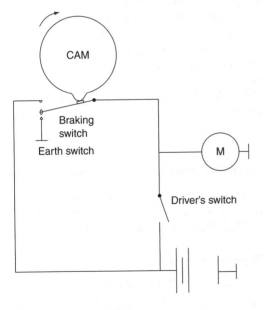

Fig. 12.5 Regenerative braking

Intermittent operation

Most modern wiper systems make provision for the driver to select a switch position that allows the wipers to provide a short spell of wiper

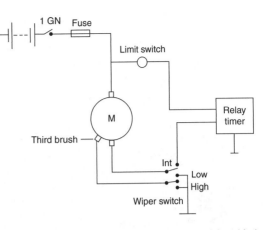

Fig. 12.6 Two speed and intermittent wiper control (simplified circuit)

action at set intervals of time. This is achieved by means of a timer circuit that controls a relay. The wipe action usually occurs at intervals of 15 to 20 seconds and is designed to help the driver cope with conditions where light rain or showers are encountered. In some designs the delay period may be adjusted at the wiper switch. In such cases, the timer circuit will contain a resistive and capacitive device in which the resistance may be altered to vary the interval between wipe actions. An outline circuit is shown in Fig. 12.6.

12.3 Screen washers

A windscreen washer on the driver's side of the vehicle is an essential piece of equipment and it must be maintained in working order. In addition to this basic requirement, most modern vehicles are equipped with screen wash facilities on the passenger's side and the rear screen, and many vehicles are now equipped with headlamp washers. Figure 12.7 shows the essential parts of a screen washer systems.

The pump is electrically operated from the screen wiper switch and it is normally mounted at the base of washer fluid reservoir where it is gravity fed. The inlet to the pump is provided with a filter in order to exclude foreign matter and thus prevent blockages and damage to the pump. The fluid, under a pressure of approximately 0.7 bar from the pump, is conveyed to the washer jets through plastic piping. The jets are arranged so that a maximum washing effect is gained from the wash-wipe action. The washer should only be

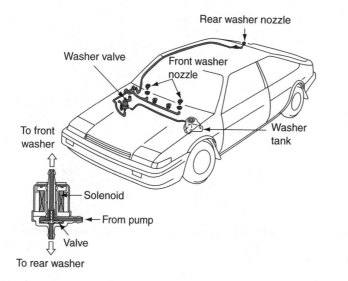

Fig. 12.7 Front and rear washers screen washers

operated for a few seconds at a time as long periods of continuous use may result in damage to the pump motor. A few seconds of use normally suffices for effective cleaning of the screen. The fluid used is water to which is added an anti-freeze agent, a detergent and an anti-corrosion agent. The strength of the washer fluid that is added to the water in the washer reservoir is varied according to the season of the year.

Rear screen washer

Figure 12.7 shows how the screen washer reservoir and pump may be connected so that it supplies both front and rear screens.

A solenoid operated washer valve is placed in the pipeline and operation of the washer switch permits the solenoid valve to direct the fluid to the required washer jets.

Maintenance and repair

It should be evident that preventive maintenance of wipers and washers is important because there may be quite long intervals when they are not in use and, if it only becomes clear that the wipe/wash system is not working properly during a journey, an inconvenient and possibly dangerous situation may arise. Maintenance of screen wipers is largely a matter of taking preventive measures, such as examining the condition of wiper blades to ensure that they are not worn, perished or hardened and replacing them

as necessary. The action of wipers is readily observed by the simple expedient of operating them in their various modes of operation, for example, slow speed, intermittent, etc.

Screen washer fluid levels must be checked at regular intervals and the strength of the washer fluid mixture that is added to the system must be adjusted to suit weather conditions. The wash action may readily be examined by operating the system and observing the pattern of fluid sprayed on to the screen through the washer jets. The condition of the fluid reservoir and its fixings, and the condition of the plastic tubes that connect the pump outlet to the jets, should be checked to ensure that they are secure and in good condition.

Repairs and fault tracing

For current purposes, screen wiper defects may usefully be grouped into two categories namely, 1) Electrical, and 2) Mechanical.

Electrical defects

Problem – electrical overload
Possible effect – fuse blown or circuit breaker triggered

Some thermal circuit breakers operate by breaking the circuit at intervals, in which case the sporadic behaviour of the wipers will indicate that there is something wrong. Persistent operation of the circuit breaker means that there is a problem that requires attention because the wiper circuit is being overloaded. In the case of a blown fuse, it may be that the fuse has blown because of a temporary overload such as ice on the screen or excessively dirty conditions. In this case it would probably be in order to replace the fuse and perform a full test on the wipers, ensuring a clean and wet screen before attempting any test. Should the fuse blow again, or the circuit breaker come into operation again, the system must be examined to see whether it is an electrical fault or a mechanical fault.

Electrical overload may arise from a short circuit where some part of the wiring has become damaged, or it may arise from friction in the motor bearing, damage in the linkage, or excessive pressure on the wiper blades. Most manufacturers' dedicated test equipment makes provision for electrical tests on wiper circuits. In general, the wiper current may be checked by placing an ammeter in series with the motor and then conducting a test to observe the current flow. On a 12 volt system, the wiper motor operating current is approximately 1.5 A to 4 A. The diagrams, in Fig. 12.8, show the positions of the ammeter for conducting current tests on Lucas-type wiper motors in the low and high speed modes. The procedure for performing such tests varies for different makes of vehicle and electrical equipment and they should only be attempted by skilled personnel and in accordance with the manufacturer's instructions.

Motor does not operate or motion is weak
Motor

Weak motion of the wipers may arise from a voltage drop in the circuit. Causes may be dirty or loose connections which may usually be detected by means of voltage tests at the motor and the switch and suitable intermediate points. The brushes in the motor may be worn and, in some cases, these can be replaced. However, the overall efficacy and cost of such work must be compared with the alternative of a replacement motor.

Switch

The control switch may fail completely, or it may fail to operate some of the wipe/wash functions. In the case of complete failure, circuit tests should be carried out in order to ascertain that it is switch failure and not some other electrical problem, such as a broken or disconnected wire or a loose connection. Once the cause of the malfunction has been confirmed as a defective switch, the remedy will probably be to replace the switch.

Mechanical defects
Blades

Poor wiping action may result from hardened, cracked or perished blades. Before condemning wiper blades the screen should be thoroughly cleaned to remove any traffic film that may have stuck to the screen. The wipers should then be retested with the screen wet. Once satisfied that the wiper blades are at fault, it is usually possible to effect a cure by replacing the rubber part of the blades.

Arms

Wiper arms are spring loaded and are designed to exert sufficient blade pressure on the screen to ensure effective wiping action. If the spring tension becomes weak, the blade pressure will be lower and wiping action will be poor with the possibility of blade chattering. Too great a spring pressure may cause scratching of the glass and poor wiper action due to extra load on the motor. Most wiper arm problems may be detected by visual examination and comparison with known good examples.

Linkage

Wear in the linkage will result in lost motion at the wiper blades and excessive noise when the wipers are in operation. Looseness in the wiper mechanism may result from the blades not being properly secured to the spindle.

12.4 Anti theft devices

The coded ignition key and engine immobiliser system

The engine immobiliser system is a theft deterrent system that is designed to prevent the engine of the vehicle from being started by any other means than the uniquely coded ignition key. The principal parts of a system are shown in Fig. 12.9.

The transponder chip in the handle of the ignition key is an integrated circuit that is programmed to communicate with the antenna (transponder key coil) that surrounds the barrel of the steering lock. When electromagnetic

TEST PROCEDURE FOR SCREEN WIPER MOTOR

Note: All tests with screen wet.

CONNECTIONS		TYPE OF MOTOR	RESULT
BATT. '+'	BATT. '−'		
*TEST 1 Red/Green	Brown/Green	All types	Motor should run at normal speed
*TEST 2 Blue/Green	Brown/Green	2-speed motors only	Motor should run at high speed
TEST 3 Red/green	Green	Self-switching types only	Motor should run to park position then stop
TEST 4 Green	Red/Green	Self-parking types only	Motor should run to extended park position then stop

*Do not disconnect battery supply from plug while the wiper blades are in the parked position.

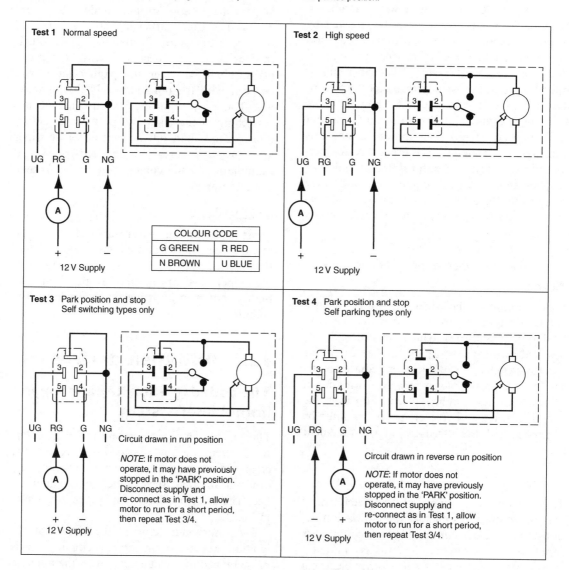

Fig. 12.8 Using an ammeter to test the operating current of a screen wiper system

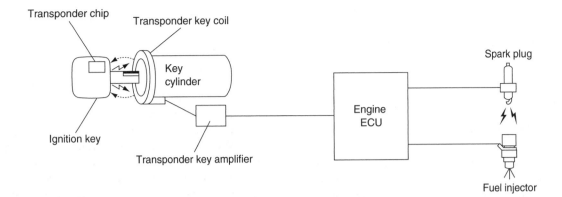

Fig. 12.9 The engine immobiliser system

communication is established between the chip in the key and the transponder key coil, the immobiliser section of the ECM recognises the key code signal that is produced and this effectively 'unlocks' the immobiliser section so that starting can proceed. Vehicle owners are provided with a back-up key that can be used to reset the system should a key be lost.

Burglar alarm

Modern vehicles are normally equipped with some form of alarm to alert people within hearing range to the fact that the vehicle is being forcibly entered. In most cases, the alarm system sounds the vehicle horn and flashes the vehicle lights for a period of time. The theft alarm system, shown in Fig. 12.10, is armed (set) whenever the ignition is switched off

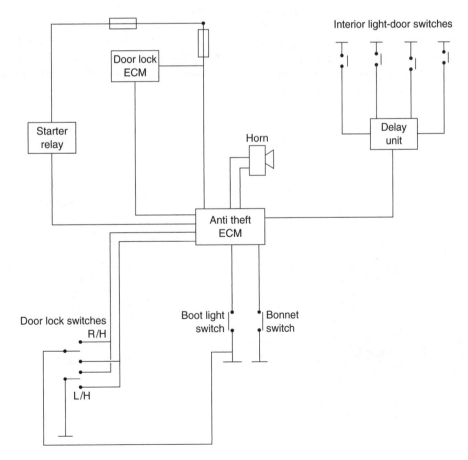

Fig. 12.10 Theft alarm circuit

and the doors and boot lid are locked. If an unauthorised entry is attempted through doors, bonnet or boot, an earth path will be made via the alarm ECM and the alarm will be triggered.

12.5 Electrically operated door mirrors

Many vehicles are equipped with door mirrors that can be adjusted by the driver, from inside the vehicle, in order to achieve the optimum rearward view. Adjustment of mirror position is achieved by means of an electric motor that is housed inside the mirror's protective moulding, as shown in Fig. 12.11 and the driver's control is situated in a suitable position close to the driving position. In many cases, the electrically operated mirrors are fitted with an electric demister facility.

Accurate setting of the mirrors, to suit an individual driver, is vital if the mirrors are to be used to full effect. Because the door mirrors protrude from the side of the vehicle they are normally retractable so that they can be folded away when necessary. Cracked mirrors can normally be replaced because the glass part of the mirror is normally attached to a mirror holder by means of adhesive pads. After fitting a replacement glass part of the mirror, it is necessary to ensure that the mirror is re-adjusted to suit the driver.

12.6 Electric horns

Motor vehicles must be equipped with an audible warning of approach and this normally takes the form of an electrically operated horn. The basic structure of a simple electric horn is shown in Fig. 12.12(a). Pressing the horn switch completes the circuit that energises the electro-magnet and this action draws the armature towards the electro-magnet causing the metal diaphragm to deflect. At the same time, the contacts open to break the circuit and this de-energises the electro-magnet so that the diaphragm springs back. Making and breaking the circuit causes the diaphragm to vibrate, thus setting up the sound. The resonator that is attached to the diaphragm acts as a tone disc in order to give the horn a suitable sound. The diaphragm is normally made from high-quality carbon or alloy steel, the thickness and hardness being factors that give the required sound properties and ensure a long working life. The contact points that make and break the circuit are made from tungsten alloy and are designed to be maintenance-free. The simple horn takes a current of approximately 4 amperes, but the wind tone and the air horn require a current of 10 amperes or more and it is common practice for the circuits for these horns to use a relay, as shown in the circuit in Fig. 12.12(d).

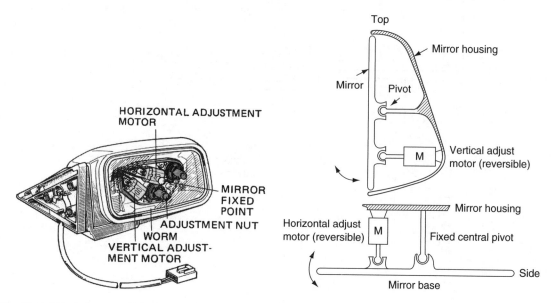

Fig. 12.11 Electrically operated door mirrors

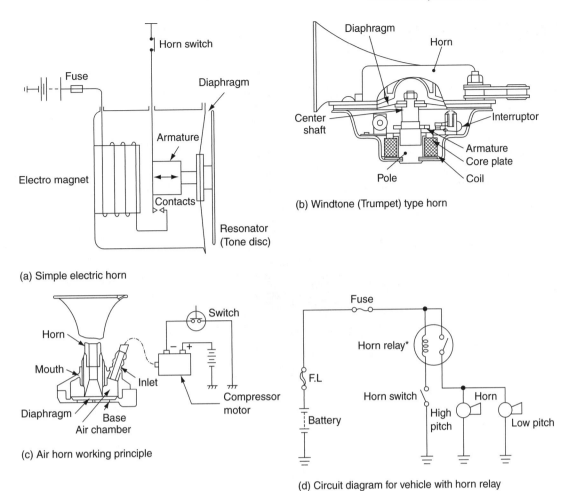

(a) Simple electric horn

(b) Windtone (Trumpet) type horn

(c) Air horn working principle

(d) Circuit diagram for vehicle with horn relay

Fig. 12.12 Basic principles of an electric horn

Wind tone horns

This type of horn also operates through the action of a vibrating diaphragm. The action is slightly different from that the simple horn because the diaphragm controls the movement of air in a trumpet-shaped tube. The tube is often spiral in form so that a relatively long 'trumpet' can be contained in a short space.

Air horns

This type of horn is operated by means of a supply of compressed air that is provided by a small electrically operated compressor that is controlled by the horn switch. Figure 12.12(c) shows the principle of operation.

Repair and maintenance

Horns are normally quite reliable. Problems are most likely to arise in the electrical circuit and the procedure for fault tracing should probably start with a visual inspection followed by checking the fuse. Horns are normally secured to the vehicle structure by means of a flexible bracket, the purpose of which is to prevent vehicle vibrations from distorting the quality of the sound emitted by the horn. This is a point to note when considering the fitment of a different type of horn.

Learning task

Inspect the horn installations on several different vehicles. Pay particular attention to the fixing bracket and the type of horn. With the aid of the notes in this book, identify the type of horn in each case and then look up the vehicle details to compare your result with the horn details provided by the vehicle maker.

12.7 Electrically operated windows

Electrically operated windows are commonly used in many types of vehicle. The two systems, illustrated in Fig. 12.13, make use of reversible electric motors. The mechanism that is used to raise and lower the windows is known as the regulator. In Fig. 12.13(a), the regulator is operated by a small gear on the motor shaft that engages with a quadrant gear on the arm of the regulator. Rotation of the motor armature causes a semi-rotary action of the quadrant. This semi-rotary action is converted into linear motion by the regulator mechanism. The system shown in Fig. 12.13(b) makes use of cables that are wound on a drum that is driven by the operating motor. In both cases, the electric motors are of the permanent magnet type. The permanent magnet motor is reversible via the

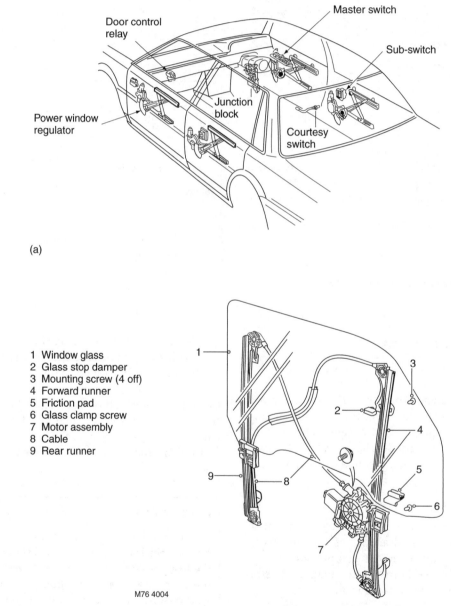

(a)

1 Window glass
2 Glass stop damper
3 Mounting screw (4 off)
4 Forward runner
5 Friction pad
6 Glass clamp screw
7 Motor assembly
8 Cable
9 Rear runner

M76 4004

LH regulator shown. RH similar

(b)

Fig. 12.13 Electrically operated windows. (Reproduced with the kind permission of Ford Motor Company Limited.)

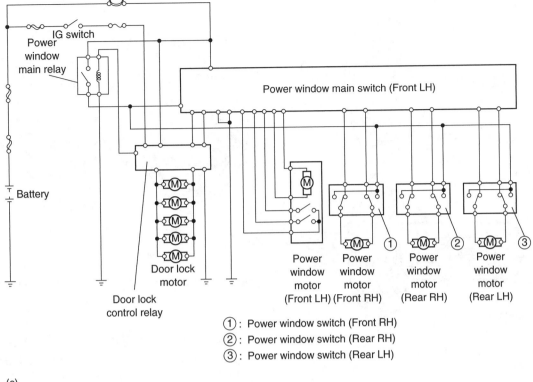

(c)

Fig. 12.13 (Continued)

switching system and this permits the windows to be operated as required. An electrically operated window circuit normally contains a thermally operated switch to protect the circuit in the event of severe icing or other condition that may lead to an excessive load on the motors and winding mechanism.

12.8 Lighting

The main purpose of lights on a vehicle is to enable the driver to see and for other road users to be able to see the vehicle, after dark and in other conditions of poor visibility. In the UK there are legal requirements for the lights that must be fitted. In summary, these requirements are:

1. Headlamps (minimum of two, one each side)
2. Side, rear and number plate lamps
3. Direction indicator lamps (flashers)
4. Stop lamps (brake lights)
5. Rear fog lamp (at least one).

All of these lamps must be maintained in working condition, including proper alignment.

Lighting regulations

The law relating to vehicle lights is quite complicated and you are advised to check the regulations to ensure that you understand them. This is particularly important when fitting extra lights to a vehicle.

Lighting circuit components

Bulbs

The main source of light for the lamps listed above is the traditional bulb. Electric current in the bulb (lamp) filament causes the filament to heat up and give out whitish light. Where other colours are required, for example, for stop and tail lights, the lamp lens is made from coloured material. Figure 12.14 shows some commonly used types of bulbs.

Ordinary twin filament headlamp bulb

The filaments are made from tungsten wire and the glass bulb is often filled with an inert gas such as argon. This permits the filament to

(a)

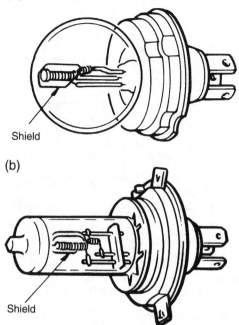

Shield

(b)

Shield

Fig. 12.14 Typical vehicle light bulbs, (a) ordinary bulb, (b) Quartz halogen blub

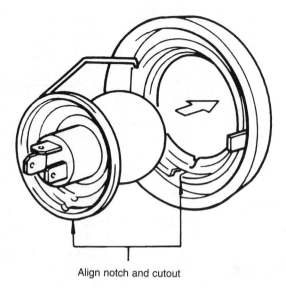

Align notch and cutout

Fig. 12.16 The location notch in the bulb base – regular type

operate at a higher temperature and increases the reliability. One of the filaments provides the main beam and the other filament, which is placed a little above centre, provides the dipped beam. The effect is shown in Fig. 12.15.

In order to locate the bulb accurately in the lamp reflector, the metal base of the bulb is equipped with a notch, as shown in Fig. 12.16.

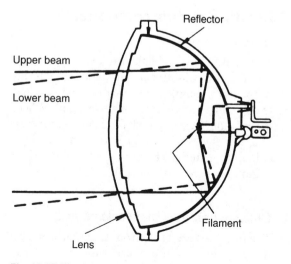

Reflector

Upper beam

Lower beam

Filament

Lens

Fig. 12.15 The dipped beam principle

Quartz halogen twin filament headlamp bulb

A problem with ordinary tungsten lamp bulbs is that, over a long period of time, the filament deteriorates (evaporates) and discolours the glass. The rather more elaborate quartz halogen bulb is designed to give brighter light and to prevent the evaporated tungsten from being deposited on the inside of the bulb. Halogens are gases, such as iodine, chlorine, etc., which react chemically inside the bulb to provide the 'halogen cycle'. The 'halogen cycle' preserves the life of the tungsten filament.

Oil, grease and salt from perspiration can damage the quartz and it is recommended that these bulbs are handled by the metal part to prevent damage. Figure 12.17 shows a typical method of locating a quartz halogen type bulb.

Stop and tail lamps

The stop/tail bulb has two filaments. They are normally 21 Watt and 5 Watt filaments. The 21 Watt filament is for the stop (brake) lights and the 5 Watt filament is for the tail lights. The metal base of the bulb is provided with off-set pins so that it cannot be fitted incorrectly.

Sealed beam units

The construction of a sealed beam unit is shown in Fig. 12.18.

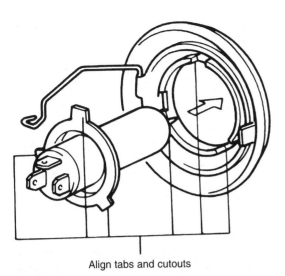

Align tabs and cutouts

Fig. 12.17 Locating a quartz halogen bulb

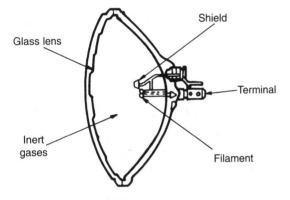

Fig. 12.18 A sealed beam headlamp unit

There is no separate bulb. The assembly – lighting filaments, reflector and lamp lens is a single unit. During manufacture, the inside of the unit is evacuated and filled with inert gas and the unit is then hermetically sealed. The idea is that the dust and other contaminants cannot enter and the accuracy of the setting of the filament in relation to the focal point of the reflector cannot be altered. A disadvantage is that, in the event of filament failure, the whole unit must be replaced.

Headlamp dipping

The reflector concentrates the light produced by the bulb and projects it in the required direction. The lens is specially patterned to 'shape' the light beam so that the beam is brighter in the centre and less bright on both sides. The intention is to reduce the risk of 'dazzling' oncoming drivers and pedestrians. Headlamps must also

be provided with the means to deflect the beam downwards and, in a two headlamp system, this is achieved by switching the lamps from the main beam filament to the dip beam filament.

Headlamp alignment

To ensure that headlamps are correctly adjusted, it is necessary to check the alignment. Special machines of the type shown in Fig. 12.19, are often used for this purpose.

If such a machine is not available it is acceptable to use a flat, vertical surface such as a wall or a door.

The wall should be marked out, as shown in Fig. 12.20.

The vehicle is placed so that the headlamps are parallel to the wall and 8 metres from the wall. The centre line of the vehicle must line up with

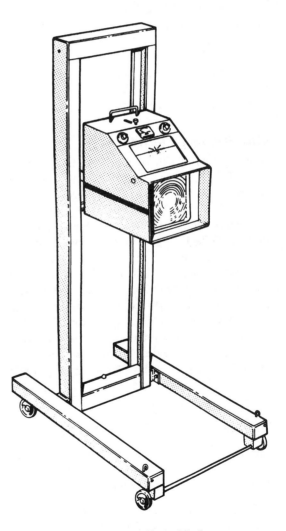

Lucas Beam Tester Mk. II

Fig. 12.19 Lucas headlamp alignment machine

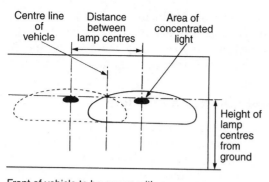

Front of vehicle to be square with screen
Vehicle to be loaded and standing on level ground
Recommended distance for setting is at least 8m
For ease of setting, one headlamp should be covered

Fig. 12.20 Using a marked wall for headlamp alignment

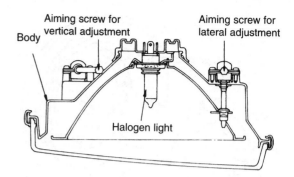

Fig. 12.21 Headlamp alignment screws

the centre line marked on the wall and, with the headlamps switched on to main beam, the area of concentrated light should be very close to that shown in Fig. 12.20. Should the settings be incorrect, it will be necessary to inspect for the cause of the inaccuracy. It may be a case of incorrect adjustment. If adjustment is required, it is probable that the headlamp unit will be equipped with screws to alter the horizontal and vertical settings, as shown in Fig. 12.21.

Gas discharge lamps

Xenon lamps operate on the gas discharge principal. The light source (bulb) contains a pair of electrodes that are encased in a special type of quartz glass bulb. The bulb is filled with Xenon gas, traces of mercury and other elements, under pressure. When a high voltage of approximately 20 kVolts is applied to the electrodes, the Xenon gas emits a bright white light and evaporates the mercury and other elements. Once illuminated, the light output of the bulb is maintained by a lower voltage. Gas discharge (Xenon) lamps produce a greater amount of light than conventional bulbs and are designed to last longer. The main features of a Xenon headlamp are shown in Fig. 12.22.

Safety

The electronic control unit for the Xenon lamp produces a very high voltage and the following precautions must be observed:

• Ensure that the lights are switched off before attempting to work on the system. This is to guard against electric shock that can arise from the high voltage.

• Gloves and protective eye wear must be used when handling Xenon bulbs and the glass part of the bulb must not be touched.

• Discarded Xenon bulbs must be treated as hazardous waste.

Vehicles that are equipped with Xenon lamps must also be equipped with a means of automatically adjusting the lamps to prevent dazzling of oncoming drivers.

Headlight and rear fog light circuit

In the circuit shown in Fig. 12.23, the rear fog lamp circuit is fed from the main light switch. The purpose of the rear fog lamp is to make the presence of the vehicle more visible in difficult driving conditions. The headlamps are connected in parallel so that failure of one lamp does not lead to failure of the others.

Direction indicators

Direction indicator lights (flashers) are required so that the driver can indicate any intended manoeuvre. Figure 12.24 shows a typical indicator lamp circuit circuit. The 'flasher' unit is designed so that the frequency of flashes is not less than 60 per minute and not more than 120 per minute. The circuit is also designed so that failure of an indicator lamp will lead to an increased frequency of flashing and the driver will notice this at the warning light.

The 'flasher' unit

The unit shown in the above circuit is marked 8FL. This is a Lucas unit that operates on the basis of thermal expansion and contraction of a metal strip. The flasher unit shown in Fig. 12.25

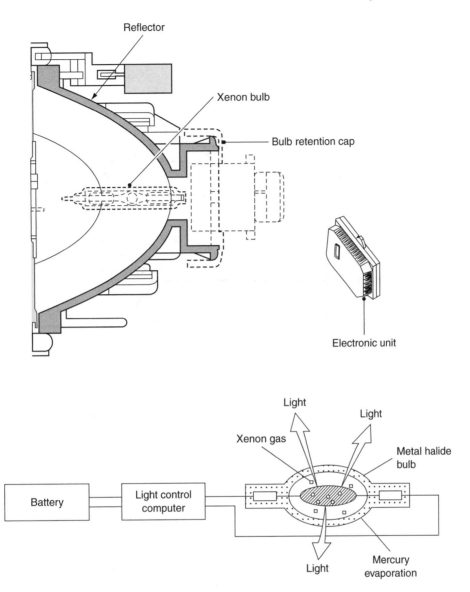

Fig. 12.22 The gas discharge (Xenon) lamp. (Reproduced with the kind permission of Ford Motor Company Limited.)

employs a capacitor and a relay to provide the flashing action. Examination of the circuit diagram shown in Fig. 12.25 shows that the contact points are normally at rest in the closed position. When the indicator switch is moved to indicate a turn, to left or right, current flows to the indicator lamps, through the upper winding of the relay. This current energises the relay and opens the contact points, the current flow is interrupted and the lamps 'go out'. As the lamps go out, the contact points close and the lights 'come on' again. This happens at a frequency of approximately 90 flashes per minute and the timing of the flash rate is controlled by the capacitor.

12.9 Circuits and circuit principles

Circuit (wiring) diagrams

A complete circuit is needed for the flow of electric current. Figure 12.26 shows two diagrams of a motor in a circuit. Should the fuse be blown, the circuit is incomplete, current will not flow and the motor will not run. In the right-hand diagram, the fuse has been replaced. The circuit is now complete and the motor will run. Basic electrical principles such as this, are fundamental to good work on electronic systems because much of the

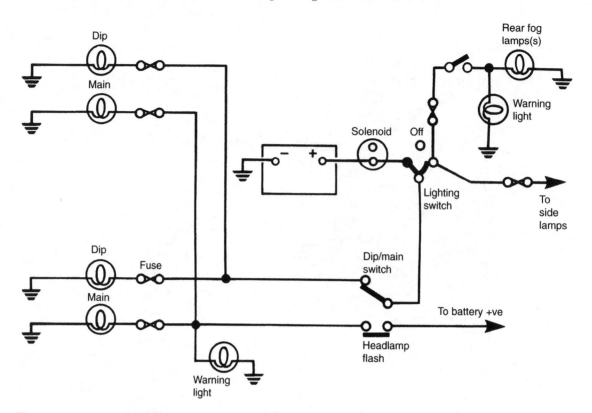

Fig. 12.23 Headlamp and rear fog lamp circuit

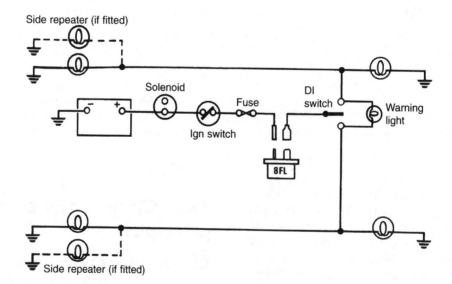

Fig. 12.24 Indicator lamp circuit

testing of electronic systems requires checking of circuits to ensure that they are complete.

It is essential to be able to understand and follow circuit diagrams and this requires a knowledge of circuit symbols. There is a set of standard circuit symbols, some of which are shown in Fig. 12.27.

However, non-standard symbols of the type shown in Fig. 12.28, are sometimes used and this can cause confusion. Fortunately, when

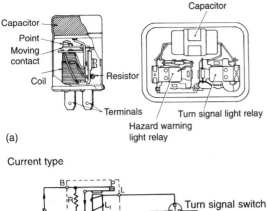

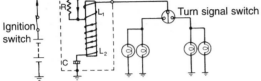

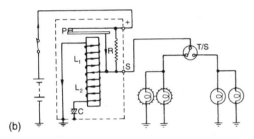

Fig. 12.25 Capacitor and relay flasher unit and circuit (a) construction of the flasher units, (b) circuits for the two types of flasher unit

non-standard symbols are used, they are normally accompanied by a descriptive list.

Figure 12.29 gives an example of the use of a descriptive list to support the wiring diagram; the injector resistors are shown as a sawtooth line, number 21 on the list, and on the diagram.

This solves the problem of decyphering the diagram.

Colour code

The wiring diagram is an essential aid to fault tracing. The colour code for wires is an important aid to fault tracing and most wiring diagrams make use of it.

A commonly used colour code:

N	= brown	Y	= yellow
P	= purple	K	= pink
W	= white	R	= red
O	= orange	LG	= light green
U	= blue	B	= black
G	= green	S	= slate

In order to assist in tracing cables, they are often provided with a second colour tracer stripe. The wiring diagram shows this by means of letters, for example, a cable on the wiring diagram, with GB written on it, is a green cable with a black tracer stripe. The first letter is the predominant colour and the second is the tracer stripe.

Note: This is not a universal colour code and, as with many other factors, it is always wise to have accurate information to hand, that relates to the product being worked on.

The predominant (main) colours frequently relate to particular circuits as follows:

Brown (N) = Main battery feeds
White (W) = Essential ignition circuits (not fused)
Light green and also Green (LG) (G) = Auxiliary ignition circuits (fused).
Blue (U) = Headlamp circuits.
Red (R) = Side and tail lamp circuits.
Black (B) = Earth connections.
Purple (P) = Auxiliary, non-ignition circuits, probably fused.

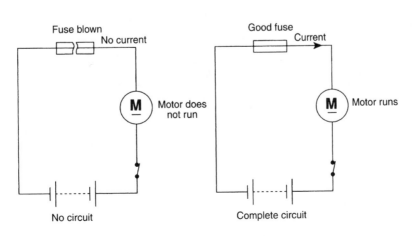

Fig. 12.26 An electric circuit

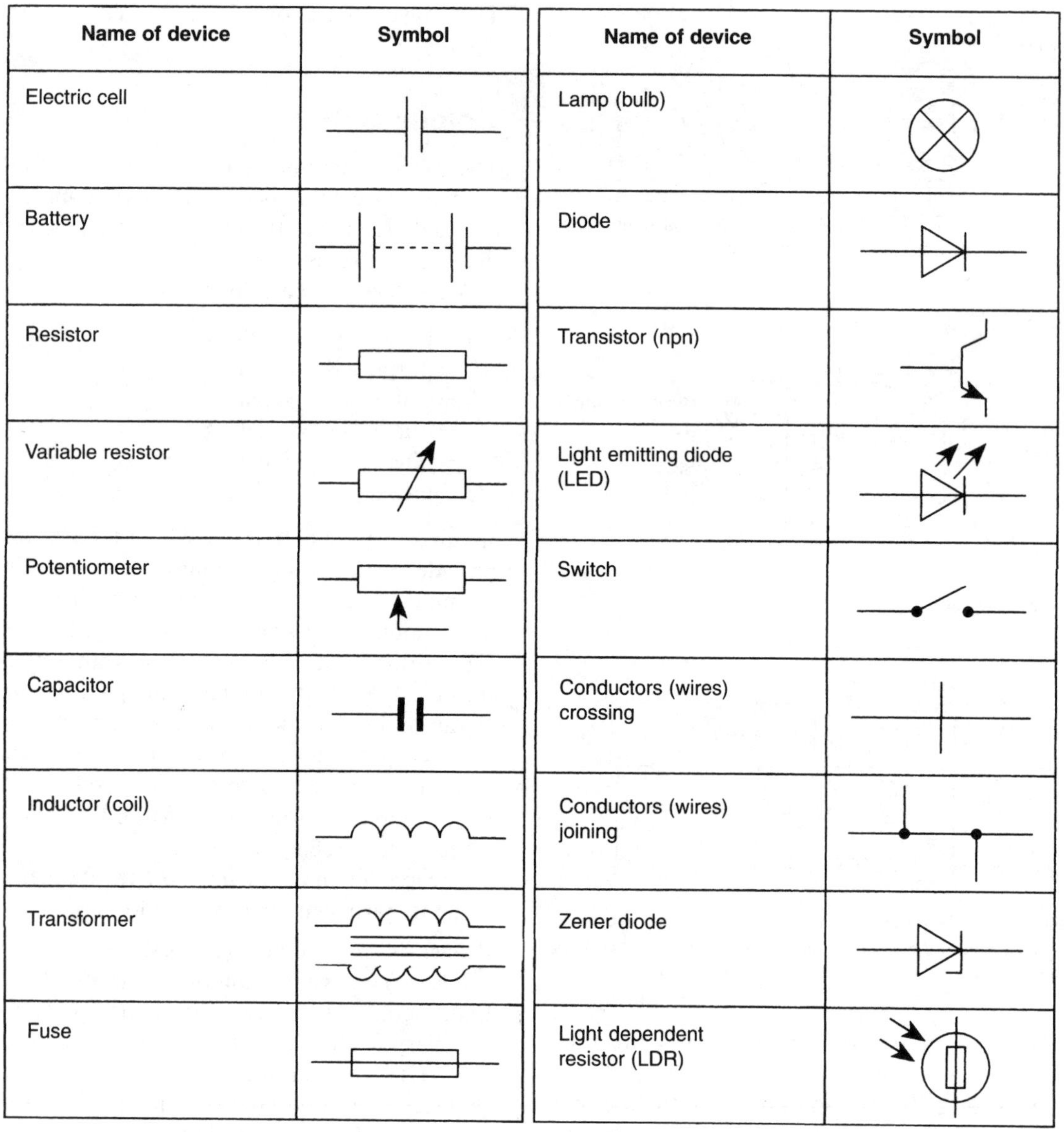

Name of device	Symbol	Name of device	Symbol
Electric cell		Lamp (bulb)	
Battery		Diode	
Resistor		Transistor (npn)	
Variable resistor		Light emitting diode (LED)	
Potentiometer		Switch	
Capacitor		Conductors (wires) crossing	
Inductor (coil)		Conductors (wires) joining	
Transformer		Zener diode	
Fuse		Light dependent resistor (LDR)	

Fig. 12.27 A selection of circuit symbols

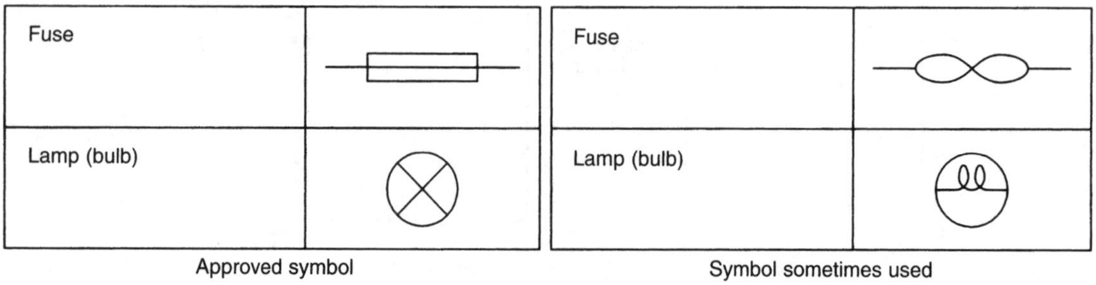

Fuse		Fuse	
Lamp (bulb)		Lamp (bulb)	

Approved symbol   Symbol sometimes used

Fig. 12.28 Non-standard circuit symbols

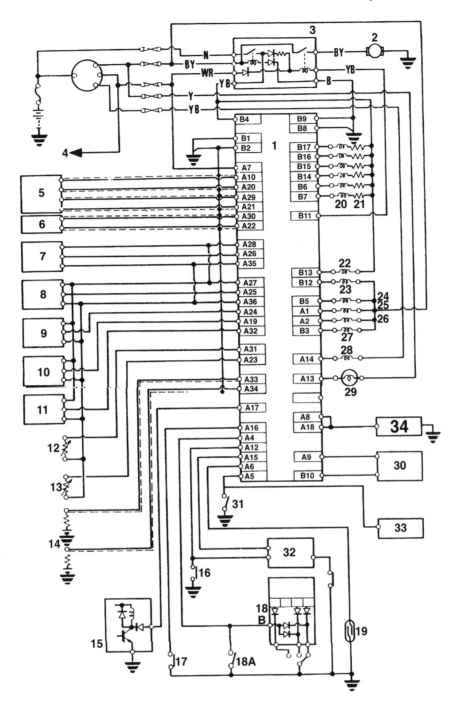

Fig. 12.29 System diagram with list describing circuit elements

1 – ECU
2 – Fuel pump
3 – Main relay
4 – To starter
5 – Cyl/crank sensor
6 – TDC sensor
7 – MAP sensor
8 – Atmospheric pressure sensor
9 – Throttle angle sensor
10 – Ignition timing adjuster
11 – EGR lift sensor
12 – Water temperature sensor

13 – Air intake temperature sensor
14 – Oxygen sensors*
15 – Alternator
16 – Cooling fan switch
17 – Power steering switch
18a – Neutral switch (MT)
18b – A/T position switch
19 – Vehicle speed sensor
20 – Injectors
21 – Injector resistors
22 – EICV
23 – Pressure regulator solenoid

24 – EGR solenoid
25 – Air suction control solenoid*
26 – By-pass solenoid B
27 – By-pass solenoid A
28 – Air-con clutch relay
29 – Check engine light
30 – EAT ECU
31 – Clutch switch M/T
32 – Radiator fan control unit
33 – Cruise control
34 – Igniter unit
*Emission vehicles only

A complete wiring diagram

Figure 12.30 shows a full wiring diagram for a vehicle. In order to make it more intelligible, this diagram uses a grid system. Numbers 1 to 4 across the page, and letters A, B, C, at the sides. This means that an area of diagram can be located. For example, the vehicle battery is in the grid area 1A.

The wiring loom, or harness

It is common practice to bind cables together to facilitate positioning them in the vehicle structure. It used to be the practice to fabricate the loom, or harness, from a woven fabric. Modern loom sheathing materials are usually a PVC type of material. Figure 12.31 shows the engine compartment part of a typical wiring loom.

Cable sizes

The resistance of a cable (wire) is affected by, among other factors, its diameter and its length. For a given material, the resistance increases with length and decreases with diameter. Doubling a given length of cable will double its resistance and doubling the diameter of the same length of cable will decrease the resistance to 1/4th of the original value. Cable sizes are, therefore, important, and only those sizes specified for a given application should be used. Most cables used on vehicles need to be quite flexible. This flexibility is provided by making the cable from a number of strands of wire and it is common practice to specify cable sizes by the number of strands and the diameter of each strand, for example, 14/0.30 means 14 strands of wire and each strand has a diameter of 0.30 mm.

Some typical current carrying capacities of cables for vehicle use are:

Size of cable	Current rating	Typical use
14/0.30	8.75 amps	Side and tail lamps
28/0.30	17.5 amps	Headlamps
120/0.30	60 amps	Alternator to battery

The choice of cable is a factor that is decided at the design stage of a vehicle. However, it sometimes affects vehicle repair work when, for example, an extra accessory is being fitted to a vehicle, or a new cable is being fitted to replace a damaged one. In such cases, the manufacturers' instructions must be observed.

Circuit protection

Fuses

The purpose of the fuse is to provide a 'weak' link in the circuit which will fail (blow) if the current exceeds a certain value and, in so doing, protect the circuit elements and the vehicle from the damage that could result from excess current.

The fuse is probably the best known circuit protection device. There are several different types of fuses and some of these are shown in Fig. 12.32. Fuses have different current ratings and this accounts for the range of types available. Care must be taken to select a correct replacement and larger rated fuses must never be used in an attempt to 'get round' a problem. Many modern vehicles are equipped with a 'fusible link' which is fitted in the main battery lead as an added safety precaution.

It is common practice to place fuses together in a reasonably accessible place on the vehicle. Another feature of the increased use of electrical/electronic circuits on vehicles is an increase in the number of fuses to be found on a vehicle. Figure 12.33 shows an engine compartment fuse box that carries fusible links in addition to 'normal' fuses.

The same vehicle also has a dashboard fuse-box. This also carries 'spare' fuses which appear to the right of the other fuses (Fig. 12.34).

Whilst, in the event of circuit failure, it is common practice to check fuses and replace any 'blown' ones, it should be remembered that something caused the fuse to blow. Recurrent fuse 'blowing' requires that circuits should be checked to ascertain the cause of the excess current that is causing the failure.

The circuit breaker

The thermal circuit breaker relies for its operation on the principle of the 'bi-metallic strip'. In Fig. 12.35, the bi-metal strip carries the current between the terminals of the circuit breaker. Excess current, above that for which the circuit is designed, will cause the temperature of the bi-metal strip to rise to a level where it will curve and cause the contacts to separate. This will open the circuit and current will cease to flow. When the temperature of the bi-metal strip falls the circuit will be re-made. This action leads to intermittent functioning of the circuit which will continue until the fault is rectified. An application may be of a 7.5 ampere circuit breaker to protect a door lock circuit. The advantage of the circuit breaker, over a fuse, is that the circuit breaker can be re-used.

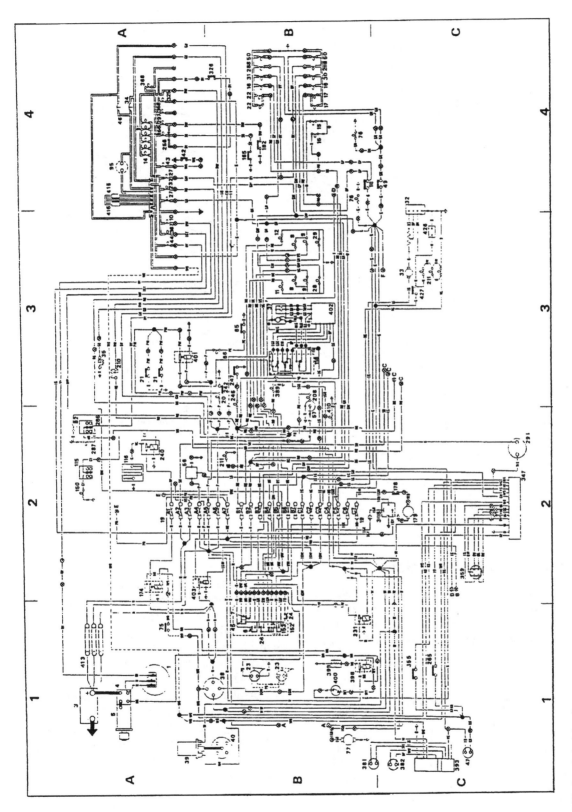

Fig. 12.30 A typical wiring diagram

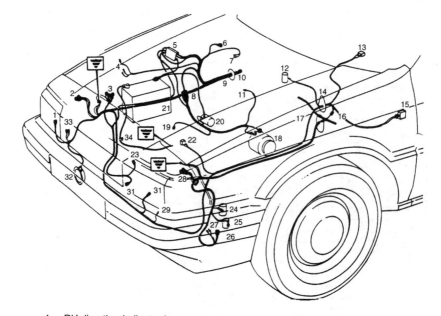

1 – RH direction indicator lamp connector
2 – RH headlamp connector
3 – Relays – air conditioning
4 – Ignition coil connector
5 – Fusible link box
6 – Windscreen wiper motor plug
7 – Handbrake warning lamp switch
    connector
8 – Main/engine harness connector
9 – Main harness
10 – Bulkhead grommet
11 – Fuel cut-off solenoid
12 – Throttle solenoid – air conditioning
13 – Switch – air conditioning
14 – Bulkhead grommets
15 – Relay – air conditioning – 'A' post
16 – Main/air conditioning harness connector
17 – Harness – air conditioning
18 – Alternator

19 – Reverse light switch
20 – Starter solenoid
21 – Battery
22 – Charging pressure switch –
    air conditioning
23 – Cooling fan connector
24 – Headlamp washer motor
25 – Windscreen washer motor
26 – LH direction indicator lamp
    connector
27 – LH side lamp
28 – LH headlamp connector
29 – Compressor clutch –
    air conditioning
30 – Fan – air conditioning
31 – Thermostatic switch
32 – Horn
33 – RH side lamp
34 – Cable clip

Fig. 12.31 Engine compartment wiring loom

Other circuit protection

Vehicle circuits are subject to 'transient' voltages which arise from several sources. Those which interest us here are load dump, alternator field decay voltage, switching of an inductive device

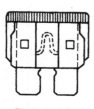

Blade-type fuse

Lug-type fuse

Cartridge-type fuse

Fig. 12.32 Fuse types

(coil, relay, etc.) and over voltage arising from incorrect use of batteries when 'jump starting'.

Load dump occurs when an alternator becomes disconnected from the vehicle battery while the alternator is charging, i.e., when the engine is running.

Figure 12.36 shows a Zener diode, as used for surge protection in an alternator circuit. The breakdown (Zener) voltage of the diode is 10 to 15 volts above the normal system voltage. Such voltages can occur if an open circuit occurs in the main alternator output lead when the engine is running. Other vehicle circuits, such as coil ignition, can also create inductive surges. Should such voltage surges occur, they could damage the alternator circuits but, with the Zener diode connected as shown, the excess voltage is 'dumped' to earth via the Zener diode.

No.	Rating	Function
G	50 amp	Radio, power amplifier, electric seats
H	50 amp	Ignition switch circuit
I	80 amp	Battery output
J	50 amp	Window lift
K	50 amp	ABS brake system
L	50 amp	Supply to fuses 4, 5 and 6 and sidelight relay

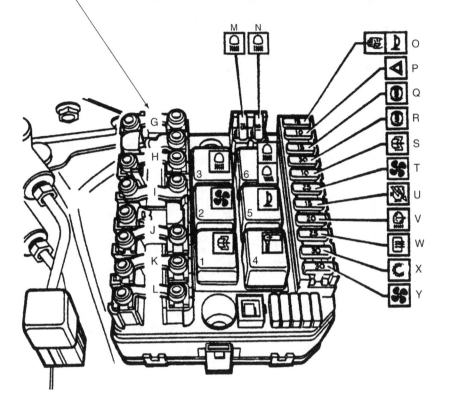

Relays
1 – Cooling fan changeover or manifold heater
2 – Cooling fan
3 – Lighting
4 – Starter
5 – Horns
6 – Main/dip beams
7 – Air conditioning changeover

Fig. 12.33 Engine compartment fusebox (Rover 800)

Should such a voltage surge occur, it may destroy the protection diode. The alternator would then cease charging. In such a case, the surge protection diode would need to be replaced, after the cause of the surge had been remedied.

Figure 12.37 shows another form of circuit protection where a diode is built into a cable connector. This reduces the risk of damage from reversed connections and it is evident that one should be aware of such uses because a continuity test on such a connector will require correct polarity at the meter leads.

Connectors

Connectors are used to connect cables to components and to join cables together. They are made in a range of different shapes and sizes for a variety of purposes, as shown in Fig. 12.38. In Fig. 12.38(b), a pair of cables are joined together

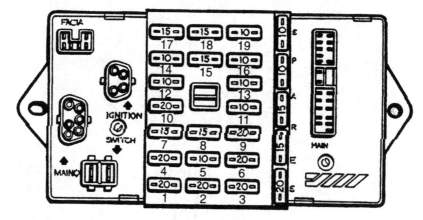

Fuse functions

Fuse No.	Rating	Wire colour	Function
1	20 amp	N/O	Sunroof, driver's seat heater
2	20 amp	S/U	RH front door window lift
3	20 amp	S/R	LH front door window lift
4	20 amp	P/N	Front and rear cigar lighters Footwell lamps
5	10 amp	P	Interior lights, boot light, interior light delay unit, map light, door open lights, trip computer memory, radio station memory, clock memory, headlight delay unit
6	20 amp	P/O	Burglar alarm ECU (optional) Central locking ECU
7	15 amp	R/O	RH number plate lamp RH side, tail and marker lights Trailer plug
8	10 amp	R/B	Cigar lighter illumination, LH licence plate lamp, sidelight warning light, glovebox light, trailer plug, dashboard illumination, LH side tail lights
9	20 amp	S/O	LH rear window lift
10	20 amp	S/G	RH rear window lift

Fig. 12.34 Dashboard fusebox (Rover 800)

by a connector and in Fig. 12.38(c), some cables are connected to a component. Figure 12.38(a) shows the principal parts of a connector. These parts are designed to be resistant to vibration and moisture. In some cases, such as the multiple pin connector for an ECU that is shown in Fig. 12.38(d), the pins are coated with gold to ensure the type of electrical connection that is needed for very small electric current.

Repair and maintenance

Some service and repair operations may require components and cables to be disconnected. In such cases, the procedures for taking connectors

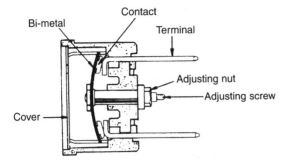

Fig. 12.35 A circuit breaker (Toyota)

apart should be observed and care taken to avoid damage to the pins. Damage to pins may cause the connection to develop a high resistance or to fail completely. Should it be necessary to probe connections, in an endeavour to trace a fault, every effort should be made to prevent damage to waterproofing and electrical insulation.

12.10 Instrumentation

Vehicles are equipped with instruments and other warning devices, in order to inform the driver about the current state of various systems on the vehicle. Some instruments, such as the speedometer, the fuel tank contents gauge, the oil pressure gauge and the ignition warning, have been features of the instrument display on vehicles for many years. Several of the more commonly used instruments are examined below, in order to assess the technology involved and to examine some approaches to fault tracing.

Thermal type gauges

These instruments make use of a bi-metal strip to provide the movement that gives an indication of the quantity that is being measured. In the example of the fuel tank contents gauge, that is shown in Fig. 12.39, the current flowing in the meter circuit is dependent on the resistance of the sensing unit at the fuel tank. As the float of the fuel tank sensing unit rises and falls with fuel level, so the resistance that it places in the meter circuit changes. Changes in resistance cause the current flowing in the heating coil that surrounds the bi-metal strip to change and this leads to changes in the temperature of the bi-metal strip. Thus, the heating effect of the current that deflects the bi-metal strip varies with fuel level and, by careful calibration, the fuel gauge is made to give a reliable guide to the quantity of fuel in the tank.

The engine coolant temperature gauge, that is also shown in Fig. 12.39, operates on the same

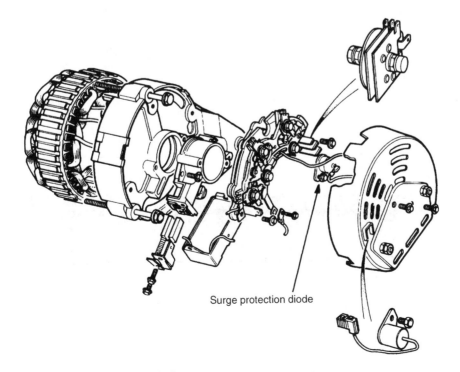

Fig. 12.36 Voltage surge protection by Zener diode

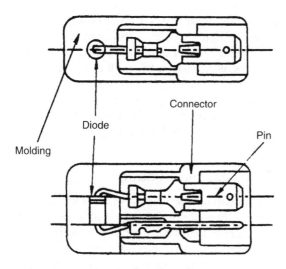

Fig. 12.37 A protection diode in a cable connector

principle as the fuel contents gauge, except that the variable resistance is provided by the thermistor. A thermistor has a negative temperature coefficient which means that the resistance of the thermistor decreases as its temperature rises. It is this variation of resistance that cause the meter current to change and thus indicate the coolant temperature.

The voltage stabiliser

Reliable operation of thermal type meters is dependent on a reasonably constant voltage supply and the voltage stabiliser is the means of providing a stable voltage. In the type of voltage stabiliser shown in Fig. 12.39, the current in the meter circuit causes the temperature of the voltage stabiliser bi-metal strip to

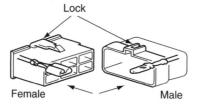

(a) Basic parts of a cable connector

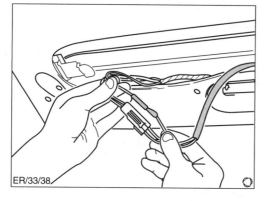

(b)

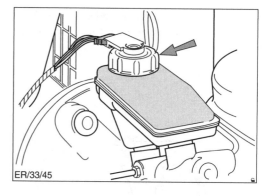

(c)

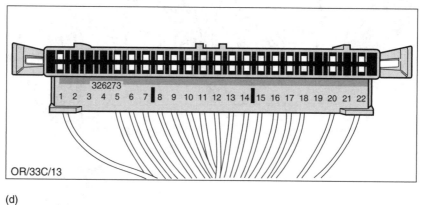

(d)

Fig. 12.38 Types of cable connectors. (Reproduced with the kind permission of Ford Motor Company Limited.)

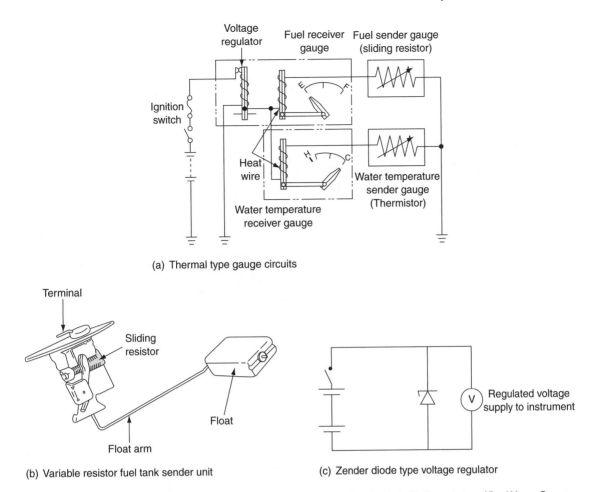

(a) Thermal type gauge circuits

(b) Variable resistor fuel tank sender unit

(c) Zender diode type voltage regulator

Fig. 12.39 A thermal type fuel gauge and variable resistor sender unit. (Reproduced with the kind permission of Ford Motor Company Limited.)

heat up and bend. The heating and bending action causes the voltage stabiliser contact to open, thus stopping the current flow. The temperature of the bi-metal strip then drops and the contacts close. This process proceeds at high frequency and provides a reasonably constant voltage level at the meter. An alternative method of voltage stabilisation is shown in Fig. 12.39(c). This type of stabiliser makes use of the breakdown voltage of an avalanche (Zener) diode. When the voltage at the terminals exceeds the breakdown value, the diode conducts and effectively 'absorbs' the excess voltage.

A vacuum fluorescent display gauge

The vacuum fluorescent display makes use of a number of segments. Each segment of the display is activated electronically through a circuit similar to that shown in Fig. 12.40(a). Figure 12.40(b) shows the main features of a single

element of a VFD. The fluorescent material, the grid and the filament are encased in a vacuum compartment. When a segment is activated by the electronic circuit, a stream of electrons bombards the fluorescent material which causes it to emit light. The filament is negatively charged and heated to approximately 600 °C. At this temperature electrons are released to pass through the grid towards the positively charged segment, where they cause the display to emit light.

Liquid crystal displays (LCDs) and light emitting diode displays (LEDs) are also used in many similar applications. They are also driven by electronic circuits similar to that shown in Fig. 12.40(a).

Vehicle condition monitoring (VCM)

Gauges, such as the oil pressure gauge, the ammeter, and warning lights (i.e., those that light up when the brake pads are worn), are all

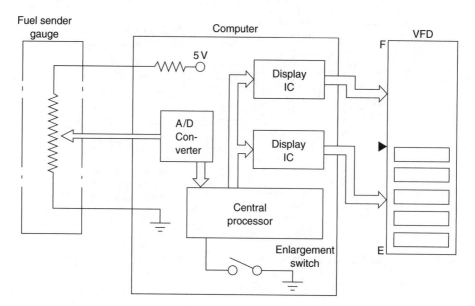

(a) A vacuum fluorescent gauge circuit

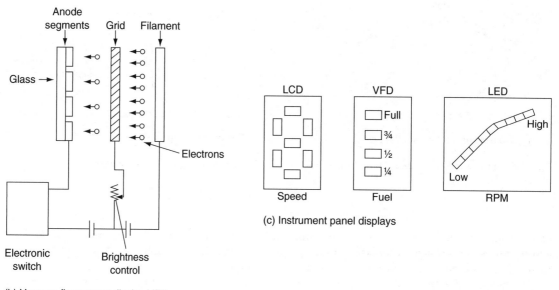

(b) Vacuum fluorescent display VFD

(c) Instrument panel displays

Fig. 12.40 Types of instrument displays. (Fig. 12.40(a) – Reproduced with the kind permission of Ford Motor Company Limited.)

examples of instruments that continuously monitor the condition of various parts of the vehicle. The extensive use of computer controlled systems on vehicles means that many of the factors that affect the use of a vehicle are constantly monitored because it is the monitoring that provides the data that allows the systems to operate. Vehicle condition monitoring has thus evolved so that it has become a feature on all modern vehicles. The most striking example is probably that provided by on board diagnostics. Modern computer

controlled systems have sufficient memory and processing capacity to constantly check a large number of variables and to take action if any variable (measurement) is wrong. This action may be to store a fault code in the memory, or it may be to illuminate a warning lamp to alert the driver. The European (EOBD) version of condition monitoring requires the use of a malfunction indicator lamp MIL. The MIL is illuminated when any item of equipment that affects vehicle emissions develops a fault.

12.11 Supplementary restraint systems (SRS)

Air bags of the type shown in Fig. 12.41, and seat belt pre-tensioners, such as that shown in Fig. 12.42, are features of a basic supplementary restraint system. In the event of a frontal impact of some severity, the air bags and seat belt pre-tensioners are deployed. The air bags are inflated to protect those provided with them from impact with parts of the vehicle. The seat belt pre-tensioners are made to operate just before the air bags are inflated and they operate by pulling about 70 mm of seat belt on to the inertia reel of the belt. This serves to pull the seat occupant back on to the seat.

The deployment of these supplementary restraint devices is initiated by the action of the collision detection sensing system. A collision detection sensing system normally utilises signals from two sensors. One sensor is a 'crash sensor' and the other is a 'safing sensor'. The safing sensor is activated at a lower deceleration than the crash sensor (about 1.5 g less) and both sensors must have been activated in order to trigger the supplementary restraint system. The safing sensor is fitted to reduce the risk of a simple error bringing the air bag into operation. Both of these sensor may be fitted inside the electronic control unit which, in some cases, is known as a diagnostic and control unit (DCU) because it contains the essential self diagnosis circuits in addition to the circuits that operate the SRS. Figure 12.43 shows the layout of a supplementary restraint system on a Rover Mini.

Air bags and seat belt pre-tensioners

Air bags are made from a durable lightweight material, such as nylon, and in Europe they have a capacity of approximately 40 litres. The pyrotechnic device that provides the inert gas to inflate the air bag contains a combustion chamber filled with fuel pellets, an electronic igniter and a filter, as shown in Fig. 12.44. Combustion of the fuel pellets produces the supply of nitrogen that inflates the air bag. The plastic cover that retains the folded air bag in place at the centre of the steering wheel is designed with built in break lines. When the air bag is inflated, the plastic cover separates at the break lines and the two flaps open out to permit unhindered inflation of the air bag.

The seat belt pre-tensioners are activated by a similar pyrotechnic device. In this case the gas is released into the cylinder of the pre-tensioner, where it drives a piston along the cylinder. The

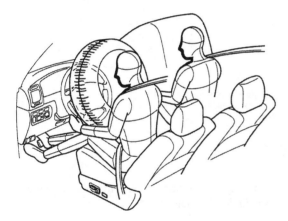

Fig. 12.41 Airbags for driver and front seat passenger

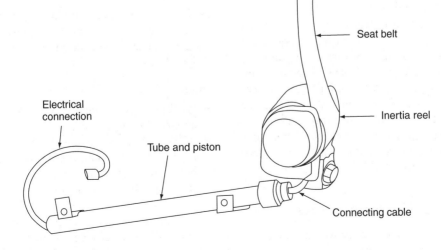

Fig. 12.42 Operation of pre-tensioner

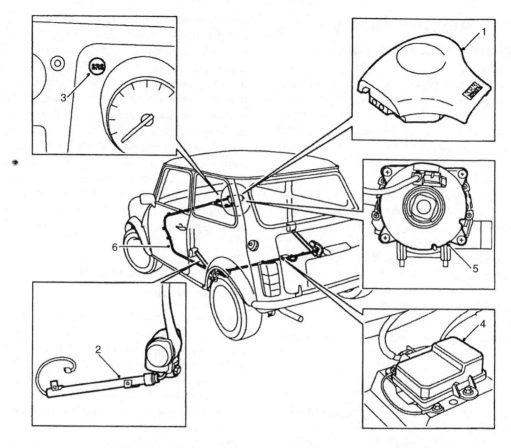

1. Driver's airbag module
2. Seatbelt pre-tensioner
3. SRS warning light
4. Diagnostic and control unit
5. Rotary coupler
6. SRS wiring harness

Fig. 12.43 The elements of a supplementary restraint system (Rover Mini)

piston is attached to a strong flexible cable that then rotates the inertia reel of the seat belt by a sufficient amount to 'reel in' the seat belt by approximately 70 mm.

The rotary coupler

This device is fitted beneath the steering wheel to provide a reliable electrical connection between the rotating steering wheel and air bag, and the static parts of the steering column. The positioning of the rotary coupler is a critical element of the air bag system and it should not be tampered with. When working on supplementary restraint systems, it is important that a technician is fully acquainted with the system and procedures for working on it.

Handling SRS components

The following notes are provided to Rover trained technicians and they are included here because they contain some valuable advice for all vehicle technicians.

Safety precautions, storage and handling

Airbags and seat belt pre-tensioners are capable of causing serious injury if abused or mishandled. The following precautions must be adhered to:

In vehicle

- ALWAYS fit genuine new parts when replacing SRS components;
- ALWAYS refer to the relevant workshop manual before commencing work on a supplementary restraint system;
- Remove the ignition key and disconnect both battery leads, earth lead first, and wait 10 minutes to allow the DCU back-up power circuits to discharge before commencing work on the SRS;

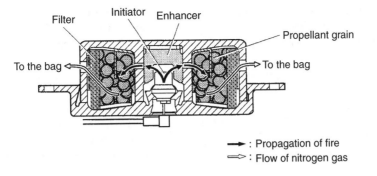

Fig. 12.44 A pyrotechnic device for inflating airbags

- DO NOT probe SRS components or harness with multi-meter probes, unless following a manufacturer's approved diagnostic routine;
- ALWAYS use the manufacturers approved equipment when diagnosing SRS faults;
- Avoid working directly in line with the air bag when connecting or disconnecting multi-plug wiring connectors;
- NEVER fit an SRS component which shows signs of damage or you suspect has been abused.

Handling

- ALWAYS carry airbag modules with the cover facing upwards;
- DO NOT carry more than one airbag module at a time;
- DO NOT drop SRS components;
- DO NOT carry airbag modules or seat belt pre-tensioners by their wires;
- DO NOT tamper with, dismantle, attempt to repair or cut any components used in the supplementary restraint system;
- DO NOT immerse SRS components in fluid;
- DO NOT attach anything to the airbag module cover;
- DO NOT transport and airbag module or seatbelt pre-tensioner in the passenger compartment of a vehicle. ALWAYS use the luggage compartment.

Storage

- ALWAYS keep SRS components dry;
- ALWAYS store a removed airbag module with the cover facing upwards;
- DO NOT allow anything to rest on the airbag module;
- ALWAYS place the airbag module or pre-tensioner in the designated storage area;
- ALWAYS store the airbag module on a flat, secure surface well away from electrical equipment or sources of high temperature.

N.B

Airbag modules and pyrotechnic seatbelt pre-tensioners are classed as explosive articles and, as such, must be stored overnight in an approved, secure steel cabinet which is registered with the Local Authority.

Disposal of airbag modules and pyrotechnic seatbelt pre-tensioners

If a vehicle that contains airbags and seatbelt pre-tensioners that have not been deployed (activated) is to be scrapped, the airbags and seatbelt pre-tensioners must be rendered inoperable by activating them manually prior to disposal. **This procedure may only be performed in accordance with the manufacturer's instruction manual**.

Self assessment questions

1. The approximate electric current used by a 12 volt wiper motor, when it is in proper operating condition, is:
   (a) 1.5 A to 4 A
   (b) 15 A to 40 A
   (c) 0.15 A to 0.4 A
   (d) 6 Watts
2. Remote controllers for central locking are infrared or radio frequency devices. The source of energy that provides the power for the remote control device is:
   (a) an electro magnet in the key
   (b) a small battery in the key
   (c) static electricity
   (d) a thermionic valve
3. Two additives that are added to water to make screen washer fluid are:
   (a) liquid paraffin and salt
   (b) anti-freeze and detergent

(c) boiling point depressant and descaler

(d) distilled water and rubber lubricant

4. Regenerative braking of a screen wiper motor is achieved by:

   (a) a clamping action on the wiper mechanism

   (b) electrically connecting the two main brushes of a permanent magnet motor

   (c) a mechanical brake on the wiper motor driving gears

   (d) the back emf from the wiper control switch

5. By what means is the self parking action of wiper blades achieved:

   (a) a cam operated trip switch at the motor switches the motor off when the park position is reached

   (b) a mechanism that permit's the driver to stop the wiper arms at any point on the screen

   (c) a rheostat that is built into the switch circuit

   (d) a sensor built into the screen

6. A wind tone horn circuit contains a relay:

   (a) to reduce power loss

   (b) because of the high current

   (c) to vibrate the diaphragm

   (d) to reduce arcing at the horn switch

7. Instrument circuits require a steady voltage that is provided by:

   (a) a relay

   (b) a voltage stabiliser

   (c) a field effect transistor

   (d) a separate battery

8. Electrically operated windows are operated by:

   (a) linear motors

   (b) semi-rotary induction motors

   (c) reversible permanent magnet motors

   (d) powerful solenoids

9. Headlamps are connected:

   (a) in series

   (b) in parallel

   (c) by a Wheatstone bridge

   (d) by a failure mode analysis circuit

10. Hazard warning lights:

    (a) make use of the direction indicator lamps circuit

    (b) operate in conjunction with the fuel circuit inertia switch

(c) operate by causing the stop lamps to flash

(d) use 5 Watt MES bulbs

11. List 3 factors that may cause scratching of the windscreen

12. Make a list of the possible causes of a weak spray of washer fluid from the jets on to the screen

13. Describe a simple procedure to test a theft alarm system

Learning tasks

1. Identify the headlamp alignment equipment that is used in your workplace. Study the instructions for use and consult with you training supervisor to arrange sessions when you will learn how to use the equipment.

2. Examine a number of vehicles and note the procedures that must be followed in order to replace headlamp bulbs and bulbs in any other external lamps on the vehicles.

3. Examine a number of screen wiper blades and, with the aid of sketches and a few notes, explain the procedure for replacing worn squeegees.

4. With the aid of workshop manuals, locate the fuse boxes on a number of vehicles. Make a note of the different types of fuses that you have seen and discuss with your supervisor the procedure for replacing blown fuses.

5. Locate the heated rear screen relay on a vehicle and write down an explanation as to why it is placed where it is. Explain this to your training supervisor.

6. Examine a number of vehicles to locate the position of the horn. Use the manual to find the position of the horn relay and fuse.

7. Describe the procedure for dealing with air-bags when a repair operation entails removal of a steering wheel.

13
Measuring instruments and measurement

Topics covered in this chapter

Micrometers
Vernier calipers
Feeler gauges
Number approximations – typical motor vehicle measurements
Electrical test meters

Accurate measurement plays an important part in vehicle maintenance and repair. For example, measurements that are routinely taken are:

- valve clearances
- steering angles and track alignment
- exhaust emissions
- compression pressures
- alternator charging rate
- tyre pressures
- brake disc 'run out'
- brake disc thickness, brake pad thickness
- shaft and bearing clearances.

Figure 13.1 shows a number of examples of the types of measurements and the measuring instruments that are used in motor vehicle maintenance and repair work.

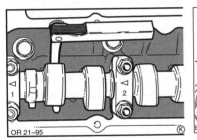

Check valve clearances.

(a) Feeler gauges

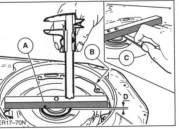

Measure drive train end float with measuring tool 17–025 and depth gauge (or feeler gauge)
A – Needle bearing No.11
B – Oil pump gasket
C – Measuring with feeler gauge
D – Distance between lower edge of measuring tool and needle bearing

(b) Straight edge & vernier gauge

(c) Micrometer – crank dimensions

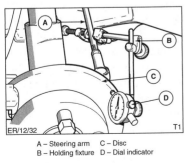

A – Steering arm   C – Disc
B – Holding fixture   D – Dial indicator

(d) Dial test indicator

(e) Vernier caliper – outside diameter of a bush

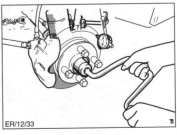

Rotating the disc using a socket and wrench.

(f) Rotating a brake disc to check run-out

Fig. 13.1 Measuring instruments and their uses

13.1 Measuring instruments

Feeler gauges

Feeler gauges are used in motor vehicle repair and maintenance for such purposes as checking valve clearances, piston ring gaps and many other clearances. The feeler gauges, shown in Fig. 13.2, are typical of the type that form a part of a technician's tool kit. They are made from high-grade steel that is tempered for hard wear and durability and the size of each blade is etched into the surface.

Figure 13.1(a) shows a feeler gauge in use, measuring valve clearance. In this example, the feeler gauge is being inserted between the heel of the cam and the cam follower, with the valve in the fully closed position. When ready to proceed with the measurement, the gauges of the required thickness are inserted in the clearance between the heel of the cam and the cam follower and a light-hand force, sufficient to slide the feeler gauge through the gap, is applied.

Non-magnetic feeler gauges

Feeler gauges made from brass or some other non-magnetic material, such as a suitable plastic, are used to check the air gap on some types of electronic ignition distributors.

Vernier calipers

Figure 13.1(b) shows the depth gauge section of a vernier caliper being used to check the drive train end-float in a final drive assembly.

Fig. 13.2 A set of feeler gauges. (Reproduced with the kind permission of Bowers Metrology Group.)

In this application the vernier gauge is being used in conjunction with an engineer's straight edge. Figure 13.3 shows a Vernier gauge in greater detail. This instrument measures to an accuracy of 0.02 mm, which is quite adequate for most vehicle repair and maintenance work. As indicated in the diagram, the instrument is versatile as it may be used for internal, external and depth measurements. The instrument should be treated carefully and kept in clean dry conditions when not in use.

Reading a vernier scale

For the following explanation please refer to the metric reading at the bottom of Fig. 13.3.

The **main scale** is graduated in millimetres and are numbered at every 10 divisions. The **vernier scale** is divided into 50 divisions over a distance of 49 mm and each division is equal to 49/50ths of a mm, that is 0.98 mm. The difference between the divisions on the two scales is $1.00 - 0.98 = 0.02$ mm.

To read the measurement, note the **main scale** reading up to the zero on the **vernier scale**. In this example, that is a reading of 21 mm. To this reading must be added the decimal reading on the **vernier scale**. This is obtained by noting the line on the **vernier scale** that exactly lines up with a line on the **main scale**. In this case the scales align at 7 large divisions and 3 small divisions of the **vernier scale**. The 7 large divisions represent 0.7 mm and the 3 small divisions represent 3×0.02 mm $= 0.06$ mm.

This gives a total reading $= 21$ mm $+ 0.7$ mm $+ 0.06$ mm $= 21.76$ mm

Micrometers

Figure 13.1(c) shows an external micrometer being used to check the diameter of a crankpin. Crankshaft measurement is normally only performed when an engine has operated for many thousands of miles, or when some failure has occurred as a result of running the engine short of oil or a related problem. The reason for checking the crank diameter is to assess its condition and thus ensure that the crank is suitable for further use.

Micrometers are available in a range of sizes, for example, 0 inch to 1 inch (0 mm to 25 mm), 1 inch to 2 inch (25 mm to 50 mm), etc.

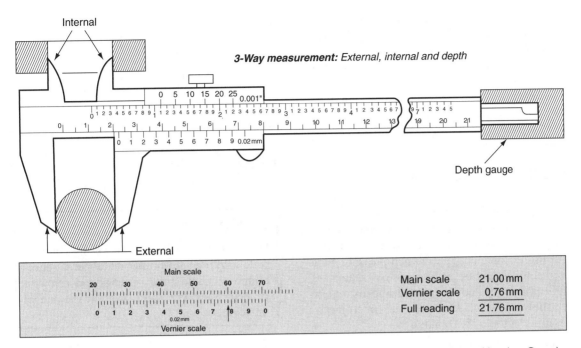

3-Way measurement: *External, internal and depth*

Internal

External

Depth gauge

Main scale	21.00 mm	
Vernier scale	0.76 mm	
Full reading	21.76 mm	

Main scale

Vernier scale

Fig. 13.3 A Vernier gauge and method of reading the scale. (Reproduced with the kind permission of Bowers Metrology Group.)

A metric external micrometer

Figure 13.4 shows the main features of a micrometer and reference should be made to this diagram when reading the following description.

The screw thread on the spindle of the metric micrometer has pitch of 0.5 mm so that one

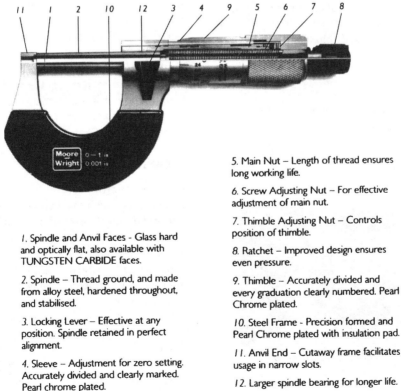

1. Spindle and Anvil Faces - Glass hard and optically flat, also available with TUNGSTEN CARBIDE faces.

2. Spindle – Thread ground, and made from alloy steel, hardened throughout, and stabilised.

3. Locking Lever – Effective at any position. Spindle retained in perfect alignment.

4. Sleeve – Adjustment for zero setting. Accurately divided and clearly marked. Pearl chrome plated.

5. Main Nut – Length of thread ensures long working life.

6. Screw Adjusting Nut – For effective adjustment of main nut.

7. Thimble Adjusting Nut – Controls position of thimble.

8. Ratchet – Improved design ensures even pressure.

9. Thimble – Accurately divided and every graduation clearly numbered. Pearl Chrome plated.

10. Steel Frame - Precision formed and Pearl Chrome plated with insulation pad.

11. Anvil End – Cutaway frame facilitates usage in narrow slots.

12. Larger spindle bearing for longer life.

Fig. 13.4 An external micrometer. (Reproduced with the kind permission of Bowers Metrology Group.)

complete turn of the thimble will move the spindle by 0.5 mm. The thimble is marked in 50 equal divisions, each of which is equal to 0.01 mm.

The sleeve is graduated in two sets of lines, one set on each side of the datum line. The set of divisions below the datum line read in 1 mm and the set above the datum line are in 0.5 mm divisions. (NOTE – on some micrometers, these positions are reversed).

Reading a micrometer

When taking measurements, the force exerted on the thimble must not be excessive and to assist in achieving this, most micrometers are equipped with a ratchet which ensures that a constant force is applied when measurements are being taken. For measuring small diameters, the micrometer should be held approximately, as shown in Fig. 13.5. Larger micrometers require the use of both hands.

Taking the reading

Please refer to Fig. 13.6.
The procedure is as follows:

- First note the whole number of millimetre divisions on the sleeve. These are called **MAJOR** divisions.
- Next observe whether there is a 0.5 mm division visible. These are called **minor** divisions.
- Finally read the thimble for 0.01 mm divisions. These are called **thimble** divisions.
- Taking the metric example shown in Fig. 13.6, the reading is:

    MAJOR divisions = 10 × 1.00 = 10 mm
    MINOR divisions = 1 × 0.50 = 0.50 mm
    THIMBLE divisions = 16 × 0.01 = 0.16 mm

- These are then added together to give a reading = 10.66 mm.

Method of holding the micrometer

Fig. 13.5 Using the micrometer on a small diameter. (Reproduced with the kind permission of Bowers Metrology Group.)

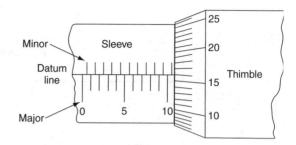

Fig. 13.6 The micrometer scales

Learning activity

Figure 13.8 shows a rage of micrometer scales. Write down the dimensions represented by these scale readings and compare your answers with those given at the end of this chapter.

Micrometers for larger diameters

A wide range of diameters may be encountered in vehicle repair work and an adjustable micrometer is useful for covering several ranges of diameters. The micrometer shown in Fig. 13.7(a) has a number of interchangeable anvils that may be screwed into place to suit the diameter being measured.

Figure 13.7(b) shows the range of diameters, from a large piston to a small shaft, that can be measured with an instrument of this type. Larger micrometers of the type shown here should be held in one hand while the thimble is turned with the other.

Internal micrometer

Internal micrometers are used to measure dimensions, such as cylinder bore diameter, internal diameter of a hole into which a bearing is to be inserted and other machined holes where the dimensions of the internal diameter are critical. Figure 13.9(a) shows an internal micrometer being used for such measurement.

Internal micrometers consist of a measuring head and extension pieces suitable for a range of measurements and the handle permits measurements to be taken in deep holes such as the cylinder of an engine. The principal features of an internal micrometer are shown in Fig. 13.9(b). The micrometer

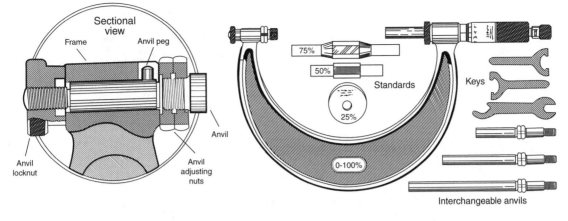

(a)

Typical applications

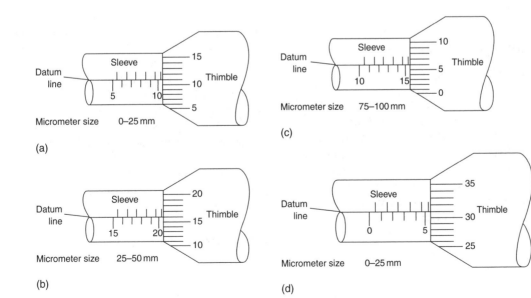

(b)

Fig. 13.7 An adjustable micrometer and its uses. (Reproduced with the kind permission of Bowers Metrology Group.)

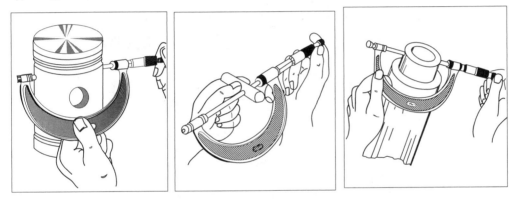

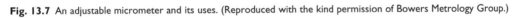

(a)

(c)

(b)

(d)

Fig. 13.8 Exercises on reading micrometer scales

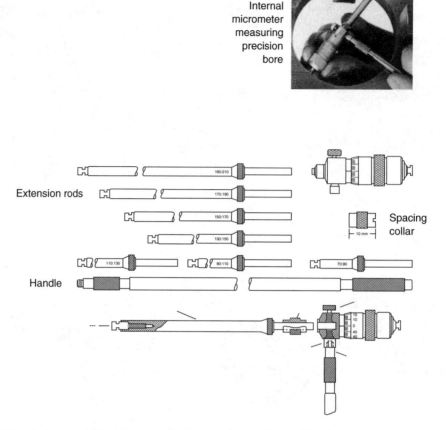

Internal
micrometer
measuring
precision
bore

Extension rods

Spacing
collar

Handle

Fig. 13.9 An internal micrometer. (Reproduced with the kind permission of Bowers Metrology Group.)

scale is read in a similar way to that shown in the external micrometer section of this chapter.

Mercer cylinder gauge

Because piston rings do not reach the top edge of the cylinder bore, when the piston is at the top of its stroke, an unworn ridge develops at the top of the cylinder and further cylinder bore wear occurs on the side where most thrust is exerted. This means that the cylinders of an engine wear unevenly from top to bottom of the stroke. In order to assess cylinder bore wear, it is necessary to measure the diameter at several points and the Mercer gauge has been developed for this purpose. Figure 13.10 shows a Mercer gauge. The gauge can be zeroed by setting it to the diameter of the unworn section at the top of the cylinder bore, or more accurately, by means of an outside micrometer.

Once zeroed, the bar gauge is inserted into the cylinder, as shown in Fig. 13.10. Measurements are taken in line with, and at right angles to, the crankshaft. The maximum difference between the standard size and gauge reading is the wear. The difference between the in-line reading and the one at right angles is the ovality. The measurements are repeated at points in the cylinder, as shown in Fig. 13.11 and differences between the diameter at the top and bottom of the cylinder are referred to as taper.

The measurements taken may then be compared with the quoted bore size and excessive wear can be remedied by reboring or, in the case of minor wear, new piston rings.

Calipers, inside and out

Figure 13.11 show internal and external calipers being set against a steel rule prior to making measurements of inside and outside diameters. These types of measurements may be found

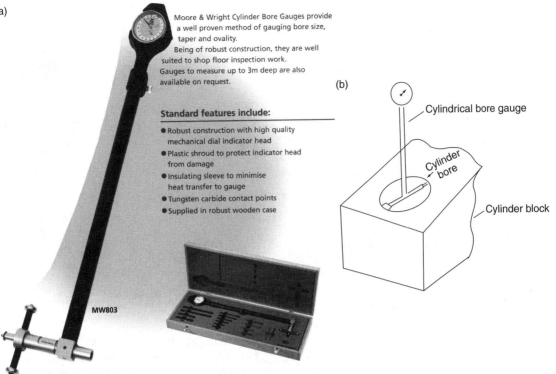

(a)

Moore & Wright Cylinder Bore Gauges provide a well proven method of gauging bore size, taper and ovality.
Being of robust construction, they are well suited to shop floor inspection work.
Gauges to measure up to 3m deep are also available on request.

Standard features include:

● Robust construction with high quality mechanical dial indicator head
● Plastic shroud to protect indicator head from damage
● Insulating sleeve to minimise heat transfer to gauge
● Tungsten carbide contact points
● Supplied in robust wooden case

MW803

(b)

Cylindrical bore gauge

Cylinder bore

Cylinder block

Fig. 13.10 A mercer cylinder gauge. (Reproduced with the kind permission of Bowers Metrology Group.)

useful in assembly work, such as fitting new parts where it may be necessary to check one diameter against another.

Dial test indicators

Dial test indicators of the type shown in Fig. 13.12 are used for measurements, such as checking run-out on brake discs and checking end float on a crankshaft.

There are many other instances where a DTI would be used, but the following two examples will suffice to explain the principles involved.

Brake disc run-out

Figure 13.1(d) shows a dial test indicator that has been set up to test the axial run-out that occurs when the disc is rotated, as shown in Fig. 13.1(f). This measurement is important because of the effect that excessive run-out may have on the operation of the brakes. Among the brake problems that may be caused are defects, such as excessive pad wear and low brake efficiency.

Excessive disc run-out may be caused by a distorted disc, dirt between the disc and the flange that it is fitted to or other problems

in the wheel hub. Run out should not exceed 0.15 mm (0.006 in) when measured in a position near the edge of the disc, but clear of any corrosion that may exist on the edge of the disc.

Crank shaft end float

End thrust on the crankshaft occurs when the clutch is operated and from the weight of the crankshaft when the vehicle is operating on an incline. The end thrust is normally taken by thrust bearings that are fitted to the centre main bearing of the crankshaft, as shown in Fig. 13.13.

There must be sufficient clearance between the machined faces on the crankshaft and the thrust faces of the thrust bearing to permit proper lubrication and allow for thermal expansion. If the clearance is too great, lubrication may fail because of loss of oil pressure, if the clearance is too small, seizure may occur. The clearance is checked by means of a DTI that measures the end float on the crankshaft, as shown in Fig. 13.14. A lever suitably positioned, is used to move the crankshaft to and fro in order to check the end float.

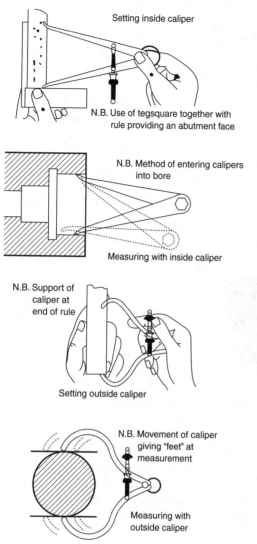

Setting inside caliper

N.B. Use of tegsquare together with rule providing an abutment face

N.B. Method of entering calipers into bore

Measuring with inside caliper

N.B. Support of caliper at end of rule

Setting outside caliper

N.B. Movement of caliper giving "feet" at measurement

Measuring with outside caliper

Fig. 13.11 Engineers calipers – checking diameters. (Reproduced with the kind permission of Bowers Metrology Group.)

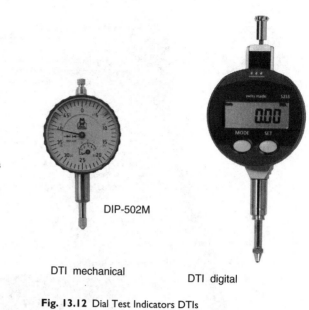

DIP-502M

DTI mechanical   DTI digital

Fig. 13.12 Dial Test Indicators DTIs

13.2 Pressure gauges

Pressure checks on various vehicle systems and components are performed as part of routine service procedures or as part of a series of fault tracing tests. The following examples show the principles involved.

Engine compression test

The compression test is performed in order to ascertain compression pressure because this is a reliable guide to engine condition. The test is performed with all sparking plugs disconnected and removed from the engine. When the compression tester adaptor is firmly pressed on to the sparking plug hole seat, as shown in Fig. 13.15, the engine throttle is opened fully and the engine is cranked over a prescribed number of times, until the maximum pressure is recorded.

In the Ford tester shown here, the pressure is recorded on the graph paper in the instrument, in other cases a note should be made of the maximum pressure recorded. This procedure is repeated for each of the cylinders. The figures are then compared with those given in the manufacturer's data. The pressure readings for each cylinder should also be compared because this information can inform the user about the possible source of a problem. With worn engines, small variations of pressure are to be expected although variations of more than 10% should be investigated.

Oil pressure checks

In addition to its function of lubricating and helping to cool the engine, the engine lubrication system plays a part in operating hydraulic tappets and variable valve-timing mechanisms, and possibly other devices on the engine. In order for the engine to function properly, the lubrication system must be able to sustain the correct oil pressure at all engine speeds. The oil pressure check that is shown in Figs 13.16(a) & 13.16(b), is carried out by temporarily removing the oil pressure detection switch and replacing it by the test type

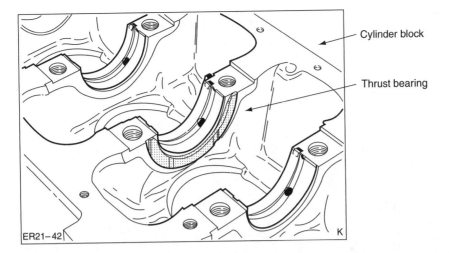

Installation of bearing shells and thrust half washers

Fig. 13.13 Crankshaft thrust bearing

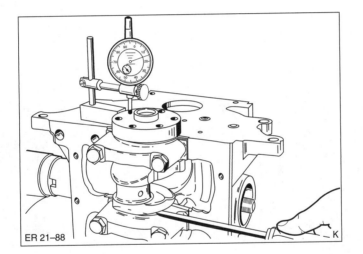

Crankshaft end float measurement

Fig. 13.14 Checking crankshaft end-float

pressure gauge. The engine is then run under the recommended conditions of speed and temperature, etc. The readings should be noted and compared with those provided by the manufacturer.

Radiator pressure cap test

Figure 13.17 shows a piece of equipment that consists of a small pump, a pressure gauge, and an adaptor to which the radiator cap is secured. The radiator cap is removed from the vehicle and then attached to the pressure tester. When the cap has been secured to the

tester, the pump is operated and the maximum pressure reached is noted. This figure should be compared with operating pressure for the cap, which is often stamped on the cap.

13.3 Belt tension gauge

In order to function correctly, the cam drive belt must be set to the correct tension. The belt tension is set at the assembly stage by means of a special gauge that is attached to the 'tight' side of the belt, as shown in

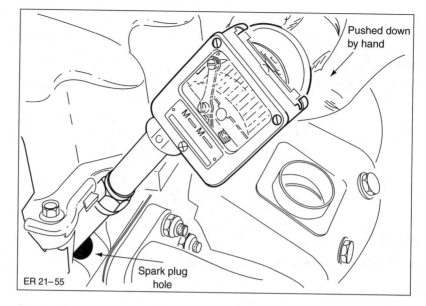

Compression tester

Fig. 13.15 Using the compression test gauge

(a) Removal of oil pressure switch

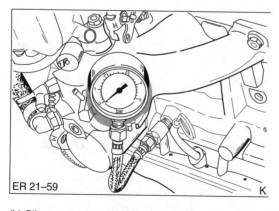

(b) Oil pressure gauge connected to engine

Fig. 13.16 An oil pressure check

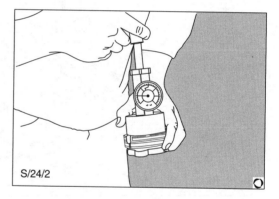

Testing pressure cap

Fig. 13.17 Pressure testing a radiator cap

Fig. 13.18. This operation is carried out whenever a new cam belt is fitted or the adjustment is disturbed for some repair operation to be performed.

13.4 Cathode ray oscilloscope

The cathode ray tube (oscilloscope), as used in domestic television sets and flat screen display, are used for a wide range of vehicle tests.

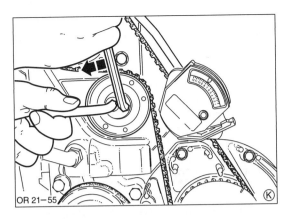

Tension toothed belt

Fig. 13.18 Cam belt tension gauge

With suitable interfaces, a range of voltages, from very small voltage values to very high ones, such as ignition secondary voltages, can be measured. Oscilloscope-based test equipment for vehicle systems normally includes a range of cables and adaptors which contain the interfaces that permit electrical signals to be collected and transferred to the oscilloscope. Because oscilloscopes measure voltage against time, the shape of the image that is shown on the screen enables a technician to observe the changes that are taking place while the system under test is in operation. Figure 13.19 shows a portable oscilloscope that is connected to the HT lead of a modern ignition system.

Figure 13.20 shows the oscilloscope pattern that is displayed on the screen during an ignition test.

Explanation of HT trace shown in Fig 13.20

1. Firing line – this represents the high voltage that is required to cause the spark to bridge the spark plug gap;
2. This is the spark line;

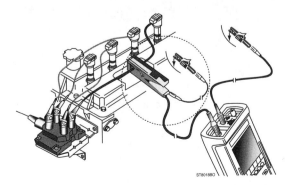

Fig. 13.19 Oscilloscope testing on an ignition system

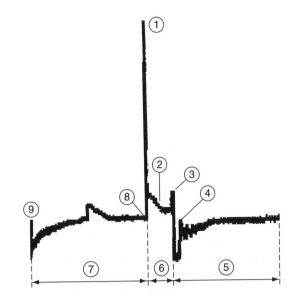

Fig. 13.20 Details of HT voltage pattern for a single cylinder

3. Spark ceases at this point;
4. Coil oscillations – caused by collapse of the magnetic field;
5. Any remaining electro magnetic energy is dissipated;
6. Firing section (represents burn time);
7. Dwell section;
8. Primary winding current of the coil is interrupted by the transistor switch that is controlled by the engine control computer (ECM);
9. Primary winding current is switched on by the ECM and this energises the primary winding of the ignition coil. The dwell period, between 9 and 8 on the trace, is important because it is the period in which primary current builds up to its maximum value.

Figure 13.20, and the accompanying explanation, give an indication of the amount of information that can be obtained from an oscilloscope test. Engine management systems on modern vehicles rely on the accuracy of factors such as burn time, etc.

Sensors

Many vehicle systems are now controlled by a computer on the vehicle.

Such control computers (ECMs) operate on low voltage electrical pulses that are interpreted as the zeroes (0) and ones (1) of digital codes. The electrical pulses originate from devices such as the crank position sensor, the coolant temperature sensor, the engine air flow sensor and many others. The sensors on a vehicle measure some quantity, such as speed or position and then convert the quantity into an electrical signal. The electrical signal that is produced is an accurate representation of the variable being measured. Sensors play a vital part in the operation of vehicle systems and the maintenance and testing of them is a major element of the work of a vehicle technician.

Figure 13.21 shows an oscilloscope screen display of the voltage pattern obtained at the output terminals of a throttle position sensor. A separate window on the screen shows an ideal pattern that is stored in test equipment memory. This is the lower trace in Fig. 13.21. The upper trace shows the voltage pattern obtained from a single throttle operation. This trace is obtained by moving the throttle from the closed position to the to fully open position.

1. show the actual test pattern showing a defective throttle position sensor trace, and
2. the voltage spikes in a downward direction indicate a short circuit to earth or a break in the resistive strip in the sensor.

The lower trace shows the pattern for a sensor in good condition. The trace between 5 and 4 is

obtained by a full operation of the throttle from closed to fully open and back to closed. The voltage at 5 represents the throttle closed position, the section marked 6 shows the voltage increasing smoothly to the throttle fully open position and the section marked 3 shows the voltage decreasing smoothly back to the throttle closed position at 4.

13.5 Digital multi-meter

Multi-meters are electrical measuring instruments (meters) that can be set to read voltage, current, resistance, frequency, etc. Digital instruments are most suited to testing modern vehicle systems because they have the high impedance (internal resistance) that is required for testing circuits that contain sensitive electronic components.

Using a multi-meter

Figure 13.22 shows a sensor that is built into an engine oil dipstick. The sensor, marked 3, is a small length of fine wire that has a high temperature coefficient of resistance which means that small changes in temperature produce large changes in resistance. The resistance of the sensing element is 7 ohms at room temperature.

When the control unit passes a current of 200 mA through the sensing element, for a period approximately 2 seconds, the sensing wire is heated and this produces an increase in resistance. This increased resistance

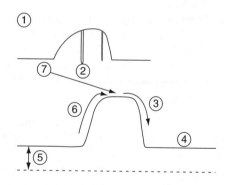

1. Defective TPS pattern.
2. Spikes in a downward direction indicate a short to ground or an intermittent open in the resistive carbon strips.
3. Voltage decrease identifies enleanment (throttle plate closing).
4. Minimum voltage indicates closed throttle plate.
5. DC offset indicates voltage at key on, throttle closed.
6. Voltage increase identifies enrichment.
7. Peak voltage indicates wide open throttle (WOT).

Analysis of voltage trace for a potentiometer-type throttle position sensor

Fig. 13.21 Testing a throttle position sensor

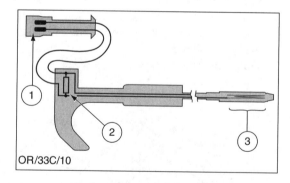

OR/33C/10

Low oil sensor circuit diagram
1 – Connector (to loom)
2 – Trimming resistor
3 – Sensor (fine wire with a high temperature coefficient of resistance)

Fig. 13.22 Low oil sensor

leads to a voltage increase across the sensor terminals.

If the oil is level is correct, the sensor is in oil and the heat is dissipated. Any voltage increase caused by the electrical pulse is small. If the oil level is low, the sensor wire will be in air. The temperature increase from the electrical pulse will be large and the voltage across the sensor terminals will be high.

The voltage across the sensor terminals is measured by the control unit. If the voltage increase is greater than the set value that is kept in the control unit memory, the 'low oil' warning light is switched on.

Testing a low oil sensor by means of a digital multi-meter

A typical use of the multi-meter is to measure the resistance of a sensor because, in many cases, this gives a guide to its electrical condition.

In Fig. 13.23(b), the ignition has been switched off and the cable connector has been detached from the sensor. The multi-meter has been set to measure resistance in ohms and the meter probes are connected to the two sensor pins. In this case, the resistance should lie between 6.4 ohms and 7.6 ohms. The test should be done quickly, because the current from the ohm-meter will heat the sensing element and change the resistance. If this test is satisfactory, the sensor should be re-connected. The multi-plug is then disconnected at the ECM and the ohm-meter probes are applied to the appropriate pins on the multi-plug, in this case, pins 10 and 11, as shown in Fig. 13.23(c). The resistance reading at this point should be between 6.4 ohms and 8.6 ohms, the higher figure being attributable to the length of cable between the sensor and the multi-plug.

Applications of number

Approximations and degrees of accuracy
Significant figures

Measurements and calculations often produce numbers that are more useful when they are rounded to a certain number of significant figures. For example, the swept volume of a certain single cylinder engine is 249.94 cc. For many purposes, this figure would be rounded to 3 significant figures, to give 250cc.

The rule for rounding to a certain number of significant figures is: "*If the next digit is 5 or more, the last digit is increased by 1. A zero in the*

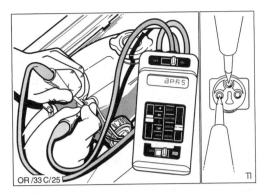

(a)

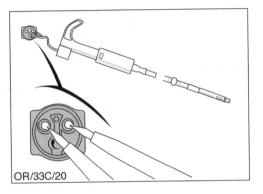

(b) Resistance at sensor 6.4 Ω to 7.6 Ω

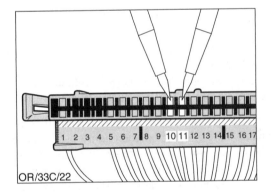

(c) Resistance at ECM 6.4 Ω to 8.6 Ω

Fig. 13.23 Resistance check of a sensor

middle of a number is counted as a significant figure, for example, 3.062 to 3 significant figures is 3.06."

Decimal places

Numbers can be rounded to a certain number of decimal places, dependent on the required degree of accuracy. For example, 3.56492, rounded to 2 decimal places, becomes 3.56.

The rule for decimal place approximations is; "*If the next number (digit) is 5 or more, the previous digit is increased by 1.*"

Examples

1. The test data for a throttle position sensor specifies a voltage of 2.50 volts for the half open throttle position. A test with a digital meter gives a voltage reading of 2.496 volts for the half open throttle position. Does this mean that the sensor is faulty?

   The answer is probably not, because the test data is given to 2 places of decimals. When corrected to 2 decimal places, the digital meter reading becomes 2.50 volts which is as it should be.

2. The maximum permitted diameter of a certain crankpin is given as 43.01 mm. When measured with a micrometer, the diameter is found to be 43.009 mm. Is this acceptable?

   Yes. The test data is given to 2 decimal places. When the measured reading of 43.009 mm is rounded to 2 decimal places, a figure of 43.01 mm is produced and this is exactly the same as the quoted figure.

Limits

Dimensions of components are normally given in the following form:

Crankshaft main bearing journals: **53.970 mm to 53.990 mm**
Crankshaft end float: **0.093 mm to 0.306 mm**
Valve length: **107.25 mm to 107.35 mm**.

In these cases, the larger figure is known as the **upper limit** and the smaller figure as the **lower limit**. The difference between the upper limit and the lower limit is known as the **tolerance**.

To calculate the tolerance, the smaller dimension is subtracted from the larger. For example, in the case of the main bearing journal, the tolerance is **53.990 mm − 53.970 mm = 0.020 mm**.

These tolerances allow for slight variations in measurements at the manufacturing stage, and also variations in dimensions that arise from wear in use.

Fits

The terminology of engineering fits is based on the dimensions of a shaft and a hole into which the shaft is inserted.

When the shaft is slightly smaller than the hole to which it is to be fitted, the result is known as a **clearance fit**. The clearance is required for lubrication and thermal expansion purposes.

When the shaft is slightly larger than the hole to which it is to be fitted the result is known as an **interference fit**. Interference fits are used where the shaft is to be firmly secured by the material surrounding the hole.

Examples of the two types of fit are shown in Fig. 13.24.

With a **floating** gudgeon pin, the pin is a **clearance fit** in the small end bush. The gudgeon pin is located in the piston by means of circlips that fit into grooves on each side of the piston.

In the **semi-floating** gudgeon pin arrangement, the gudgeon pin is an **interference fit** in the small end eye of the connecting rod. The gudgeon pin is held in place by the strain energy in the materials. At the fitting stage, the gudgeon pin is either pressed into the connecting rod using a hand operated screw press or the connecting rod is heated to make the small end eye expand so that the gudgeon pin may be inserted. When the assembly cools down, the gudgeon pin is firmly held in the small end eye.

Angular measurement

Examples of angles that occur in motor vehicle dimensions as shown in the following table:

Steering	Angle (Approximate)
Camber angle	1° 26′
Castor angle	2° 11′
King pin inclination	3°
Toe-out on turns	Outer wheel 20° Inner wheel 25° 30′
Engine	Angle (approximate)
Spark ignition – static advance	12° before TDC
Valve timing – inlet valve opens	8° before TDC
Diesel fuel pump – start of injection	15° before TDC

Specification of angles

The degree is the basic measurement of angle. As there are 360 degrees in a complete circle, for more accurate measurement, a degree is subdivided into 60 minutes.

A degree is denoted by the small zero above the number, i.e., 1°.

A minute is denoted by a small dash above the number, i.e., 5′.

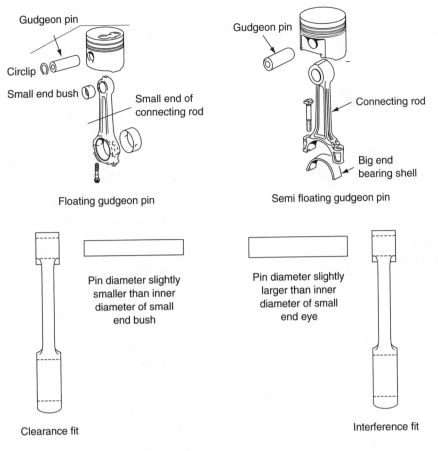

Gudgeon pin

Circlip

Small end bush

Small end of
connecting rod

Floating gudgeon pin

Gudgeon pin

Connecting rod

Big end
bearing shell

Semi floating gudgeon pin

Pin diameter slightly
smaller than inner
diameter of small
end bush

Pin diameter slightly
larger than inner
diameter of small
end eye

Clearance fit

Interference fit

Gudgeon pins – clearance & incerference

Fig. 13.24 Floating and semi-floating small ends – clearance and interference fits. (Reproduced with the kind permission of Bowers Metrology Group.)

The camber angle given in the Table is 1° 26′; this reads as one degree, twenty-six minutes.

Figure 13.25 shows two examples of angular measuring devices in use. In Fig. 13.25(a) a 360° protractor is attached to the timing gear of a diesel in order to set the injection timing. In Fig. 13.25(b), the angle is read from the protractor on the track alignment gauges.

> *Learning activity answer. Figure 13.8.*
>
> Ans. (a) 10.61 mm
> Micrometer size 0–25 mm
> Reading:
> - Major divisions $= 10 \times 1 = 10$ mm
> - Minor divisions $= 1 = 0.5$ mm
> - Thimble divisions $= 11 = 0.11$ mm
> - Total $= 10.61$ mm
>
> Ans. (b) 45.66 mm

> Micrometer size 25–50 mm
> Reading:
> - Major divisions $= 20 \times 1 = 20$ mm
> - Minor divisions $= 1 \times 0.5 = 0.5$ mm
> - Thimble divisions $= 16 = 0.16$ mm
> - Total 45.66 mm
>
> Ans. (c) 90.56 mm
> Micrometer size 75–100 mm
> Reading:
> - Major divisions $= 15 \times 1 = 15$ mm
> - Minor divisions $= 1 \times 0.5 = 0.5$ mm
> - Thimble divisions $= 6 \times 0.01 = 0.06$ mm
> - Total 90.51 mm
>
> Ans. (d) 6.31 mm
> Micrometer size 0–25 mm
> Reading:
> - Major divisions $= 5 \times 1 = 5$ mm
> - Minor divisions $= 0$
> - Thimble divisions $= 31 \times 0.01 = 0.31$
> - Total 5.31 mm

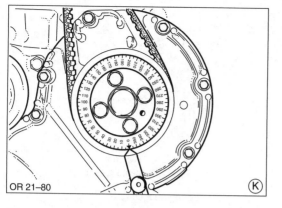

(a)

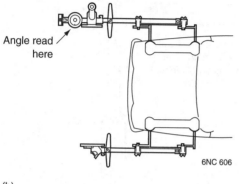

Angle read
here

6NC 606

(b)

Fig. 13.25 Angular measurement on motor vehicles

Self assessment questions

1. Correct the following figures to 2 decimal places:
   (a) 43.001
   (b) 0.768
   (c) 5.887

2. Correct the following figures to the specified number of significant figures:
   (a) 19.28 to 3 significant figures
   (b) 2.034 to 3 significant figures
   (c) 337.9 to 3 significant figures

3. The capacity of a certain engine is calculated to be 1998 cm³. For descriptive purposes, this figure is expressed in litres. Calculate the engine capacity in litres, correct to 2 significant figures.

4. 43 bolts, priced at 45.6 pence each, are to be added to an account. Calculate the total amount for the 40 bolts, correct to nearest 1 penny. Give the answer in pounds and pence.

5. Figure 13.26 shows a feeler gauge being used to check the gap in a piston ring.
   In order to perform this check:
   (a) the piston ring should just be pushed into the cylinder
   (b) the piston ring should be gently pushed into the cylinder using a piston as a guide to ensure that the ring is at right angles to the axis of the cylinder
   (c) the ring gap must not exceed 0.001 mm
   (d) it does not matter if the ring is tilted in the cylinder

6. Figure 13.27 shows a dial gauge that is set up to measure the end float on an engine crankshaft.
   End float is important because:
   (a) it allows space for lubrication of the crankshaft thrust bearings
   (b) it provides free travel at the clutch pedal
   (c) it helps to keep the timing gears in alignment
   (d) it allows oil to be sprayed onto the cylinder walls to lubricate them

7. A crankpin has a diameter of 58.20 mm and the big end bearing that is fitted to it has an internal diameter of 58.28 mm. The clearance between the crankpin and the big end bearing is:
   (a) 1.0013 mm
   (b) 0.008 mm
   (c) 0.08 mm
   (d) 0.48 mm

8. Figure 13.28 shows a feeler gauge marked 5 being used to check the clearance between the cam marked 4 and the cam follower.
   This clearance is known as the valve clearance and it is checked when:
   (a) the cam is in the fully closed position
   (b) when the engine is at the recommended temperature and the valve is in the fully closed position

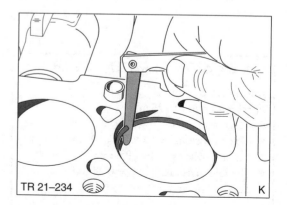

Fig. 13.26 Checking a piston ring gap

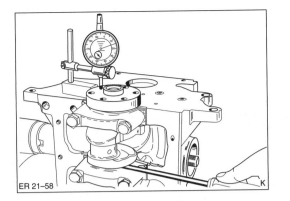

Fig. 13.27 Checking crankshaft end float

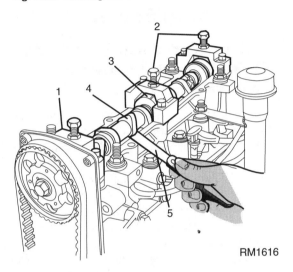

Fig. 13.28 Checking valve clearances

(c) the valve is fully open

(d) the cam is at any position between valve fully open and valve fully closed

9. The crankpin shown in Fig. 13.29 is measured at A, B, C and D.

Dimensions A and B are taken to determine taper, and dimensions C and D to determine ovality. In this case the dimensions are:

- A = 42.99 mm
- B = 42.75 mm
- C = 42.99 mm
- D = 42.60 mm.

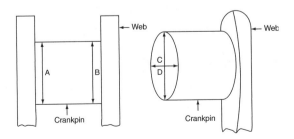

Fig. 13.29 Crankshaft measurement

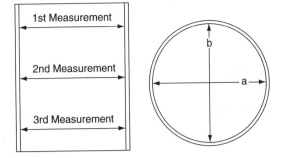

Fig. 13.30 Cylinder bore measurement

The taper on this crankpin is:

(a) 0.39 mm

(b) 0.24 mm

(c) 0.15 mm

(d) 0.01 mm

10. The ovality on this crankpin is:

(a) 0.24 mm

(b) 0.00 mm

(c) 0.39 mm

(d) 0.15 mm

11. Figure 13.30 shows the positions in an engine cylinder at which measurements of diameter are taken in order to assess wear. The measurements are taken in line and at right-angles to the crankshaft, at the three levels shown. The third measurement is taken at the lowest extent of piston ring travel. The terms ovality and taper are used to describe the condition of the cylinder bore. Ovality is the difference betweeen a and b, and taper is the difference between the largest diameter at the first measurement and the smallest diameter at the third measurement.

In an example, the following dimensions were recorded:

- First measurement, a = 74.22 mm, b = 73.95 mm,
- Second measurement, a = 74.19 mm, b = 74.00 mm,
- Third measurement, a = 74.00 mm, b = 73.97 mm.

The maximum ovality is:

(a) 0.22 mm

(b) 0.19 mm

(c) 0.27 mm

(d) 0.03 mm

12. The taper is:

(a) 0.25 mm

(b) 0.03 mm

(c) 0.19 mm

(d) 0.27 mm

14

Organisations and qualifications

Topics covered in this chapter

Organisations in the motor trade
Qualifications and careers

14.1 Qualifications

It is unfortunate that training and education tend to be treated as separate activities because this leads to difficulty when one is attempting to choose a course of study or training. However, there are many fundamentals of motor vehicle engineering where the distinction between training and education cannot sensibly be made. With valve timing of an engine, it is essential to understand cycles of engine operation, for example, the four stroke cycle. The knowledge required is known as engine technology and is generally considered to be education and therefore is part of the syllabus for courses such as City and Guilds Progression Award, BTEC First Diploma and Certificate and similar qualifications. The actual setting of the valve timing is a practical task and learning how to do it is considered to be training. The instruction and practice that leads to becoming proficient at the task is also considered to be training and is considered to be part of the National Vocational Qualification (NVQ). The outcome of this rather artificial distinction between education and training is that educational courses contain much that contributes to ones ability to be a vehicle technician, while practical training courses for NVQ also contain much that is of educational value and, as a result, there are several routes that people can follow in order to qualify as a motor vehicle technician.

14.2 Modern motor vehicle qualifications

National Vocational Qualification NVQ – Scottish Vocational Qualification-SVQ

A list of skills that are required for competent performance of the tasks involved in maintenance and repair of vehicles has been devised and these skills are known as National Occupational Standards, a typical example being, *Carry out routine vehicle maintenance*. These Occupational Standards are used to design qualifications such as the NVQ. Because the work of a motor car technician differs in some respects from that of a heavy vehicle technician, the training programme and course content is not the same and NVQs have been designed to cater for these differences. The same is true of other motor vehicle occupations, such as Auto Electrical Technician, Vehicle Body Repair Technician, Parts Department technician, etc.

Levels

In the school years, levels of attainment tend to be expressed in terms of year 2, year 6, etc., but in post school training and education, the use of years as a description of a degree of difficulty involved in some learning step, is not appropriate and the training and education courses are classified in Levels. The result is that NVQs are available at several Levels, the main ones being at Levels 1, 2 and 3.

Studying for NVQ

To qualify for NVQ, it is necessary to demonstrate competence, namely ability to perform a given task in a normal work situation. However, as already explained in the example of valve timing, the performance of a practical task often requires an underpinning knowledge of technology. The result is that courses of study for NVQ consist of a balanced program that takes place on the job (in a workshop), for the practical and off the job (in a classroom), for the underpinning knowledge. NVQs in Motor Vehicle Maintenance and Repair certainly require a good deal of underpinning technical knowledge together with practical training and courses of education and training are designed to cater for these factors.

Modern apprenticeships

Modern apprenticeships provide paid employment, practical training and further education. During the apprenticeship, trainees must study for NVQ and Key Skills and a Technical Certificate, such as City and Guilds of BTEC.

There are two levels of apprenticeships; foundation for beginners and the training and education programme normally lasts for approximately 1 year. The other level is the advanced apprenticeship, which lasts for approximately 2 years.

Foundation modern apprenticeship

This first level of apprenticeship trains apprentices for Level 2 NVQ and the associated Key Skills. In addition, a Technical Certificate is studied and this may be a BTEC First, City&Guilds or Institute Of motor Industry certificate.

Advanced modern apprenticeship

Advanced Apprenticeships are designed to last for 2 years. They prepare trainees for Level 3 NVQ and Key Skills and a Technical Certificate, such as BTEC National or similar educational qualification is studied. Apprenticeships should, in future, provide the most attractive and efficient route to qualification and employment in the vehicle maintenance and repair industry.

Courses in colleges

Many Further Education Colleges provide courses of Motor Vehicle Studies. The courses vary from those for beginners, such as the City & Guilds Progression Award, to main stream courses for BTEC National Certificates and Diplomas. Many of these courses are linked to employers so that the students gain work experience and this practical work, plus work done at college, may be taken into account when making a later claim for NVQ. For students who opt for study in a Further Education College, as opposed to School 6th form, the BTEC National and Higher National programmes may provide a route into University education.

Courses in secondary schools, technology colleges, etc.

Over the past few years, a number of secondary schools have offered motor vehicle courses and, with the renewed emphasis on vocational education, such courses are likely to become more widely available.

This book is designed for Levels 1 & 2 of current training and education programmes but there are elements that extend into the Level 3 region of some topics. The text shows a large number of engineering concepts and it should help readers to make links between practical work and other subjects, such as science and number work, and thus assist in providing students with a balanced educational experience.

14.3 Career Prospects

When embarking on a career in the vehicle repair and maintenance industry, thoughts about future jobs may not be uppermost in one's mind. As one's career progresses this may change and it is useful to have an insight into the types of jobs that may be open to people who have trained as technicians. The following list gives an indication of the types of jobs that are often taken by people who started work as trainees or apprentices and who, through gaining experience and qualifications, equipped themselves for promotion:

- Foreman
- Technical receptionist
- Service manager
- Branch manager
- Training instructor
- Teacher/lecturer
- Insurance engineers and assessors
- Transport managers.

How to get there

There are approximately 30 000 businesses, large and small, in the UK, that carry out vehicle repair and maintenance. There are, therefore, many opportunities for trainees technicians to advance to more senior positions, if they so desire. Holders of the jobs in the above list need management skills and these are often taught through specialised management courses. Management studies are also included in higher-level qualifications such as Higher National Certificates and Diplomas. A well presented curriculum vitae (CV) is likely to be an asset when applying for more senior positions. Apprenticeships and other opportunities for beginners are likely to be advertised and promoted through the career service. College courses that contain an element of work experience often lead to employment – it is probably a good idea to shop around in your area to see what courses are on offer.

14.4 Motor industry organisations

The Institute of the Motor Industry (IMI), and the Institute of Road Transport Engineers (IRTE), are two of the engineering/management institutes that many people engaged in motor vehicle maintenance and repair belong to. These institutes have regional centres that arrange lectures and other events relating to the industry. These events provide an opportunity to meet people from many different companies in an informal setting, in addition to allowing an opportunity to help people keep abreast of developments. Two magazines, *Motor Industry Management* published by the IMI and *Road Transport Engineer*, published by the IRTE, advertise posts nationally, and the local press normally carries advertisements about job vacancies in their areas.

Answers to self assessment questions

Chapter 1

1. (b)
2. (a)
3. (c)
4. (c)
5. (a)
6. (b)
7. (b)
8. (d)
9. (b)
10. (c)

Chapter 2

1. (b)
2. (c)
3. (a)
4. (c)
5. (b)
6. (c)
7. (d)
8. (b)

Chapter 3

1. (b)
2. (a)
3. (b)
4. (c)
5. (c)
6. (c)
7. (a)
8. (b)
9. (b)

Chapter 3

1. (b)
2. (b)
3. (c)
4. (c)
5. (b)

6. (b)
7. (a)
8. (b)
9. (c)
10. (a)

Chapter 4

1. (b)
2. (b)
3. (b)
4. (a)
5. (b)
6. (a)
7. (a)
8. (b)
9. (b)
10. (c)

Chapter 5

1. (c)
2. (d)
3. (c)
4. (b)
5. (a)
6. (a)
7. (c)
8. (c)
9. (c)
10. (c)

Chapter 6

1. (a)
2. (a)
3. (b)
4. (a)
5. (c)
6. (b)
7. (b)
8. (a)

Chapter 7

1. (b)
2. (c)
3. (b)
4. (a)
5. (c)
6. (b)
7. (b)

Chapter 8

1. (a) Over inflation.
   (b) Under inflation.
   (c) Incorrect tyre, tyre very old.
   (d) Incorrect camber angle.
   (e) Incorrect track setting.
   (f) Wheel and tyre out of balance, defective brake discs.
2. (a)
3. (b)
4. (b)
5. (b)
6. (c)
7. (a)

Chapter 9

1. (c)
2. (b)
3. (c)
4. (c)
5. (b)
6. (b)
7. (c)
8. (a)
9. (a)
10. (b)

Chapter 10

1. (a)
2. (b)
3. (c)
4. (b)
5. (c)
6. (b)
7. (a)
8. (c)

9. (c)
10. (b)

Chapter 11

1. (a)
2. (b)
3. (d)
4. (a)
5. (b)
6. (c)
7. (a)
8. (a)
9. (b)
10. (b)
11. (c)

Chapter 12

1. (a)
2. (b)
3. (b)
4. (b)
5. (a)
6. (b)
7. (b)
8. (c)
9. (b)
10. (a)
11. Using the wipers without washer fluid. Hardened or broken wiper blades. Worn wiper blades causing metal parts to rub on the screen. Too much blade pressure.
12. Blocked jets. Leaking hose connections. Blocked filter. Worn pump. Poor electrical connections to the pump.

Chapter 13

1. (a) 43.00, (b) 0.77, (c) 5.89.
2. (a) 19.3, (b) 2.03, (c) 338.
3. 2.0 litres.
4. $43 \times 45.6 = 1960.8 = £19$ and 61 pence.
5. (b)
6. (a)
7. (c)
8. (b)
9. (b)
10. (c)
11. Ovality (c)
12. (a) Taper $= 0.25$ mm.

Index